04T

Lecture Notes in Applied and Computational Mechanics

Volume 16

Series Editors

Prof. Dr.-Ing. Friedrich Pfeiffer
Prof. Dr.-Ing. Peter Wriggers

Springer

Berlin
Heidelberg
New York
Hong Kong
London
Milan
Paris
Tokyo

Theories of Plates and Shells

Critical Review and New Applications

Reinhold Kienzler
Holm Altenbach
Ingrid Ott (Editors)

Springer

Professor Dr. Reinhold Kienzler
Dipl.-Ing. Ingrid Ott
Universität Bremen
Fachbereich 4, Produktionstechnik
Technische Mechanik, Strukturmechanik
28334 Bremen
Germany

Professor Holm Altenbach
Martin-Luther-Universität
Halle-Wittenberg
Fachbereich Ingenieurwissenschaften
06099 Halle (Saale)
Germany

With 75 Figures

Cataloging-in-Publication Data applied for
Bibliographic information published by Die Deutsche Bibliothek
Die Deutsche Bibliothek lists this publication in the Deutsche Nationalbibliografie; detailed bibliographic data is available in the Internet at <http://dnb.de>

ISBN 3-540-20997-2 Springer Verlag Berlin Heidelberg New York

Springer-Verlag is a part of Springer Science+Business Media

springeronline.com

Printed in Germany

Cover design: design & production GmbH, Heidelberg
Typesetting: Digital data supplied by editors

Printed on acid-free paper 62/3020Rw -5 4 3 2 1 0

EUROMECH COLLOQUIUM 444

Critical Review of the Theories of Plates and Shells and New Application

EUROMECH 444

September 22-25, 2002

Bremen / Germany

organized by

R. Kienzler, Universität Bremen

H. Altenbach, Martin-Luther-Universität, Halle-Wittenberg

Universität Bremen

Martin-Luther-Universität
Halle-Wittenberg

Preface

Theories of plates and shells were initiated in the 19th century with the pioneering publications of Kirchhoff and Aron/Love. In the 20th century these theories have been further developed and refined. The effect of shear deformations, the influence of the change of thickness and cross-sectional warping have been explored as well as non-linear theories - geometrically and physically - were essential objects of thorough scientific investigations. Plate and shell theories are inherently approximative in character since the in-fact three-dimensional state of stress and deformation is described by quantities, which "live" on a two-dimensional surface. Therefore the establishment of consistent theories attracted broad attention.

After a period of a reduced publication activity in the field, plate and shell theories experienced a "renaissance" in recent years. The attention of the plate and shell community was brought back by new applications. The potentials of smart materials, the challenges of adaptive structures, the demands of thin-film technologies etc. on the one hand and the availability of newly developed mathematical tools, the tremendous increase in computer facilities and the improvement of commercial software packages on the other caused a reanimation of the scientific interest.

The aim of the EUROMECH Colloquium 444 "Critical Review of the Theories of Plates and Shells and New Applications" was to bring together active scientists in the field to discuss the common roots of different plate and shell approaches, to review the current state-of-the-art and to develop future lines of research. 50 participants were attracted from 15 countries with civil and mechanical engineering as well as mathematical and physical background. The Colloquium was held at the Atlantic Hotel Universum in Bremen, Germany. During three days 26 oral lectures were presented and 16 poster contributions were thoroughly discussed in an extended poster session. Not only the scientific activities but also the social part of the program furthered fruitful interactions between the participants and contributed to friendship and mutual understanding.

In the present book of lecture notes, the contributions of the participants have been collected and ordered alphabetically by the names of the presenting authors. The responsibility for the contents of the paper lies solely with each author.

The organizers of the EUROMECH Colloquium express their sincere thanks to the sponsors of the symposium: EUROMECH, Gesellschaft für Angewandte Mathematik und Mechanik, Deutsche Froschungsgemeinschaft, Universität Bremen, HVG Hanseatische Veranstaltungs GmbH, Naberterm GmbH & Co., Deutsche Airbus, Bremer Straßenbahn AG. Without their generous support the realization of the conference would not have been possible. The warm hospitality of the Atlantic Hotel Universum is also gratefully acknowledged. They made the participants feel like being at home. Assistants and students of the Fachgruppe Technische Mechanik - Strukturmechanik formed the local organizing committee leaded open-eyed and omnipresent by Dipl.-Ing. Ingrid Ott. Their enthusiasm, readiness and helpfulness made the conference smooth-running and pleasant. Especially to mention are Ralf Neuhaus for carefully editing and layouting the book of abstracts and Sebastian Holweg for spending much time and patience in editing the lecture notes. Finally the support by Dr. Merkle of Springer Verlag is much appreciated.

Bremen, August 2003

R. Kienzler H. Altenbach

Fig. 1. Participants of the EUROMECH Colloquium 444 'Critical Review of the Theories of Plates and Shells' held in Bremen, Germany, September 22-25, 2002

Contents

Authors and Co-authors

Altenbach, Holm, Prof. Dr.-Ing.
Technische Mechanik, Fachbereich Ingenieurwissenschaften
Martin-Luther-Universitaet Halle-Wittenberg
D-06099 Halle (Saale)
Germany
e-mail: holm.altenbach@iw.uni-halle.de

Audoly, Basile, Dr.
Laboratoire de Modélisation en Méchanique, UMR 7607 du CNRS
Université Pierre et Marie Curie
4 place Jussiue, F- 75252 Paris Cedex 05
France
e-mail: audoly@lmm.jussieu.fr

Chróścielewski, Jacek, Prof.
Department of Civil Engineering, Technical University of Gdańsk
ul. Narutowicza 11/12
80-952 Gdansk
Poland
e-mail: jchrost@pg.gda.pl

Eisenberger, Moshe, Associate Prof.
Faculty of Civil Engineering, Technion – Israel, Inst. of Technology
Technion City, 32000
Israel
e-mail: cvrmosh@tx.technion.ac.il

Ganczarski, Artur, Dr.
Institute of Mechanics and Machine Design, Cracow University of Technology
al. Jana Pawla II 37, Krakow
Poland
e-mail: artur@cut1.mech.pk.edu.pl

Gavrylenko, G. D., Prof.
S.P. Timoshenko Institute of Mechanics, National Academy of Science
Nesterov 3, 03057 Kiev,
Ukraine
e-mail: GAVR@inmech.kiev.ua

Golze, Mark, Dipl.-Ing.
Chair of structural mechanics and vehicle vibrations
Brandenburg University of Technology Cottbus
Universitätsplatz 3-4, D-03044 Cottbus
Germany

Hackl, Klaus, Prof. Dr.-Ing.
Lehrstuhl für Allgemeine Mechanik, Ruhr-Universität Bochum
D-44780 Bochum
Germany
e-mail: hackl@am.bi.ruhr-uni-bochum.de

Hohe, Jörg, Dr.-Ing.
Fraunhofer Institut für Werkstoffmechanik
Wöhlerstr. 11
79108 Freiburg/Brsg.
Germany
e-mail: hohe@iwm.fhg.de

Hoppe, Ulrich, Dr.-Ing.
Lehrstuhl für Allgemeine Mechanik, Ruhr-Universität Bochum
D-44780 Bochum
e-mail: ulrich.hoppe@ruhr-uni-bochum.de

Jański, Łukasz, Dipl.-Ing.
Institut für Festkörpermechanik, Technische Universität Dresden
Mommsenstr. 13, D-01062 Dresden
Germany
e-mail: janski@mkf.mw.tu-dresden.de

Kaplunov, Julius, Prof. Dr.
Department of Mathematics, The University of Manchester
Oxford Road, Manchester, M13 9 PL
United Kingdom
e-mail: kaplunov@maths.man.ac.uk

Kienzler, Reinhold, Prof. Dr.-Ing.
University of Bremen
PO Box 330440, D- 28334 Bremen
Germany
e-mail: rkienzler@uni-bremen.de

Kirillova, Irina V., Dr.
Department of Mathematical Theory of Elasticity and Biomechanics,
Saratov State University
Astrakhanskaya Str. 83, Saratov, 410012
Russia
e-mail: KirillovaIV@yandex.ru

Kolpakov, Alexander G., Prof. Dr. Sci
Russian Academy of Sciences, Novosibirsk, Russia
324, Bld.95, 9th Novemberstr., Novosibirsk, 630009
Russia
e-mail: agk@neic.nsk.su

Kühhorn, Arnold, Prof. Dr.-Ing.
Chair of structural mechanics and vehicle vibrations
Brandenburg University of Technology Cottbus
Universitätsplatz 3-4, D-03044 Cottbus
Germany
e-mail: kuehhorn@tu-cottbus.de

Lammering, Rolf, Prof. Dr.-Ing.
Institute of Mechanics, University of the Federal Armed Forces Hamburg
Holstenhofweg 85, D-22043 Hamburg
Germany
e-mail: rolf.lammering@unibw-hamburg.de

Lancioni, Giovanni, PhD Student
Dipartimento di Ingegneria Civile, Università di Roma "Tor Vergata"
Viale del Politecnico 1,00133, Roma
Italy
e-mail: giovannilancioni@tin.it

Lellep, Jaan, Prof.
Institute of Applied Mathematics, Tartu University
2 Liivi str., Tartu, EE2400
Estonia
e-mail: lellep@math.ut.ee

Librescu, Liviu, Prof. Dr.
Virginia Polytechnic Institute and State University
Department of Engineering Science and Mechanics
Blacksburg, VA, 24061-0219
USA
e-mail: librescu@vt.edu

Lubowiecka, Izabela, Dr.
Department of Civil Engineering, Technical University of Gdańsk
ul. Narutowicza 11/12
80-952 Gdansk
Poland
e-mail: lubow@pg.gda.pl

Matheas, Jan, Dipl.-Ing
Institut für Baumechanik und Bauinformatik, Technische Universität Dresden
D-01062 Dresden
Germany
e-mail: Jan.Matheas@mailbox.tu-dresden.de

Mikhailovskii, Evgeny I., Prof. Dr. Sci.
Matematicheskii Fakultet, Syktyvkarskii Universitet
Oktjabrskii pr. 55, 167001 Syktyvkar, Komi
Russia

Meenen, Johannes, Dr.-Ing.
BASF Aktiengesellschaft, WLF/FF Materials Engineering
D- 67056 Ludwigshafen
Germany
e-mail: johannes.meenen@t-online.de

Mesecke-Rischmann, Simone, Dipl.-Ing.
Institute of Mechanics, University of the Federal Armed Forces Hamburg
Holstenhofweg 85, D-22043 Hamburg
Germany
e-mail: simone.mesecke@UniBw-Hamburg.de

Mestrovic, Mladen, M.Sc., Civ.Eng., Math.Eng.
Faculty of Civil Engineering, University of Zagreb
Kaciceva 26, 10000 Zagreb
Croatia
e-mail: mestar@master.grad.hr

Mürk, Annely, PhD Student
Institute of Applied Mathematics, Tartu University
2 Liivi str., Tartu, EE2400
Estonia
e-mail: annely@math.ut.ee

Naumenko, Konstantin, Dr.-Ing.
Technische Mechanik, Fachbereich Ingenieurwissenschaften
Martin-Luther-Universität Halle-Wittenberg
D-06099 Halle (Saale)
Germany
e-mail: Konstantin.naumenko@iw.uni-halle.de

Parfenova, Ya. A., Dr.
Department of Mathematical Theory of Elasticity and Biomechanics,
Saratov State University
Astrakhanskaya Str. 83, Saratov, 410012
Russia

Pichugin, Aleksey V., Dr.
Department of Mathematics, University of Manchester
Manchester, M13 9PL
United Kingdom
e-mail: pichugin@maths.man.ac.uk

Pietraszkiewicz, Wojciech, Prof. Dr. zw. hab. Inż.
Institute of Fluid-Flow Machinery of the Polish Academy of Science
Ul.Fiszera 14, 80-952 Gdańsk
Poland
e-mail: pietrasz@imp.gda.pl

Rogacheva, Nellya N., Prof.
Institute for Economy Communal and Engineering
Srednaya Kalitnikovskaya 30, 109807 Moscow
Russia
e-mail: rogachev@mech.titech.ac.jp

Rogerson, Graham A., Prof.
Department of Computer and Mathematical Sciences, University of Salford
Salford M5 4WT
United Kingdom
e-mail: G.A.Rogerson@salford.ac.uk

Sargsyan, S. H., Prof.
Head of "Mathematical Analysis and Differential Equations" Chair
Gyumri State Pedagogical Institute
Gyumri, 377501, Armenia
e-mail: evrica@shirak.am

Schlebusch, Rainer, Dipl.-Ing
Institut für Baumechanik und Bauinformatik, Technische Universität Dresden
D-01062 Dresden
Germany
e-mail: Rainer.Schlebusch@mailbox.tu-dresden.de

Selezov, Igor T., Prof.
Department of Wave Processes, Institute of Hydromechanics
National Academy of Science
8/4 Sheliabov Str., Kiev 03680
Ukraine
e-mail: selezov@uninet.kiev.ua

Sheremet, I.G.
324, Bld.95, 9th Novemberstr.
Novosibirsk, 630009
Russia

Skrzypek, Jacek, Prof.
Institute of Mechanics and Machine Design, Cracow University of Technology
al. Jana Pawla II 37, Krakow
Poland
e-mail: skrzyp@cut1.mech.pk.edu.pl

Ulbricht, Volker, Prof.
Institut für Festkörpermechanik, Technische Universität Dresden
Mommsenstr. 13, D-01062 Dresden
Germany
e-mail: ulbricht@mfk1.mw.tu-dresden.de

Wierzbicki, Ewaryst, Prof.
Institute of Mathematics and Informatics
Częstochowa University of Technology
ul. Dabrowskiego 73, 42-200 Częstochowa
Poland
e-mail: wierzbicki@matinf.pcz.czest.pl

Woźniak, Czesław, Prof.
Institute of Mathematics and Informatics
Częstochowa University of Technology
ul. Dabrowskiego 73, 42-200 Częstochowa
Poland
e-mail: wozniak@matinf.pcz.czest.pl

Yermolenko, Andrey V., assistant professor
Matematicheskii Fakultet, Syktyvkarskii Universitet
Oktjabrskii pr. 55, 167001 Syktyvkar, Komi
Russia
e-mail: emich@email.ruemich@ssu.komi.com

Zarutsky, V. A., Prof.
S. P. Timoshenko Institute of Mechanics, National Academy of Sciences
Nesterov str. 3, 03057, Kiev
Ukraine
e-mail: string@inmech.kiev.ua

Zastrau, Bernd W., Prof. Dr.-Ing.
Institut für Baumechanik und Bauinformatik, Technische Universität Dresden
D-01062 Dresden
Germany
e-mail: Bernd.W.Zastrau@mailbox.tu-dresden.de

Zhilin, Pavel A., Prof. Dr.
Department of Dynamics of Mechanical Systems, Institute for Problems in Mechanical Engineering of Russian Academy of Sciences
V.O.Bolshoy pr. 61, RUS-195225 St. Petersburg
Russia
e-mail: zhilin@mail.admiral.ru

The Theory of Simple Elastic Shells

H. Altenbach[1] and P. A. Zhilin[2]

[1]Martin-Luther-University Halle-Wittenberg, Department of Engineering Sciences, Halle (Saale), Germany
[2]Department of Dynamics of Mechanical Systems, Institute for Problems in Mechanical Engineering of Russian Academy of Sciences, St. Petersburg, Russia

1 Introduction

At present the shell theory find out new branches of applications. Biological membranes, thin polymeric films and thin structures made from shape memory materials may be pointed out as examples . In addition, the manufacturing technology of shells leads to significant changes of the material properties. As a result the conventional variants of the shell theory, based on the derivation of the basic equations from the 3D-theory of elasticity, cannot be used. The effective elastic moduli of the shell must be found directly for the shell structure. That means that we have to use the direct method for the formulation of the shell theory. The main idea of the direct approach is the introduction of a 2D-continuum with some physical properties. The basic laws of mechanics and thermodynamics are applied directly to this 2D-continuum. The main advantage of the direct approach is the possibility to obtain quite strict equations.

Today many variants of the shell theory exist. Most of them can be characterized by two principal features: a) they are based on two-dimensional equations and b) they operate with forces and moments (higher order moments are ignored). These two facts may be used for the following definition [8]:

> *A simple shell is a 2D-continuum in which the neighboring parts interact by means of forces and moments.*

A simple shell is a model for the description of the mechanical behavior of shell-type structures. The theory of simple shells allows to make a correct plane photo of three-dimensional phenomena. The advantage of such a theory is that it can be applied for shells with a complex internal structure - for multilayered, for stiffened, etc. In addition, such a theory can be used in the analysis of biological membranes, soft shells, etc. In this sense the theory of simple shells allows the formalization of an old engineering problem - the formulation of a shell theory with effective stiffness.

2 General nonlinear theory

2.1 Kinematics of simple shells

The kinematical model of a simple shell is based on the introduction of a *directed material surface*, i.e. the carrying surface each point of which is connected with an orthonormal triad of vectors. In what follows the direct tensor notation will be used. In the reference configuration ($t = 0$) the directed surface is determined by

$$\{\mathbf{r}(q^1, q^2); \mathbf{d}_1(q^1, q^2), \mathbf{d}_2(q^1, q^2), \mathbf{d}_3(q^1, q^2)\} \equiv \{\mathbf{r}(q); \mathbf{d}_k(q)\},$$

where $\mathbf{r}(q) \equiv \mathbf{r}(q^1, q^2)$ is the position vector defining the geometry of the surface. $q^1, q^2 \in \Omega$, $\mathbf{d}_k(q)$ with $k = 1, 2, 3$ denote a triad of orthonormal vectors obeying the condition $\mathbf{d}_k \cdot \mathbf{d}_m = \delta_{km}$. In the actual configuration ($t \neq 0$) we have

$$\{\mathbf{R}(q, t); \mathbf{D}_k(q, t)\}, \qquad \mathbf{D}_k \cdot \mathbf{D}_m = \delta_{km}.$$

Note that $\mathbf{R}(q, 0) = \mathbf{r}(q), \mathbf{D}_k(q, 0) = \mathbf{d}_k(q)$. In addition, a natural basis can be introduced

$$\mathbf{R}_\alpha(q, t) = \frac{\partial \mathbf{R}}{\partial q^\alpha} \equiv \partial_\alpha \mathbf{R}, \quad \mathbf{R}_3 \equiv \mathbf{N} : \quad \mathbf{N} \cdot \mathbf{R}_\alpha = 0, \quad \mathbf{N} \cdot \mathbf{N} = 1.$$

In what follows we use the relation $\mathbf{d}_3 = \mathbf{n}$. For further derivations we introduce the dual basis

$$\mathbf{R}^i : \quad \mathbf{R}^i \cdot \mathbf{R}_k = \delta^i_k, \quad \mathbf{R}^3 \equiv \mathbf{N}.$$

Then we can define the following two-dimensional Hamilton operators

$$\tilde{\boldsymbol{\nabla}} \equiv \mathbf{R}^\alpha(q, t) \frac{\partial}{\partial q^\alpha}, \quad \boldsymbol{\nabla} = \mathbf{r}^\alpha(q) \frac{\partial}{\partial q^\alpha}.$$

Let us introduce the first ($\mathbf{A}$) and second ($\mathbf{B}$) fundamental tensors of the surface

$$\mathbf{A} = \tilde{\boldsymbol{\nabla}} \mathbf{R} = \mathbf{R}^\alpha(q, t) \otimes \mathbf{R}_\alpha(q, t) = \mathbf{R}_\alpha \otimes \mathbf{R}^\alpha = A^{\alpha\beta} \mathbf{R}_\alpha \otimes \mathbf{R}_\beta = A_{\alpha\beta} \mathbf{R}^\alpha \otimes \mathbf{R}^\beta,$$
$$\mathbf{1} = \mathbf{A} + \mathbf{N} \otimes \mathbf{N}, \quad \mathbf{B} = -\tilde{\boldsymbol{\nabla}} \mathbf{N} = -\mathbf{R}^\alpha \otimes \partial_\alpha \mathbf{N} = B^{\alpha\beta} \mathbf{R}_\alpha \otimes \mathbf{R}_\beta = B_{\alpha\beta} \mathbf{R}^\alpha \otimes \mathbf{R}^\beta,$$

where $\mathbf{1}$ is the second rank unit tensor. In addition, we accept $\mathbf{a} = \mathbf{A}(q, 0) = \boldsymbol{\nabla} \mathbf{r}$, $\mathbf{b} = \mathbf{B}(q, 0) = -\boldsymbol{\nabla} \mathbf{n}$ and $\mathbf{n} = \mathbf{N}(q, 0)$. Note that all tensors here and below are introduced as quantities of the 3D-space defined on the surface.

The motion of the directed surface can be defined by

$$\mathbf{R}(q, t), \quad \mathbf{P}(q, t) \equiv \mathbf{D}^k(q, t) \otimes \mathbf{d}_k(q), \quad \mathbf{D}^k = \delta^{km} \mathbf{D}_m.$$

Here $\mathbf{P}(q, t)$ is an orthogonal turn-tensor, $\operatorname{Det} \mathbf{P} = +1$, $\mathbf{P}(q, 0) = \mathbf{1}$. Let us introduce the linear and the angular velocities $\mathbf{v}$, $\boldsymbol{\omega}$ of the points of the surface which behave like rigid bodies

$$\mathbf{v}(q, t) = \dot{\mathbf{R}}(q, t), \quad \dot{\mathbf{P}}(q, t) = \boldsymbol{\omega}(q, t) \times \mathbf{P}(q, t), \quad \mathbf{P}(q, 0) = \mathbf{1}, \quad \dot{f} \equiv \frac{df}{dt}.$$

For further derivations we need a vector $\mathbf{\Phi}_\alpha$ characterizing the change of $\mathbf{P}(q,t)$ along the surface

$$\partial_\alpha \mathbf{P} = \mathbf{\Phi}_\alpha(q,t) \times \mathbf{P}(q,t), \quad \Rightarrow \quad \mathbf{\Phi}_\alpha = -\frac{1}{2}[\partial_\alpha \mathbf{P} \cdot \mathbf{P}^T]_\times.$$

It is easy to show that

$$\dot{\mathbf{\Phi}}_\alpha = \partial_\alpha \boldsymbol{\omega} + \boldsymbol{\omega} \times \mathbf{\Phi}_\alpha, \quad \partial_\alpha \mathbf{\Phi}_\beta - \partial_\beta \mathbf{\Phi}_\alpha - \mathbf{\Phi}_\alpha \times \mathbf{\Phi}_\beta = \mathbf{0}.$$

Note that $\mathbf{\Phi}_\alpha = \mathbf{0}$ only for a rigid body motion.

2.2 Equations of motion

The local form of equations of motion can be written as

$$\begin{aligned} \widetilde{\boldsymbol{\nabla}} \cdot \mathbf{T} + \rho \mathbf{F}_* &= \rho(\mathbf{v} + \mathbf{\Theta}_1^T \cdot \boldsymbol{\omega})^{\cdot}, \\ \widetilde{\boldsymbol{\nabla}} \cdot \mathbf{M} + \mathbf{T}_\times + \rho \mathbf{L} &= \rho(\mathbf{\Theta}_1 \cdot \mathbf{v} + \mathbf{\Theta}_2 \cdot \boldsymbol{\omega})^{\cdot} + \rho \mathbf{v} \times \mathbf{\Theta}_1^T \cdot \boldsymbol{\omega}, \end{aligned} \qquad (1)$$

where $\mathbf{T} = \mathbf{R}_\alpha \otimes \mathbf{T}^\alpha$ denotes the force tensor, $\mathbf{M} = \mathbf{R}_\alpha \otimes \mathbf{M}^\alpha$ – the moment tensor and $\mathbf{T}_\times \equiv \mathbf{R}_\alpha \times \mathbf{T}^\alpha$. The vectors $\mathbf{F}_*$ and $\mathbf{L}$ are the mass density of the external forces and moments, respectively.

Let $\mathbf{T}_{(\nu)}$ and $\mathbf{M}_{(\nu)}$ be the vectors of external force and moment acting on the boundary curve with the external unit normal $\boldsymbol{\nu}$. Then the Cauchy's formulae are

$$\mathbf{T}_{(\nu)} = \boldsymbol{\nu} \cdot \mathbf{T}, \quad \mathbf{M}_{(\nu)} = \boldsymbol{\nu} \cdot \mathbf{M}.$$

Introducing the Piola-Kirchhoff tensors

$$\mathbf{T}_\Pi = \sqrt{\frac{A}{a}} (\widetilde{\boldsymbol{\nabla}} \mathbf{r})^T \cdot \mathbf{T}, \quad \mathbf{M}_\Pi = \sqrt{\frac{A}{a}} (\widetilde{\boldsymbol{\nabla}} \mathbf{r})^T \cdot \mathbf{M},$$

(1) can be rewritten in the reference configuration

$$\begin{aligned} \boldsymbol{\nabla} \cdot \mathbf{T}_\Pi + \rho_0 \mathbf{F} &= \rho_0 (\mathbf{v} + \mathbf{\Theta}_1^T \cdot \boldsymbol{\omega})^{\cdot}, \\ \boldsymbol{\nabla} \cdot \mathbf{M}_\Pi + (\boldsymbol{\nabla} \mathbf{R}^T \cdot \mathbf{T}_\Pi)_\times + \rho_0 \mathbf{L} &= \rho_0 (\mathbf{\Theta}_1 \cdot \mathbf{v} + \mathbf{\Theta}_2 \cdot \boldsymbol{\omega})^{\cdot} + \rho_0 \mathbf{v} \times \mathbf{\Theta}_1^T \cdot \boldsymbol{\omega}. \end{aligned}$$

The last one equations are convenient for shell stability problems [1].

2.3 Energy balance equation

Let us formulate the balance of energy for the two-dimensional continuum

$$\frac{d}{dt} \int\limits_{(\Delta\Sigma)} \rho(\mathcal{K} + \mathcal{U}) d\Sigma = \int\limits_{(\Delta\Sigma)} \rho(\mathbf{F} \cdot \mathbf{v} + \mathbf{L} \cdot \boldsymbol{\omega}) d\Sigma + \int\limits_C (\mathbf{T}_{(\nu)} \cdot \mathbf{v} + \mathbf{M}_{(\nu)} \cdot \boldsymbol{\omega}) dC,$$

where $\mathcal{U}$ is the mass density of the internal energy. For isothermal processes $\mathcal{U}$ is called the deformation energy. The local form can be expressed as

$$\rho \dot{\mathcal{U}} = \mathbf{T}^{\mathrm{T}} \cdot\cdot \widetilde{\boldsymbol{\nabla}} \mathbf{v} - (\mathbf{R}_{\alpha} \times \mathbf{T}^{\alpha}) \cdot \boldsymbol{\omega} - \mathbf{M}^{\mathrm{T}} \cdot\cdot \widetilde{\boldsymbol{\nabla}} \boldsymbol{\omega}.$$

Introducing the energetic tensors

$$\mathbf{T}_e = (\widetilde{\boldsymbol{\nabla}} \mathbf{r})^{\mathrm{T}} \cdot \mathbf{T} \cdot \mathbf{P}, \quad \mathbf{M}_e = (\widetilde{\boldsymbol{\nabla}} \mathbf{r})^{\mathrm{T}} \cdot \mathbf{M} \cdot \mathbf{P},$$

we obtain

$$\rho \dot{\mathcal{U}} = \mathbf{T}_e^{\mathrm{T}} \cdot\cdot \dot{\mathbf{E}} + \mathbf{M}_e^{\mathrm{T}} \cdot\cdot \dot{\mathbf{F}}, \tag{2}$$

where $\mathbf{E}, \mathbf{F}$ denote the first and the second deformation tensors

$$\mathbf{E} = \boldsymbol{\nabla} \mathbf{R} \cdot \mathbf{P} - \mathbf{a}, \quad \mathbf{F} = (\boldsymbol{\Phi}_{\alpha} \cdot \mathbf{D}_k) \mathbf{r}^{\alpha} \otimes \mathbf{d}^k. \tag{3}$$

For elastic simple shells from the energy balance equation (3) follow the Cauchy-Green relations (with $A = \operatorname{Det} \mathbf{A}$ and $a = \operatorname{Det} \mathbf{a}$)

$$\mathbf{T}_e = \sqrt{\frac{A}{a}} \frac{\partial \rho_0 \mathcal{U}}{\partial \mathbf{E}}, \quad \mathbf{M}_e = \sqrt{\frac{A}{a}} \frac{\partial \rho_0 \mathcal{U}}{\partial \mathbf{F}} \tag{4}$$

2.4 Definition of the tensors of inertia. Restrictions on the tensors of forces and moments

The tensors of inertia $\rho \boldsymbol{\Theta}_{\alpha}$ define the distribution of the mass in the material body-points in the actual configuration. The following equation describes the relation between the tensors of inertia in the actual and the initial configurations

$$\rho \boldsymbol{\Theta}_{\alpha}(q, t) = \mathbf{P}(q, t) \cdot \rho_0 \boldsymbol{\Theta}_{\alpha}^0(q) \cdot \mathbf{P}^{\mathrm{T}}(q, t).$$

The following representations may be proved for the density, the first and the second tensor of inertia of the directed surface

$$\rho_0 = \langle \tilde{\rho}_0 \rangle, \quad \rho_0 \boldsymbol{\Theta}_1^0 = -\langle \tilde{\rho}_0 z \rangle \mathbf{c}, \quad \rho_0 \boldsymbol{\Theta}_2^0 = -\langle \tilde{\rho}_0 z^2 \rangle \mathbf{a}, \quad \langle \mathbf{f} \rangle = \int\limits_{-h_1}^{h_2} \mathbf{f} \mu \, d\Sigma$$

with $\mathbf{c} = -\mathbf{a} \times \mathbf{n}$, $\mu = 1 - 2zH + z^2 G$, H and G are the mean and Gaussian curvatures, $\tilde{\rho}_0(q, z)$ denotes the 3D-density of mass.

It can be shown that the following relations are valid

$$\rho \boldsymbol{\Theta}_2 = \frac{\rho}{\rho_0} \langle \tilde{\rho}_0 z^2 \rangle (\mathbf{1} - \mathbf{D}_3 \mathbf{D}_3), \quad \rho \boldsymbol{\Theta}_1^{\mathrm{T}} = \frac{\rho}{\rho_0} \langle \tilde{\rho}_0 z \rangle (\mathbf{1} \times \mathbf{D}_3). \tag{5}$$

Besides, for simple shells of constant thickness and made from non-polar materials the following restrictions are valid

$$\mathbf{L} \cdot \mathbf{D}_3 = \mathbf{0}, \quad \mathbf{M} \cdot \mathbf{D}_3 = \mathbf{0}, \quad \mathbf{M}_e^{\mathrm{T}} \cdot\cdot [(\mathbf{F} - \mathbf{b} \cdot \mathbf{c}) \cdot \mathbf{c}] + \mathbf{T}_e^{\mathrm{T}} \cdot\cdot [(\mathbf{E} + \mathbf{a}) \cdot \mathbf{c}] = \mathbf{0}. \tag{6}$$

2.5 Reduced deformation tensors

The specific elastic deformation energy $\mathcal{U} = \mathcal{U}(\mathbf{E}, \mathbf{F})$ contains 12 scalar arguments. The number of arguments can be reduced because we have to satisfy the restrictions (6). Making use of (4) and (6) one may obtain the following system of equations for the specific energy $\mathcal{U}(\mathbf{E}, \mathbf{F})$

$$\left(\frac{\partial \mathcal{U}}{\partial \mathbf{E}}\right)^{\mathsf{T}} \cdot\cdot [(\mathbf{E} + \mathbf{a}) \cdot \mathbf{c}] + \left(\frac{\partial \mathcal{U}}{\partial \mathbf{F}}\right)^{\mathsf{T}} \cdot\cdot [(\mathbf{F} - \mathbf{b} \cdot \mathbf{c}) \cdot \mathbf{c}] = 0, \qquad \frac{\partial \rho_0 \mathcal{U}}{\partial (\mathbf{F} \cdot \mathbf{n})} = \mathbf{0}. \quad (7)$$

For the first equation of (7) we have the characteristic system

$$\frac{\mathrm{d}}{\mathrm{d}s}\mathbf{E} = (\mathbf{E} + \mathbf{a}) \cdot \mathbf{c}, \quad \frac{\mathrm{d}}{\mathrm{d}s}\mathbf{F} = (\mathbf{F} - \mathbf{b} \cdot \mathbf{c}) \cdot \mathbf{c}. \quad (8)$$

This is a 12th order system, which has only 11 independent integrals. One can select the following 11 integrals

$$\begin{aligned} \mathcal{E} &= \frac{1}{2}\left[(\mathbf{E} + \mathbf{a}) \cdot \mathbf{a} \cdot (\mathbf{E} + \mathbf{a})^{\mathsf{T}} - \mathbf{a}\right], \\ \mathbf{\Phi} &= (\mathbf{F} - \mathbf{b} \cdot \mathbf{c}) \cdot \mathbf{a} \cdot (\mathbf{E} + \mathbf{a})^{\mathsf{T}} + \mathbf{b} \cdot \mathbf{c} \cdot \mathcal{E} + \mathbf{b} \cdot \mathbf{c}, \\ \boldsymbol{\gamma} &= \mathbf{E} \cdot \mathbf{n}, \quad \boldsymbol{\gamma}_* = \mathbf{F} \cdot \mathbf{n}. \end{aligned} \quad (9)$$

Of course, it is possible to introduce another set of integrals instead of those in (9), but all of them may be expressed in terms of integrals given in (9). The arbitrary function $\mathcal{U}(\mathcal{E}, \mathbf{\Phi}, \boldsymbol{\gamma}, \boldsymbol{\gamma}_*)$ of the integrals (9) satisfies the first equation of the system (7). However the second equation in (7) shows that the specific energy does not depend on the vector $\boldsymbol{\gamma}_*$. Thus we finally have $\mathcal{U} = \mathcal{U}(\mathcal{E}, \mathbf{\Phi}, \boldsymbol{\gamma})$. Tensors $\mathcal{E}, \mathbf{\Phi}, \boldsymbol{\gamma}$ are the reduced deformation tensors. Here $\mathcal{E}$ is the tensor of plane tensile and shear strains, $\mathbf{\Phi}$ - the tensor of bending and torsional strains and $\boldsymbol{\gamma}$ the vector of transverse shear strains.

Up to here all results are the exact ones. They are valid for shells made from arbitrary materials. Note that all physical properties of a shell are contained in the specific deformation energy. The above described theory of shells is a ***Reissner-type theory***.

2.6 Special cases

Let us briefly discuss some variants of the shell theory.

Love-type theory. This is the mostly used case in applied mechanics. In this case the deformation of the transverse shear is supposed to be zero and $\mathbf{D}_3 = \mathbf{P} \cdot \mathbf{n}$

$$\boldsymbol{\gamma} \equiv (\mathbf{E} + \mathbf{a}) \cdot \mathbf{n} = \boldsymbol{\nabla}\mathbf{R} \cdot \mathbf{D}_3 = \mathbf{r}^{\alpha}(\mathbf{R}_{\alpha} \cdot \mathbf{D}_3) = \mathbf{0} \quad \Rightarrow \quad \mathbf{D}_3 = \mathbf{N}.$$

The inertia tensors (5) and the deformation tensor $\mathbf{\Phi}$ must be replaced by

$$\begin{aligned} \rho\mathbf{\Theta}_2 = \frac{\rho}{\rho_0}\langle\tilde{\rho}_0 z^2\rangle\mathbf{A}, \quad \rho\mathbf{\Theta}_1^{\mathsf{T}} = -\frac{\rho}{\rho_0}\langle\tilde{\rho}_0 z\rangle\mathbf{C}, \quad \mathbf{C} = -\mathbf{A} \times \mathbf{N}, \\ \mathbf{\Phi} = -\boldsymbol{\nabla}\mathbf{R} \cdot \mathbf{B} \cdot \mathbf{C} \cdot \boldsymbol{\nabla}\mathbf{R}^{\mathsf{T}} + \mathbf{b} \cdot \mathbf{c} \cdot (\mathbf{1} + \mathcal{E}). \end{aligned} \quad (10)$$

Moment-free (membrane) shell theory. This case follows from the Love-type theory when the specific deformation energy depends on tensor $\mathcal{E}$ only. We have

$$\mathbf{T}_e = \sqrt{\frac{A}{a}}\frac{\partial \rho_0 \mathcal{U}}{\partial \mathcal{E}}, \quad \mathbf{M}_e = \mathbf{0}, \quad \mathbf{\Theta}_2 = \mathbf{\Theta}_1 = \mathbf{0} \quad \Rightarrow \quad \mathbf{T} = \mathbf{T}^{\mathrm{T}}, \ \mathbf{T} \cdot \mathbf{N} = \mathbf{0}.$$

Soft shells. Soft shells are made from a material like textile. In addition to the previous case, we have to accept

$$\mathbf{a} \cdot \frac{\partial \mathcal{U}}{\partial \mathcal{E}} \cdot \mathbf{a} \geq 0 \quad \forall \mathbf{a} : |\mathbf{a}| \neq 0, \ \mathbf{a} \cdot \mathbf{n} = 0.$$

Besides, some additional restrictions must be introduced. As far as we know a general theory of soft shells is not presented in the literature.

2.7 Deformation energy of simple shells

For a shell made from an elastic material the deformations can be assumed to be relatively small while the displacements and rotations can be relatively large. In such a case the following quadratic approximation can be introduced

$$\begin{aligned} \rho_0 \mathcal{U} = &\mathbf{T}_0 \cdot\cdot \mathbf{E} + \mathbf{M}_0^{\mathrm{T}} \cdot\cdot \mathbf{\Phi} + \mathbf{N}_0 \cdot \boldsymbol{\gamma} + \frac{1}{2}\mathcal{E} \cdot\cdot {}^{(4)}\mathbf{C}_1 \cdot\cdot \mathcal{E} + \mathcal{E} \cdot\cdot {}^{(4)}\mathbf{C}_2 \cdot\cdot \mathbf{\Phi} \\ &+ \frac{1}{2}\mathbf{\Phi} \cdot\cdot {}^{(4)}\mathbf{C}_3 \cdot\cdot \mathbf{\Phi} + \frac{1}{2}\boldsymbol{\gamma} \cdot \mathbf{\Gamma} \cdot \boldsymbol{\gamma} + \boldsymbol{\gamma} \cdot ({}^{(3)}\mathbf{\Gamma}_1 \cdot\cdot \mathcal{E} + {}^{(3)}\mathbf{\Gamma}_2 \cdot\cdot \mathbf{\Phi}). \end{aligned} \tag{11}$$

Here $\mathbf{T}_0, \mathbf{M}_0, \mathbf{N}_0, {}^{(4)}\mathbf{C}_1, {}^{(4)}\mathbf{C}_2, {}^{(4)}\mathbf{C}_3, {}^{(3)}\mathbf{\Gamma}_1, {}^{(3)}\mathbf{\Gamma}_2, \mathbf{\Gamma}$ denote stiffness tensors of different rank. They express the effective elastic properties of the simple shell. The differences between various classes of simple shells are expressed in the stiffness tensors, the tensors of inertia and the two-dimensional density. The stiffness tensors in (11) do not depend on the deformations. Thus they may be found from tests based on the linear shell theory.

2.8 Linearized basic equations

In the linear theory there are no differences between the actual and the reference configurations. This is equivalent to the condition that the energetic and the true tensors of forces and moments are the same. The displacements and rotations are supposed to be small. In such a case instead of (9) we have

$$\mathcal{E} \simeq \boldsymbol{\varepsilon} \equiv \frac{1}{2}(\mathbf{e} \cdot \mathbf{a} + \mathbf{a} \cdot \mathbf{e}^{\mathrm{T}}), \quad \mathbf{\Phi} \simeq \mathbf{k} \equiv \boldsymbol{\kappa} \cdot \mathbf{a} + \frac{1}{2}(\mathbf{e} \cdot\cdot \mathbf{c})\mathbf{b}, \quad \boldsymbol{\gamma} = \mathbf{e} \cdot \mathbf{n} = \nabla \mathbf{u} \cdot \mathbf{n} + \mathbf{c} \cdot \boldsymbol{\varphi} \tag{12}$$

with

$$\mathbf{e} = \nabla \mathbf{u} + \mathbf{a} \times \boldsymbol{\varphi}, \quad \boldsymbol{\kappa} = \nabla \boldsymbol{\varphi}. \tag{13}$$

The linearized equations of motion (1) take the form

$$\nabla \cdot \mathbf{T} + \rho \mathbf{F}_* = \rho(\ddot{\mathbf{u}} + \mathbf{\Theta}_1^{\mathrm{T}} \cdot \ddot{\boldsymbol{\varphi}}), \quad \nabla \cdot \mathbf{M} + \mathbf{T}_{\times} + \rho \mathbf{L} = \rho(\mathbf{\Theta}_1 \cdot \ddot{\mathbf{u}} + \mathbf{\Theta}_2 \cdot \ddot{\boldsymbol{\varphi}}). \tag{14}$$

The Cauchy–Green relations can be rewritten as

$$\mathbf{T}\cdot\mathbf{a}+\frac{1}{2}(\mathbf{M}\cdot\cdot\,\mathbf{b})\mathbf{c}=\frac{\partial\rho\mathfrak{U}}{\partial\varepsilon},\quad \mathbf{T}\cdot\mathbf{n}=\frac{\partial\rho\mathfrak{U}}{\partial\gamma},\quad \mathbf{M}=\frac{\partial\rho\mathfrak{U}}{\partial\mathbf{k}}.$$

Then from (11) we get

$$\begin{aligned}\mathbf{T}\cdot\mathbf{a}+\frac{1}{2}(\mathbf{M}\cdot\cdot\,\mathbf{b})\mathbf{c}&=\mathbf{T}_0+{}^{(4)}\mathbf{C}_1\cdot\cdot\,\varepsilon+{}^{(4)}\mathbf{C}_2\cdot\cdot\,\kappa+\gamma\cdot{}^{(3)}\boldsymbol{\Gamma}_1,\\ \mathbf{T}\cdot\mathbf{n}&=\mathbf{N}_0+\boldsymbol{\Gamma}\cdot\gamma+{}^{(3)}\boldsymbol{\Gamma}_1\cdot\cdot\,\varepsilon+{}^{(3)}\boldsymbol{\Gamma}_2\cdot\cdot\,\kappa,\\ \mathbf{M}^{\mathrm{T}}&=\mathbf{M}_0^{\mathrm{T}}+\varepsilon\cdot\cdot\,{}^{(4)}\mathbf{C}_2+{}^{(4)}\mathbf{C}_3\cdot\cdot\,\kappa+\gamma\cdot{}^{(3)}\boldsymbol{\Gamma}_2.\end{aligned}\tag{15}$$

3 Determination of the elastic stiffness tensors

Now we have to discuss the most complicated part of the direct approach - the construction of the stiffness tensors. The solution of the problem was given by [9]. It is clear that we have to use the symmetry properties. Therefore we have to solve two problems. First one: the classical theory of symmetry cannot be used because it is valid for Euclidean tensors only. In the shell theory we operate with non-Euclidean tensors. Second one: the stiffness tensors depend on the symmetry of the material of the shell, symmetry of the surface shape at the point under consideration and symmetry of the intrinsic structure of the shell.

3.1 General remarks and restrictions

It is necessary to specify the quantities: $\mathbf{T}_0$, $\mathbf{M}_0$, $\mathbf{N}_0$, ${}^{(4)}\mathbf{C}_1$, ${}^{(4)}\mathbf{C}_2$, ${}^{(4)}\mathbf{C}_3$, $\boldsymbol{\Gamma}$, ${}^{3}\boldsymbol{\Gamma}_1$, ${}^{3}\boldsymbol{\Gamma}_2$. The following constraints are obvious: $\mathbf{d}\cdot\cdot\,{}^{(4)}\mathbf{C}_1={}^{(4)}\mathbf{C}_1\cdot\cdot\,\mathbf{d}$, $\mathbf{d}\cdot\cdot\,{}^{(4)}\mathbf{C}_3={}^{(4)}\mathbf{C}_3\cdot\cdot\,\mathbf{d}$, $\mathbf{c}\cdot\cdot\,\boldsymbol{\Gamma}=0$, $\mathbf{c}\cdot\cdot\,{}^{(4)}\mathbf{C}_1=\mathbf{0}$, $\mathbf{c}\cdot\cdot\,{}^{(4)}\mathbf{C}_2=\mathbf{0}$, ${}^{(3)}\boldsymbol{\Gamma}\cdot\cdot\,\mathbf{c}=\mathbf{0}$, $\mathbf{T}_0\cdot\cdot\,\mathbf{c}=0$, where $\mathbf{d}$ is an arbitrary tensor and $\mathbf{c}$ denotes an antisymmetric tensor (both of second rank). In the Euclidean space $\mathbb{R}^3$ we have only polar vectors. In the oriented Euclidean space $\mathbb{R}_0^3$ - polar and axial vectors. For the shell theory it is convenient to use the representation $\mathbb{R}_{0,\mathfrak{n}}^3=\mathbb{R}_0^2\oplus\mathbb{R}_{\mathfrak{n}}^1$. Here we have three orientations, but only two of them are independent. We will use the orientation of the 3D-space and the orientation of the line spanned on the normal $\mathbf{n}$. In the space $\mathbb{R}_{0,\mathfrak{n}}^3$ four types of tensors can be introduced:

1. polar tensors, which are independent from the orientation in $\mathbb{R}^3$ and in the subspaces,
2. axial tensors, which change the sign if the orientation in $\mathbb{R}^3$ is changing, but not if the orientation changes in $\mathbb{R}_{\mathfrak{n}}^1$,
3. $\mathfrak{n}$-oriented tensors, which change the sign if the orientation in $\mathbb{R}_{\mathfrak{n}}^1$ changes, but independent of the orientation in $\mathbb{R}^3$, and
4. axial $\mathfrak{n}$-oriented tensors, which change the sign if the orientation in $\mathbb{R}^3$ is changing, and if the orientation changes in $\mathbb{R}_{\mathfrak{n}}^1$.

In the shell theory polar tensors $\rho, \mathcal{U}, \mathcal{W}, \mathbf{u}, \dot{\mathbf{u}}, \mathbf{E}, \mathbf{T}_0, \mathbf{T}, \mathbf{a}, {}^{(4)}\mathbf{C}_1, {}^{(4)}\mathbf{C}_3, \boldsymbol{\Gamma}, \rho\boldsymbol{\Theta}_2$, axial tensors $\rho\boldsymbol{\Theta}_1, \boldsymbol{\varphi}, \boldsymbol{\omega}, \mathbf{F}, \boldsymbol{\Phi}, \mathbf{b}\cdot\mathbf{c}, \mathbf{M}_0, {}^{(4)}\mathbf{C}_2$, the tensors $\mathbf{b}, \mathbf{B}, \boldsymbol{\gamma}, \mathbf{Q} = \mathbf{T}\cdot\mathbf{n}, {}^{(3)}\boldsymbol{\Gamma}_1, \mathbf{Q}_0$ are n-oriented objects, and axial n-oriented tensors $\mathbf{c} = -\mathbf{a} \times \mathbf{n}, {}^{(3)}\boldsymbol{\Gamma}_2$ are introduced.

Let $\mathbf{Q}$ be an orthogonal tensor. Considering the orthogonal transformation ${}^{(p)}\mathbf{S} = S^{i_1 i_2 \ldots i_p}\, \mathbf{g}_{i_1} \otimes \mathbf{g}_{i_2} \otimes \ldots \otimes \mathbf{g}_{i_p}$ with the basis $\mathbf{g}_i$ in $\mathbb{R}^3$. We shall use the notation

$$\otimes_1^p \mathbf{Q} \cdot {}^{(p)}\mathbf{S} \equiv S^{i_1 i_2 \ldots i_p}\, \mathbf{Q}\cdot\mathbf{g}_{i_1} \otimes \mathbf{Q}\cdot\mathbf{g}_{i_2} \otimes \ldots \otimes \mathbf{Q}\cdot\mathbf{g}_{i_p},$$
$${}^{(p)}\mathbf{S}' \equiv (\mathbf{n}\cdot\mathbf{Q}\cdot\mathbf{n})^{\beta} (\mathrm{Det}\mathbf{Q})^{\alpha} \otimes_1^p \mathbf{Q} \cdot {}^{(p)}\mathbf{S},$$

where $\alpha = \beta = 0$, if ${}^{(p)}\mathbf{S}$ is polar; $\alpha = 1, \beta = 0$, if ${}^{(p)}\mathbf{S}$ is axial; $\alpha = 0, \beta = 1$, if ${}^{(p)}\mathbf{S}$ is n-oriented; $\alpha = \beta = 1$, if ${}^{(p)}\mathbf{S}$ - is axial n-oriented. Note that $\mathbf{Q}\cdot\mathbf{n} = \pm\mathbf{n}$. The symmetry group (SG) for a tensor ${}^{(p)}\mathbf{S}$ is called a set of orthogonal solutions of the equation ${}^{(p)}\mathbf{S}' = {}^{(p)}\mathbf{S}$, where ${}^{(p)}\mathbf{S}$ is given and $\mathbf{Q}$ must be found.

In what follows we shall use the conventional relation

$$\mathbf{T} = \langle \mu^{-1} \cdot \boldsymbol{\sigma} \rangle, \quad \mathbf{M} = \langle \mu^{-1} \cdot \boldsymbol{\sigma} \cdot \mathbf{c}z \rangle,$$

where $\boldsymbol{\sigma}$ is the stress tensor of the classical theory of elasticity. The relations

$$\rho_0(\dot{\mathbf{u}} + \boldsymbol{\Theta}_1^T \cdot \dot{\boldsymbol{\varphi}}) = \langle \tilde{\rho}_0 \dot{\mathbf{u}}_* \rangle, \quad \rho_0(\boldsymbol{\Theta}_1 \cdot \dot{\mathbf{u}} + \boldsymbol{\Theta}_2^T \cdot \dot{\boldsymbol{\varphi}}) = \langle \tilde{\rho}_0 \dot{\mathbf{u}}_* \cdot \mathbf{c}z \rangle$$

may be obtained in order to find the displacements and rotations in terms of the three-dimensional displacement vector $\mathbf{u}_*$. The external force $\rho\mathbf{F}_*$ and moment $\rho_0\mathbf{L}$ may be found as

$$\rho_0\mathbf{F}_* = \langle \tilde{\rho}_0 \tilde{\mathbf{F}} \rangle + \mu^+ \boldsymbol{\sigma}_n^+ + \mu^- \boldsymbol{\sigma}_n^-, \quad \rho_0\mathbf{L} = \mathbf{n} \times \langle \tilde{\rho}_0 \tilde{\mathbf{F}} z \rangle + (h/2)\mathbf{n} \times (\mu^+ \boldsymbol{\sigma}_n^+ - \mu^- \boldsymbol{\sigma}_n^-),$$

where $\mu^{+(-)} = 1 - (+)hH + (h^2/4)G$, $\boldsymbol{\sigma}_n^{+(-)}$ are the stress vectors on the upper and lower face surfaces of the shell.

3.2 Local symmetry groups of simple shells

The local symmetry group (LSG) is a set of orthogonal solutions of the following system

$$\otimes_1^4 \mathbf{Q}\cdot\mathbf{C}_1 = \mathbf{C}_1, \quad (\mathrm{Det})\mathbf{Q} \otimes_1^4 \mathbf{Q}\cdot\mathbf{C}_2 = \mathbf{C}_2, \quad \otimes_1^4 \mathbf{Q}\cdot\mathbf{C}_3 = \mathbf{C}_3, \quad \mathbf{Q}\cdot\boldsymbol{\Gamma}\cdot\mathbf{Q}^T = \boldsymbol{\Gamma},$$
$$(\mathbf{n}\cdot\mathbf{Q}\cdot\mathbf{n}) \otimes_1^3 \mathbf{Q}\cdot\boldsymbol{\Gamma}_1 = \boldsymbol{\Gamma}_1, \quad (\mathbf{n}\cdot\mathbf{Q}\cdot\mathbf{n})(\mathrm{Det}\mathbf{Q}) \otimes_1^3 \mathbf{Q}\cdot\boldsymbol{\Gamma}_2 = \boldsymbol{\Gamma}_2.$$

If we know the stiffness tensors, then we are able to find the LSG of the shell. However, it is much more important to solve the inverse problem and to find the structure of the stiffness tensors, if we know some elements of the shell symmetry. For this purpose let us introduce the Curie-Neumann's Principle:

SG *of the consequence contains* SG *of the reason.*

The SG of the reason is the intersection of the following symmetry groups: 1. SG of the material at given point of the shell, 2. LSG of the surface, and 3. LSG of the intrinsic structure of the shell. For the surface the LSG is determined as a set of orthogonal solutions of the system

$$\mathbf{Q} \cdot \mathbf{a} \cdot \mathbf{Q}^{\mathsf{T}} = \mathbf{a}, \quad (\mathbf{n} \cdot \mathbf{Q} \cdot \mathbf{n})\mathbf{Q} \cdot \mathbf{b} \cdot \mathbf{Q}^{\mathsf{T}} = \mathbf{b}. \tag{16}$$

It is easy to see that LSG of the surface in the general case contains only three irreducible elements: $\mathbf{1}$, $\mathbf{n} \otimes \mathbf{n} - \mathbf{e}_1 \otimes \mathbf{e}_1 + \mathbf{e}_2 \otimes \mathbf{e}_2$, $\mathbf{n} \otimes \mathbf{n} + \mathbf{e}_1 \otimes \mathbf{e}_1 - \mathbf{e}_2 \otimes \mathbf{e}_2$, where $\mathbf{e}_\alpha$ are the eigenvectors of $\mathbf{b}$. For plates ($\mathbf{b} = \mathbf{0}$) the LSG is much more rich. In the general case the SG of a shell can not be richer than (16). That means that we are not able to simplify the structure of the stiffness tensors without additional assumptions. But up to here we do not use the fact that the shell has a small thickness.

3.3 Dimension analysis

Let the material of a shell be isotropic. In such a case the stiffness tensors depend on the following quantities: $h(q)$ - thickness of the shell, $E(q)$, $\nu(q)$ - isotropic elastic properties, $\mathbf{a}$, $\mathbf{b}$ - first and second metric tensors. Making use of standard analysis of dimensions one may prove the following representations

$$\begin{aligned} \mathbf{C}_1 &= \frac{Eh}{12(1-\nu^2)}\mathbf{C}_1^*(h\mathbf{b} \cdot \mathbf{c}, \nu), \quad \mathbf{C}_2 = \frac{Eh^2}{12(1-\nu^2)}\mathbf{C}_2^*(h\mathbf{b} \cdot \mathbf{c}, \nu), \\ \mathbf{C}_3 &= \frac{Eh^3}{12(1-\nu^2)}\mathbf{C}_3^*(h\mathbf{b} \cdot \mathbf{c}, \nu), \quad \boldsymbol{\Gamma} = Gh\boldsymbol{\Gamma}^*(h\mathbf{b} \cdot \mathbf{c}, \nu) \end{aligned} \tag{17}$$

with $G = E/[2(1+\nu)]$. Here all quantities with a star depend on the dimensionless tensor which is assumed to be small

$$||h\mathbf{b} \cdot \mathbf{c}||^2 = (h\mathbf{b} \cdot \mathbf{c}) \cdot\cdot (h\mathbf{b} \cdot \mathbf{c})^{\mathsf{T}} = \frac{h^2}{R_1^2} + \frac{h^2}{R_2^2} \ll 1.$$

Thus, one can use the following representations

$$\begin{aligned} \mathbf{C}_s &= \frac{Eh^s}{12(1-\nu^2)}\left[\mathbf{C}_s^{(0)} + \mathbf{C}_s^{(1)} \cdot\cdot (h\mathbf{b} \cdot \mathbf{c}) + (h\mathbf{b} \cdot \mathbf{c}) \cdot\cdot \mathbf{C}_s^{(2)} \cdot\cdot (h\mathbf{b} \cdot \mathbf{c}) + 0(h^3)\right], \\ \boldsymbol{\Gamma} &= Gh\left[\boldsymbol{\Gamma}^{(0)} + \boldsymbol{\Gamma}^{(1)} \cdot\cdot (h\mathbf{b} \cdot \mathbf{c}) + 0(h^2)\right] \end{aligned}$$

with $0(h^p) \equiv 0(||h\mathbf{b} \cdot \mathbf{c}||^p)$, $s = 1, 2, 3$. In what follows we ignore the terms of an order $0(h^2)$ with respect to 1. Only in some situations it is necessary to take into account higher order terms, e.g. in order to provide the positive definiteness of the deformation energy. Instead of the tensors $\mathbf{C}_i$ and $\boldsymbol{\Gamma}$ we have to consider the tensors $\mathbf{C}_i^{(p)}$ and $\boldsymbol{\Gamma}^{(p)}$. It seems that the representations for the stiffness tensors does not simplify our discussions. But the tensors $\mathbf{C}_i^{(p)}$ do not depend on the geometrical shape of the surface. In this case the symmetry group of the tensors $\mathbf{C}_i^{(p)}$ does not include the SG of the second metric tensor $\mathbf{b}$.

3.4 Homogeneous thin shells

Let us assume that the shell is made from transversally isotropic material. If $\mathbf{n}$ is the axis of isotropy, and the structure of the shell is of such kind that for $\mathbf{b} \to \mathbf{0}$ we have instead of the shell a plate with a reference (middle) surface which is a surface of symmetry. In this case the tensors $\mathbf{C}_i^{(k)}$, $\boldsymbol{\Gamma}^{(k)}$ (not the tensors $\mathbf{C}_i$, $\boldsymbol{\Gamma}$) contain the following elements of symmetry

$$\mathbf{Q} = \pm\mathbf{n}\otimes\mathbf{n} + \mathbf{q}, \quad \mathbf{q}\cdot\mathbf{q}^{\mathsf{T}} = \mathbf{a}, \quad \mathbf{q}\cdot\mathbf{n} = \mathbf{n}\cdot\mathbf{q} = \mathbf{0}. \tag{18}$$

In such a case it is easy to show that

$$\mathbf{C}_1^{(1)} = \mathbf{0}, \quad \mathbf{C}_2^{(0)} = \mathbf{0}, \quad \mathbf{C}_2^{(2)} = \mathbf{0}, \quad \mathbf{C}_3^{(1)} = \mathbf{0}, \quad \boldsymbol{\Gamma}^{(1)} = \mathbf{0}. \tag{19}$$

We see that the tensors $\mathbf{C}_1, \mathbf{C}_3$ and $\boldsymbol{\Gamma}$ can be found from plate tests with an error of $0(h^2)$. The tensor $\mathbf{C}_2$ may be found only from shell tests. Let the tensor (18) belongs to SG of $\mathbf{C}_i^{(k)}, \boldsymbol{\Gamma}^{(k)}$. That means that these tensors must be transversally isotropic. It is not difficult to find such tensors

$$\begin{aligned}
\mathbf{C}_1 &= \frac{Eh}{1-\nu^2}\left[A_1\mathbf{a}\otimes\mathbf{a} + A_2(\mathbf{r}^\alpha\otimes\mathbf{r}^\beta\otimes\mathbf{r}_\alpha\otimes\mathbf{r}_\beta + \mathbf{r}^\alpha\otimes\mathbf{a}\otimes\mathbf{r}_\alpha - \mathbf{a}\otimes\mathbf{a})\right],\\
\mathbf{C}_2 &= \frac{Eh^2}{12(1-\nu^2)}\left[B_1 hH\mathbf{a}\otimes\mathbf{c} + B_2 hH(\mathbf{r}^\alpha\otimes\mathbf{c}\otimes\mathbf{r}_\alpha + c^{\alpha\beta}\mathbf{r}_\alpha\otimes\mathbf{a}\otimes\mathbf{r}_\beta)\right.\\
&\quad \left. + B_3\mathbf{a}\otimes(h\mathbf{b}\cdot\mathbf{c} - hH\mathbf{c}) + B_4 h(\mathbf{b}\cdot\mathbf{c} - H\mathbf{c})\otimes\mathbf{a} + B_5 h(\mathbf{b} - H\mathbf{a})\otimes\mathbf{c}\right],\\
\mathbf{C}_3 &= \frac{Eh^3}{12(1-\nu^2)}\left[C_1\mathbf{c}\otimes\mathbf{c} + C_2(\mathbf{r}^\alpha\otimes\mathbf{r}^\beta\otimes\mathbf{r}_\alpha\otimes\mathbf{r}_\beta + \mathbf{r}^\alpha\otimes\mathbf{a}\otimes\mathbf{r}_\alpha - \mathbf{a}\otimes\mathbf{a})\right.\\
&\quad \left. + h^2H_1^2C_4\mathbf{a}\otimes\mathbf{a}\right],\\
\boldsymbol{\Gamma} &= Gh\Gamma_0\mathbf{a}
\end{aligned}$$

with $2H_1 = -(1/R_1) + (1/R_2)$. All results are valid for non-polar materials and have an error $0(h^2)$. The moduli $A_1, A_2, C_1, C_2, C_4, \Gamma_0, B_1, \ldots, B_5$ depend only on the Poisson's ratio.

Making use of the solutions of some test problems one can obtain the following elastic moduli

$$\begin{gathered}
A_1 = C_1 = \frac{1+\nu}{2}, \quad A_2 = C_2 = \frac{1-\nu}{2}, \quad \Gamma_0 = \frac{\pi^2}{12}, \quad C_4 = \frac{1-\nu}{24},\\
B_1 = \frac{\nu(1+\nu)}{2(1-\nu)}, \quad B_2 = 0, \quad B_3 = \frac{1+\nu}{2}, \quad B_4 = -\frac{1-\nu}{4}, \quad B_5 = -\frac{1}{2}.
\end{gathered} \tag{20}$$

All moduli in (20), except C_4, were found from the model problems in which they determine the leading terms of asymptotic expansions. The modulus C_4 is needed for the positive definiteness of the deformation energy. Some comments must be made with respect to the coefficient of transverse shear Γ_0. It may be shown that the inequality $\pi^2/12 \leq \Gamma_0 < 1$ must be valid always. If we are not interested in the high frequencies, then it would be better to accept $\Gamma_0 = 5/(6-\nu)$. In such a case the low frequencies can be found more exactly.

Finally, the tensors $\mathbf{T}_0$, $\mathbf{M}_0$, $\mathbf{N}_0$ are defined by the expressions

$$\begin{aligned}
\mathbf{T}_0 &= \frac{\nu h}{2(1-\nu)}\mathbf{a}\left[\mathbf{n}\cdot(\sigma_n^+ - \sigma_n^-)\right], \\
\mathbf{M}_0 &= \frac{\nu h^2}{12(1-\nu)}\mathbf{c}\left[\mathbf{n}\cdot(\sigma_n^+ - \sigma_n^-)\right], \\
\mathbf{N}_0 &= h(1-\Gamma_0)\mathbf{a}\left[\mathbf{n}\cdot(\sigma_n^+ - \sigma_n^-)\right].
\end{aligned} \tag{21}$$

3.5 Simplest shell theory

In the general case the theory given above cannot be simplified. If someone makes it, then an error in leading terms may appear for some problems. However, there are many practical problems where it is possible to use much more simple theory. The most popular theory of such a kind was given in [3] and [6]. Practically the same theory was presented in [5] and [7]. It was shown in [4] that the Koiter-Sanders theory differs from the Balabukh-Novozhilov theory in small terms. Let us note that there exist problems for arbitrary thin shells when the Balabukh-Novozhilov-Koiter-Sanders theory gives the errors in the leading terms. Thus in general this theory cannot be called a first-approximation theory. A shell theory may be called simplest if it is described by means of minimal number of the elastic moduli. However, the deformation energy in the simplest theory must be positive definit for any type of deformations. In such a case we have to ignore the tensors $\mathbf{T}_0$, $\mathbf{M}_0$ and $\mathbf{Q}_0$. Besides, we have to accept $\Gamma_0 \to \infty \Rightarrow \boldsymbol{\gamma} = \mathbf{0}$ and $B_1 = B_2 = B_3 = B_4 = B_5 = C_4 = 0$. From the restriction that the deformation energy must be positive, we obtain $A_1 > 0, A_2 > 0, C_1 > 0, C_2 > 0, \Gamma_0 > 0$. The deformation energy takes the form

$$\begin{aligned}
\rho_0 \mathcal{U} &= \frac{Gh}{2}\left[2\mathcal{E}\cdot\cdot\mathcal{E} + \frac{2\nu}{1-\nu}(\mathrm{tr}\mathcal{E})^2\right] \\
&+ \frac{Gh^3}{24}\left[\frac{1}{2}(\boldsymbol{\Phi}+\boldsymbol{\Phi}^T)\cdot\cdot(\boldsymbol{\Phi}+\boldsymbol{\Phi}^T) - (\mathrm{tr}\,\boldsymbol{\Phi})^2 + \frac{1+\nu}{1-\nu}(\mathbf{c}\cdot\cdot\boldsymbol{\Phi})^2\right],
\end{aligned} \tag{22}$$

where tensors $\mathcal{E}$ and $\boldsymbol{\Phi}$ are defined by expressions (10). The deformation energy (22) may be used for small deformations and large rotations. For small rotations one can use the linear shell theory. The vector of small rotation $\boldsymbol{\varphi}$ may be found in terms of the displacement vector

$$\Gamma_0 \to \infty, \quad |\mathbf{T}\cdot\mathbf{n}| < \infty \quad \Rightarrow \quad \boldsymbol{\gamma} = \mathbf{0}, \quad \Rightarrow \quad \mathbf{a}\cdot\boldsymbol{\varphi} = \mathbf{c}\cdot(\boldsymbol{\nabla} w + \mathbf{b}\cdot\mathbf{u}),$$

where $w = \mathbf{u}\cdot\mathbf{n}$. The Cauchy–Green relations (15) take the form

$$\begin{aligned}
\mathbf{T}\cdot\mathbf{a} + \frac{1}{2}(\mathbf{M}\cdot\cdot\mathbf{b})\mathbf{c} &= \frac{Eh}{1-\nu^2}\left[(1-\nu)\varepsilon + \nu(\mathrm{tr}\,\varepsilon)\mathbf{a}\right], \\
\mathbf{M} &= \frac{Gh^3}{12}\left[\mathbf{k} + \mathbf{k}^T - (\mathrm{tr}\,\mathbf{k})\mathbf{a} - \frac{1+\nu}{1-\nu}(\mathbf{c}\cdot\cdot\mathbf{k})\mathbf{c}\right].
\end{aligned}$$

The vector $\mathbf{T}\cdot\mathbf{n}$ of the transverse force is defined by the equations of motion.

4 Conclusion

It was shown that the direct approach to the shell theory is based on some new ideas of physical nature. As a result, this approach allows to build up a theory which cannot be improved without introduction of new dynamic quantities like moments of higher orders. It is important that the theory does not need any hypothesis and may be applied for all possible cases. Of course, the elastic moduli must be found for these cases by means of special considerations. For example, the elastic transverse shear moduli of three-layered shells may be found from some transcendental equations [2] which do not contain any small parameter and therefore cannot be solved by means of asymptotic methods. The moduli may be found for the case when the friction between layers is present. In such a case the constitutive equations must be modified.

Acknowledgement

This work was partly supported by RFBR grant No. 02-01-00514.

References

1. Altenbach H, Zhilin PA (1982) Eine nichtlineare Theorie dünner Dreischichtschalen und ihre Anwendung auf die Stabilitätsuntersuchung eines dreischichtigen Streifens. Techn Mech 3(2): 23-30
2. Altenbach H, Zhilin PA (1988) A general theory of elastic simple shells (in Russ.). Uspekhi Mekhaniki 11(4): 107-148
3. Balabukh LI (1946) Bending and twisting of conical shells (in Russ.). Trudy TsAGI, 577
4. Chernykh KF (1964) Linear Theory of Shell, Part II. LGU Publishing, Leningrad
5. Koiter WT (1960) A consistent first approximation in the general theory of thin elatic shells, Proc. IUTAM Symp. on the theory of thin elastic shells (Delft, 1959). North-Holland, Amsterdam, pp 12-33
6. Novozilov VV (1946) New method of the shell analysis (in Russ.). Izvestia of AN USSR. OTN (1): 35-48
7. Sanders JL (1959) An improved first-approximation theory for thin shells. NASA Tech. Report R-24, pp 1-11
8. Zhilin PA (1976) Mechanics of deformable directed surfaces. Int. J. Solids Struct. 12: 635-648
9. Zhilin PA (1982) Basic equations of the non-classical shell theory (in Russ.). Trudy Leningradskogo Politekhnicheskogo instituta (386): 29-46

Geometric Boundary Layers in Shells With Mixed Type

B. Audoly

Laboratoire de modélisation en mécanique, Université Pierre et Marie Curie, Paris, France

1 Introduction

In problems involving thin elastic shells, a common approximation is to neglect flexural effects in front of stretching effects. Indeed, the former are associated with terms in the equations of equilibrium that are higher order in the small thickness h. Neglecting flexural effects constitutes to the so-called membrane approximation.

In some situations, however, the membrane approximation is unsuited. Shells of revolution with mixed type is perhaps the simplest situation where membrane approximation breaks down. Shells with mixed type are defined as shells containing both elliptic regions (as the outer part of a torus) and hyperbolic ones (as the inner part of a torus). For such shells, membrane approximation indeed leads to divergences of the stresses and/or displacements near the parallels where the type of the surface changes — in the case of a torus, this is near the topmost and bottommost parallels. This region with inconsistencies in the membrane approximation will be called the transition region after its fluid dynamics equivalent.

These inconsistencies mean that the membrane theory is based on approximations which are not justified in the transition region. In order to address the mechanical behaviour of the shell there, one has to consider the full shell theory. There is a so-called boundary layer in this region where new effects have to be taken into account. Regularizing effects that will be considered in this paper are (a) flexural effects and (b) geometric nonlinearities present in the definition of strain. They correspond to two main approximations used to derive the membrane equations, whose failure explains the apparent inconsistencies. In this paper, we present a general theory of such boundary layers. For sake of definiteness, we consider the case of a toroidal shell with circular section, but the present results can easily be extend to arbitrary shells of revolution with mixed type.

Several partial study of these boundary layers have been undertaken in the past: Rossettos and Sanders [6], for instance, considered the case of a boundary layer transmitting no shear stress, as happens in the particular case of a torus with circular section loaded by a uniform surface pressure. Steele [8] studied the case of nonvanishing shear stresses, but restricted his investigations to purely flexural regularizion. In the present paper, we present a more general analysis that brings together all these partial results while significantly extending them. For instance, we show that the boundary layer obeys an original nonlinear boundary layer equation in the generic case and we discuss the possibility of double boundary layers when flexural effects are very weak.

2 Governing equations

In this section, we derive the mechanical equations for thin shells that are the basis of the boundary layer analysis. Rather than starting from the full – and intricate – shell equations found in the literature and later simplifying them using approximations suited to the boundary layers, it is more instructive to re-derive the equations for shells from the elementary mechanical principles, performing these approximations from the beginning.

Let (s,θ) be a parameterization of the toroidal shell, as shown in Fig. 1. In the rest state and in cylindrical coordinates, (s,θ) is mapped to a point that is located at $(r(s)e_r + q(s)e_z)$, where $\phi = s/R$, ϕ $r(s) = R(a+\sin\phi)$, $q(s) = R\cos\phi$, R is the radius of the torus section, aR is the radius of the generating circle ($a > 1$), and e_r and e_z are the unit vectors in the radial and vertical directions respectively. Let $(u(s), w(s))$ be the radial and vertical of the displacement field (stresses hence displacement are assumed to have cylindrical symmetry) and $\psi(s)$ be the change of orientation of the local tangent to the shell in the section due to deformation. Assuming that the displacement leads to small strains (lengths are almost conserved), one computes: $\psi(s) = -w' \cos(s/R) - u'(s)\sin(s/r)$. This comes from $\tan(\phi+\psi) = -(q' + w')/(r' + u')$ expanded for small u and w.

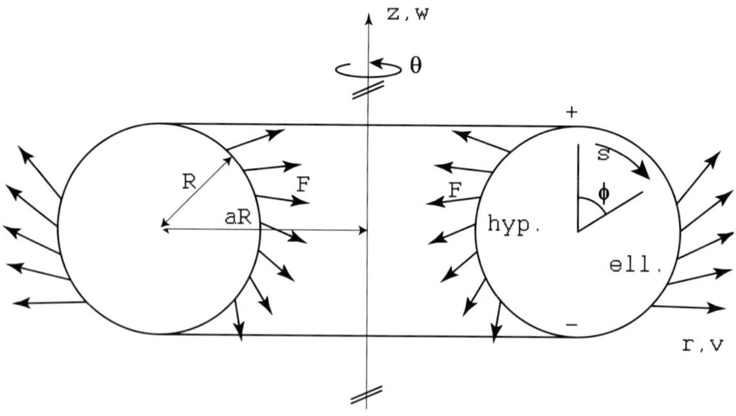

Fig. 1. A toroidal shell deforms under surface loads

The strains can be computed as:

$$\varepsilon_s = u' \cos\phi - w' \sin\phi + \varepsilon w'^2 / 2 \quad \text{and} \quad \varepsilon_\theta = u / r \tag{1}$$

Due to cylindrical symmetry, there is no shear strain. In the first equation above for ε_s, we have neglected terms of order $\varepsilon u'^2$ but not terms of order $\varepsilon w'^2/2$ as the latter nonlinearity plays a crucial role in the following. In ε_θ, we have neglected a term of order $\varepsilon u^2 / r^2$. There, the number ε measures the importance of geometrical nonlinearities: it is defined as $\varepsilon = p\ R/E\ h$, where p is the typical pressure applied on the shell and E its Young modulus.

The constitutive relations of the material, derived from Hooke law, read:

$$\varepsilon_s = N_s - \nu N_\theta \quad \text{and} \quad \varepsilon_\theta = N_\theta - \nu N_s \tag{2}$$

where N_s and N_θ are the (ss) and $(\theta\theta)$ components of the stress. Finally, these equations are supplemented by equilibrium equations along the axis:

$$\frac{d}{ds}\left[r N_s (\sin\phi + \varepsilon\psi \cos\phi) - Qr\cos\phi \right] = r(f_z - \varepsilon\psi f_r) \tag{3a}$$

and in the radial direction:

$$\frac{d}{ds}\left[r N_s (\cos\phi - \varepsilon\psi \sin\phi) + Qr\sin\phi \right] = -r(f_r + \varepsilon\psi f_z) \tag{3b}$$

In these equations, f_z and f_r are the vertical and radial components of the stresses applied on the shell surface, and Q is the bending shear stress $Q = \mu^2\psi''$, where

$$\mu = (12(1-\nu^2))^{1/2}\ R/h \tag{4}$$

is a large dimensionless number measuring the importance of flexural effects.

In these equations, there are two dimensionless numbers ε and μ, measuring the intensity of geometric nonlinearities and flexural effects respectively. ε is very small and μ is very large.

3 Membrane divergence near crowns

This system of equations is singular for $s = 0$ and $s = \pi$, which corresponds to the topmost and bottommost parallel circles, as explained in the introduction. We shall focus on the boundary layer near $s = 0$, the other being similar up to a horizontal reflexion.

The origin of the singularity in the transition region is twofold. First, it is related to a simple mechanical explanation: a membrane cannot transmit axial forces from the outer to the inner part of the torus, because it cannot sustain shear stresses. This has been noted earlier in the literature. There is second, perhaps more subtile, reason for the occurence of singularities. Pure bending modes of the shell (also called isometric deformations) are diverging near the transition circles. This has to do with deep geometrical results [1]. Stated differently, the problem of determining pure bending modes is singular in the transition region. Now, this problem is underlaid by the computation of displacements from strain, since computing pure bending deformations is just a particular case of computing displacements in the absence of any strain. Therefore, due to the particular geometrical properties of the transition circles, there might exist smooth solution of the membrane equations for the stresses, but these solutions give rise to singular displacements fields.

These two sources of singularities indeed lead to two distinct types of singularities among membrane solutions. This does not seem to have clearly been recognized in earlier works. As we show, two classes of surface loading should be distinguished. On one hand, there are loadings such that no axial force is transmitted from the outer to the inner part of the torus (later called the $\Xi = 0$ case). In this case, no shear stress is present in the transition region and only the second mechanism generates singularities. This leads to relatively weak singularities. On the other hand, when shear stress is transmitted accross the transition region ($\Xi \neq 0$), both mechanisms come into play and the singularities are stronger. The mathematical structure of the boundary layers is different depending on whether $\Xi = 0$ or $\Xi \neq 0$. We refer to these cases as shearless transition region ($\Xi \neq 0$) and transition region under shear ($\Xi \neq 0$).

Let us first analyze the divergence implied by the governing equations above in the membrane approximation $(Q = 0)$. Integration of the equilibrium equation along axis yields, for small s, to:

$$N_s = \Xi\ R/(as) + \text{smooth part} \tag{5}$$

where Ξ is a constant of integration which can be interpreted as the axial force transmitted from the outer half of the torus to the inner half.

3.1 Crowns sustaining shear ($\Xi \neq 0$)

At the dominant order for small s, one finds in the membrane approximation, starting from Eq. (5) above:

$$N_\theta = -\ \Xi R\ /\ s^2 \tag{6}$$

hence by integration of the definitions of strain (1):

$$u = -a\ R\ \Xi / s^2 \text{ and } w = -\ 2/3\ a\ R^2\ \Xi / s^3 \tag{7}$$

the change of direction of the tangent being given by:

$$\psi = -\ 2\ a\ R^2\ \Xi / s^4 \tag{8}$$

Because ψ diverges, both nonlinear and flexural effects become important and must be considered in this region.

3.2 Shearless crowns ($\Xi = 0$)

As explained above, this leads to smooth stresses in the transition region by (5) but owing to singularities in the geometric problem of the existence of isometric deformations, displacement is singular. After some algebra, one computes a logarithmic divergence for w and a divergence in $1/s$ for ψ. These weaker singularities can be removed by considering nonlinearities or flexural effect near the transition region, but the boundary layer equations are different than when shear is present.

4 Analysis of the boundary layer with shear ($\Xi \neq 0$)

According to the general technique of boundary layer analysis, the shell equations can be simplified in the transition region based on the different orders of magnitudes of the various quantities provided by the outer solution on one hand, and on geometric simplifications (the angle ϕ is small) on the other hand. This leads to a set equations which by later rescaling shall yield the boundary layer equations:

$$a^2 N_s'' + \phi\,\psi + \varepsilon\,\psi^2/2 = 0 \tag{9a}$$

and

$$a\,N_s\,(\phi + \varepsilon\,\psi) - a\,\mu^{-2}\,\psi'' = \Xi \tag{9b}$$

This set of coupled nonlinear ordinary differential equations is for the unknown functions $N_s(s)$ and $\psi(s)$. For large s, it should match with the asymptotics of the (diverging) membrane solutions. In these equations, there are two small parameters, ε and $1/\mu$. Depending on the relative smallness of ε and $1/\mu$, that is to say depending on the relative importance of nonlinear versus flexural effects, the boundary layer can have different structures.

4.1 Single boundary layer (mixed regularization)

When ε is of order $\mu^{-5/3}$, flexural and nonlinear effects are of the same order of magnitude in the boundary layer. To show this, let us indeed introduce a modified Reissner parameter as $\rho = |\Xi|\,a^{-2/3}\,|\varepsilon|\,R^{4/3}\,\mu^{5/3}$. This definition is inspired from that introduced by Reissner [5]: $\rho_R = a^{2/3}\,|\varepsilon|\,\mu^{4/3}$ in the case $\Xi = 0$. With the following definition for x, f and g, the rescaled s, ψ and N_s respectively:

$$s = a^{1/3}\,R^{1/3}\,\mu^{-1/3}\,x,\quad \psi = \rho\,a^{1/3} R^{-2/3}\,\varepsilon^{-1}\,\mu^{-1/3}\,f \tag{10}$$

$$\text{and } N_S = \rho R^{-2/3} \varepsilon^{-1} a^{-2/3} \mu^{-4/3} g$$

Equation (9) are transformed into the boundary layer equations:

$$g''(x) + x f(x) + 1/2 \rho f(x)^2 = 0$$
$$\text{and } -f''(x) + x\, g(x) + \rho f(x)\, g(x) = 1 \qquad (11)$$

The boundary conditions allowing matching with the outer solution read:

$$g(x) \sim 1/x \text{ and } f(x) \sim -2/x^4 \text{ for } x \to \pm\infty \qquad (12)$$

While earlier analyses of the transition region were made in particular cases such that the boundary layer equations were linear [6], the present one is nonlinear.
A simple counting argument proves that solutions $f(x)$, $g(x)$ of these equations are at most discrete for any value of ρ. In Fig. 1, a numerical solution of this set of boundary layer equations has been used to plot the deformed torus when $\rho = 5/2$. The numerical procedure is based on a shooting method for the nonlinear boundary layer Eqs. (11) with asymptotic boundary conditions (12).

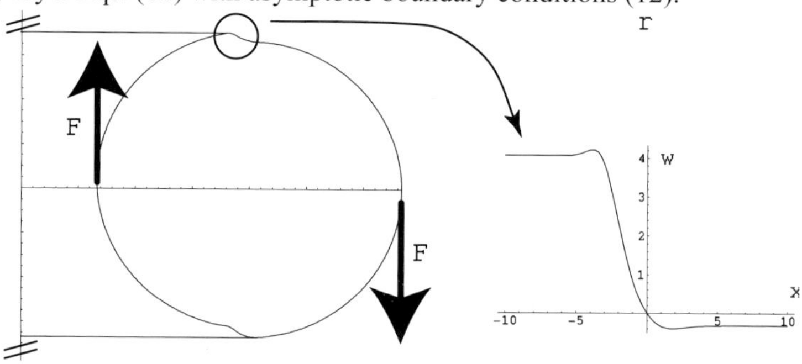

Fig. 2. Deformation of torus with modified Reissner parameter $\rho = 5/2$. Deformation is concentrated in two boundary layers near the transition circles. This plot uses a numerical solution of the boundary layer equations shown on the right, obtained by numerical integration and shooting of Eqs. (11) so as to satisfy boundary conditions (12).

4.2 Double boundary layer (strong nonlinearities)

The limit $\rho \to \infty$ is studied here. It corresponds to a loading and a shell thickness such that geometric nonlinearities are the dominant regularazing mechanism. Interestingly, the nonlinearities are not sufficient by themselves to fully regularize the divergent membrane solution. As a result, a tiny secondary boundary layer forms within the nonlinear boundary layer. In this inner boundary layer, flexural effects ultimately become important and serve to regularize the still divergent displacement of the first, nonlinear boundary layer. Defining a rescaled stress N_s by $G = \rho^{-1/5} g$ and a rescaled change of normal orientation ψ by $F = \rho^{-4/5} f$ and a rescaled coordinate $X = \rho^{1/5} x$, one obtains the following boundary layer equations for the outer, purely nonlinear, boundary layer:

$$G'' + 1/(2\,G^2) = X^2/2 \qquad (13)$$

with asymptotic conditions for $X \to \infty$ similar to those in Eq. (12). F is now a function of G by the relation $G\,(F{+}X) = 1$. These equations remain valid as long as ($\rho^{-6/5}\,|F''| << 1$). This condition is violated somewhere within the boundary layer, meaning that a secondary, flexural boundary layer must be considered. For details, see reference [3].

4.3 Flexural boundary layer

When $\rho << 1$, regularization is mostly by flexural effects. The above general equations can be used to recover the boundary layer equation $T''(x) - ixT = 1$, where $T(x) = -f(x) + i\,g(x)$. This particular case has been already been studied, by Clark [4] and Steele [8, 9] independently.

5 Shearless boundary layer ($\Xi = 0$)

The general equations derived above can be used to study shearless boundary layers. It turns out that this particular case has already been rather extensively studied in the past. Indeed, calculations of the mechanical response of a toroidal shell was motivated by the project of a toroidal space station. The main source of loading is from internal pressurization, and it turns out that in the case of uniform pressure loading on a toroidal shell with circular section, no axial force is transmitted across the transition region, hence $\Xi = 0$. We stress, however, that for more general loading, or even for a uniform pressure applied on a toroidal shell with an arbitrary (that is non circular) section, the transition region becomes under shear: the case $\Xi = 0$ is not generic by any means.

5.1 Intermediate regime

We refer to Rossettos and Sanders' analysis [6] which can be recovered perhaps in a more natural way starting from the present set of Eqs. [2].

5.2 Boundary layer for geometric nonlinearities

The limit $\mu = \infty$, ε being small but finite has been studied in paper by Sanders and Liepins [7]. Surprisingly, the resulting boundary layer is described by *linear* boundary layer equations, while it accounts for geometric nonlinearities.

5.3 Flexural boundary layer

The other limit, namely $\varepsilon = 0$ while μ is large but finite has been first studied by Clark [4] and Steele [8, 9]. This yields the same analysis as $\rho \to 0$ when $\Xi \neq 0$ discussed above.

6 Conclusion

We have shown the existence of geometric inner boundary layers in shells of revolution with mixed type. These boundary layers were called geometric because their location is dictated by a purely geometric constraint (the Gauss curvature of the surface changes sign, leading to equations similar to those describing transsonic flows in fluid mechanics). We have derived general boundary layer equations for such shells of revolution. Although we focused on the particular case of a torus with a circular section, our main results are easily extended to other shells of revolution with mixed type. An original nonlinear boundary layer equation was derived and solved numerically, which describes the deformations in the transition region in the most generic case. In various limits, results reported earlier in the literature were recovered.

We would like to thank Prof. J. G. Simmonds to pointing out useful references to us.

References

1. Audoly B (1999) Courbes rigidifiant les surfaces. C R Acad Sci Paris 328 I: 313-316
2. Audoly B, Pomeau Y (2002) The elastic torus: Anomalous stiffness in shells with mixed type. C R Mécanique 330: 425-432
3. Audoly B, Pomeau Y, (2003) in preparation
4. Clark A (1950) On the theory of thin elastic toroidal shells. J Math Phys 29: 146-178
5. Reissner E (1963) On stresses and deformations in toroidal shells of circular cross section which are acted upon by uniform normal pressure. Quart App Math 21: 177-187
6. Rossettos JN, Sanders JL (1965) Toroidal Shells under Internal Pressure in the Transition Range. AIAA Journal 3: 1901-1909
7. Sanders JL, Liepins AA (1963) Toroidal Membrane under Internal Pressure. AIAA Journal 1: 2105-2110
8. Steele CR (1964) Orthotropic pressure vessels with axial constraint. AIAA Journal 2: 703-709
9. Steele CR (1965) Toroidal pressure vessels. J Spacecraft 2: 937-943

FEM and Time Stepping Procedures in Non-Linear Dynamics of Flexible Branched Shell Structures

J. Chróścielewski[1], I. Lubowiecka[1] and W. Pietraszkiewicz[2]

[1] Gdańsk University of Technology, Poland
[2] Institute of Fluid-Flow Machinery, PASci., Gdańsk, Poland

1 Introduction

Non-linear dynamic behaviour of flexible irregular shell structures was discussed recently by Chróścielewski et al. [3,4], Lubowiecka [7], and Lubowiecka and Chróścielewski [8], where references to other related papers are given. The shell evolution in time was described by two fields: the vector $\boldsymbol{u}$ of the translatory motion of the shell base surface, and the proper orthogonal tensor $\boldsymbol{Q}$ of the mean rotation of the shell cross sections. As a result, the rotation group $SO(3)$ entered the definition of the configuration space. In such problems of structural mechanics finite-dimensional approximations, like the finite element method or the time-stepping algorithm, require non-standard approaches.

We worked out in [3,4,7,8] the extended time-stepping algorithm for the non-linear dynamic analysis on the configuration space containing the rotation group $SO(3)$. Applying the algorithm, several FE numerical simulations of 3D large overall motion of flexible elastic shell structures have been performed. In particular, we have successfully simulated the dynamic behaviour of branched shell structures forced to free flight in space by the pair of forces initially applied during a short time period. The structures perform many turns in space, exhibiting large relative deformations as well.

In this report we want to draw attention to two critical aspects of the numerical analyses performed in [3,4,7,8] which can considerably influence the conservation of total potential energy of the system. These are: a) the influence of the time step size, and b) the influence of the FE mesh condensation.

For any time step size below a critical one, the numerical tests show that values of potential and kinetic energies may oscillate in time, but the total energy remains conserved in a considered time period. The rapid increase of the total energy indicates the appearance of numerical instability of the solution. It has been noted that the numerical instability in the analysis of shell structures undergoing free motion in space is preceded by a rising trend of the potential energy. This fact makes it possible to predict the moment of numerical instability and to appropriately reduce in advance the time step size. This allows one to prolong considerably the time duration of stable numerical simulations without any special procedures.

By refining the FE mesh, or by using more complex elements, we increase the number of degrees of freedom (dof), which dramatically raises the highest eigenfrequency of the discrete system. This, in turn, makes the system more sensitive to the time step size as well.

2 Weak formulation of non-linear shell dynamics

Motion of the irregular shell structure can in general be described by the displacement vector field $\boldsymbol{u}(\boldsymbol{x},t)=\boldsymbol{y}(\boldsymbol{x},t)-\boldsymbol{x}$, where $\boldsymbol{x}\in M$ and $\boldsymbol{y}(\boldsymbol{x},t)$ are position vectors of the undeformed shell base surface and the base surface at time t, respectively, together with the independent proper orthogonal tensor field $\boldsymbol{Q}(\boldsymbol{x},t)$ representing the mean rotary motion of the shell cross sections. The fields $\boldsymbol{y}(\boldsymbol{x},t)$ and $\boldsymbol{Q}(\boldsymbol{x},t)$ are assumed to be continuous during the motion, and on any stationary singular curve $\Gamma\subset M$ determining an irregularity (fold, intersection, branching, etc., see [5,6]), i.e. $\boldsymbol{y}_\Gamma(\boldsymbol{x}_\Gamma,t)=\boldsymbol{y}(\boldsymbol{x},t)|_\Gamma$, $\boldsymbol{Q}_\Gamma(\boldsymbol{x}_\Gamma,t)=\boldsymbol{Q}(\boldsymbol{x},t)|_\Gamma$.

For the translational $\boldsymbol{p}(\boldsymbol{x},t)$ and rotational $\boldsymbol{m}(\boldsymbol{x},t)$ momentum surface vectors we take simple kinetic constitutive relations $\boldsymbol{p}(\boldsymbol{x},t)=m_0\boldsymbol{\upsilon}=\rho_0 h_0\boldsymbol{\upsilon}$ and $\boldsymbol{m}(\boldsymbol{x},t)=I_0\boldsymbol{\omega}=(\rho_0 h_0^3/12)\boldsymbol{\omega}$ (see Libai and Simmonds [6]), where $\rho_0(\boldsymbol{x})$ is the initial shell mass density, $h_0(\boldsymbol{x})$ the initial shell thickness, $\boldsymbol{\upsilon}(\mathbf{x},t)=\dot{\boldsymbol{y}}(\boldsymbol{x},t)=\dot{\boldsymbol{u}}(\boldsymbol{x},t)$ the linear velocity vector, and $\boldsymbol{\omega}(\boldsymbol{x},t)$ the angular velocity vector in the spatial representation, with $\mathrm{ad}\,\boldsymbol{\omega}=\dot{\boldsymbol{Q}}\boldsymbol{Q}^T$, $\mathrm{ad}:E^3\rightarrow SO(3)$.

We confine our considerations to hyper-elastic shells for which there exists a 2D strain energy density $W(\boldsymbol{\varepsilon}_\beta,\boldsymbol{\kappa}_\beta;\boldsymbol{x})$ of the shell strain vectors $\boldsymbol{\varepsilon}_\beta=\boldsymbol{y},_\beta-\boldsymbol{Q}\boldsymbol{x},_\beta$ and $\boldsymbol{\kappa}_\beta=\mathrm{ad}^{-1}(\boldsymbol{Q},_\beta\boldsymbol{Q}^{\mathrm{T}})$. Then the constitutive relations of the shell material are given by $\boldsymbol{n}^\beta=\partial W/\partial\boldsymbol{\varepsilon}_\beta$, $\boldsymbol{m}^\beta=\partial W/\partial\boldsymbol{\kappa}_\beta$, where $\boldsymbol{n}^\beta(\boldsymbol{x},t)$ and $\boldsymbol{m}^\beta(\boldsymbol{x},t)$ are the internal stress and couple resultant vectors, respectively.

When expressed in the weak form, the initial-boundary value problem for the branched shell-like structure can be formulated as follows. Given the external resultant force and couple vector fields $\boldsymbol{f}(\boldsymbol{x},t)$ and $\boldsymbol{c}(\boldsymbol{x},t)$ on $\boldsymbol{x}\in M\setminus\Gamma$, $\boldsymbol{n}^*(\boldsymbol{x},t)$ and $\boldsymbol{m}^*(\boldsymbol{x},t)$ along ∂M_f, $\boldsymbol{f}_\Gamma(\boldsymbol{x}_\Gamma,t)$ and $\boldsymbol{c}_\Gamma(\boldsymbol{x}_\Gamma,t)$ along the singular curve $\Gamma\subset M$, and the initial values $\boldsymbol{u}_0(\boldsymbol{x})$, $\boldsymbol{Q}_0(\boldsymbol{x})$, $\dot{\boldsymbol{u}}_0(\boldsymbol{x})$, $\dot{\boldsymbol{Q}}_0(\boldsymbol{x})$ at $t=0$, find a curve $\mathrm{u}(\boldsymbol{x},t)=(\boldsymbol{u}(\boldsymbol{x},t),\boldsymbol{Q}(\boldsymbol{x},t))$ on the configuration space $C(M,E^3\times SO(3))$ such that for any continuous kinematically admissible virtual vector field $\mathrm{w}(\boldsymbol{x})\equiv(\boldsymbol{v}(\boldsymbol{x}),\boldsymbol{w}(\boldsymbol{x}))$ the following principle of virtual work is satisfied:

$$G[\mathrm{u};\mathrm{w}] = \iint\limits_{M\setminus\Gamma} [m_0 \dot{\boldsymbol{\upsilon}} \boldsymbol{\cdot} \boldsymbol{v} + I_0 \dot{\boldsymbol{\omega}} \boldsymbol{\cdot} \boldsymbol{w}]\, da$$
$$+ \iint\limits_{M\setminus\Gamma} [\boldsymbol{n}^\beta \boldsymbol{\cdot} (\boldsymbol{v},_\beta + \boldsymbol{y},_\beta \times \boldsymbol{w}) + \boldsymbol{m}^\beta \boldsymbol{\cdot} \boldsymbol{w},_\beta]\, da - \iint\limits_{M\setminus\Gamma} (\boldsymbol{f} \boldsymbol{\cdot} \boldsymbol{v} + \boldsymbol{c} \boldsymbol{\cdot} \boldsymbol{w})\, da \tag{1}$$
$$- \int\limits_{\partial M_f} (\boldsymbol{n}^* \boldsymbol{\cdot} \boldsymbol{v} + \boldsymbol{m}^* \boldsymbol{\cdot} \boldsymbol{w})\, ds - \int\limits_{\Gamma} (\boldsymbol{f}_\Gamma \boldsymbol{\cdot} \boldsymbol{v}_\Gamma + \boldsymbol{c}_\Gamma \boldsymbol{\cdot} \boldsymbol{w}_\Gamma)\, ds \quad = 0\,.$$

In (1) it is implicitly assumed that the kinematic boundary conditions $\boldsymbol{u}(\boldsymbol{x},t) = \boldsymbol{u}^*(\boldsymbol{x},t)$ and $\boldsymbol{Q}(\boldsymbol{x},t) = \boldsymbol{Q}^*(\boldsymbol{x},t)$ are satisfied along the complementary part $\partial M_d = \partial M \setminus \partial M_f$ of the shell boundary, and the virtual vector fields are kinematically admissible if $\boldsymbol{v}(\boldsymbol{x}) = \mathbf{0}$ and $\boldsymbol{w}(\boldsymbol{x}) = \mathbf{0}$ along ∂M_d

3 Temporal and spatial discretisation in SO(3)

The solution of (1) is achieved by an iterative procedure which reduces the problem to a sequence of solutions of linearised problems. Each linearised problem is formulated at discrete values of both temporal and spatial variables. The main difficulty of the solution procedure is associated with the structure of the configuration space involving the rotation group $SO(3)$.

The Newmark type algorithm extended to the rotation group was developed in [3,4,7,8]. It is based on ideas suggested among others by Simo and Vu-Quoc [10], and Cardona and Geradin [1]. In the algorithm the linearised equations of shell dynamics are written in the spatial representation relative to the last instantaneous configuration, while the angular velocity and acceleration vectors are approximated in time in the material representation.

With the time-stepping algorithm, the solution of the non-linear problem (1) is constructed by the incremental-iterative procedure based on the Newton-Kantorovich method applied in the configuration space $C(M, E^3 \times SO(3))$. Let an i-th approximation $\mathrm{u}_{n+1}^{(i)}$ to the solution u_{n+1} has been found. In order to calculate the correction $\delta\mathrm{u}_{n+1}^{(i)}$, which would allow us to find the successive approximation $\mathrm{u}_{n+1}^{(i+1)}$ to an unknown solution u_{n+1}, we linearise G_{n+1} at the approximation $\mathrm{u}_{n+1}^{(i)}$:

$$G[\mathrm{u}_{n+1}^{(i)}, t_{n+1}; \mathrm{w}] + \delta G[\mathrm{u}_{n+1}^{(i)}, t_{n+1}; \mathrm{w}, \delta\mathrm{u}_{n+1}^{(i)}] = 0 \tag{2}$$

The second term in (2) denotes directional derivative of the functional G_{n+1}, taken at the point $\mathrm{u}_{n+1}^{(i)} \in C$ in the direction $\delta\mathrm{u}_{n+1}^{(i)} \in T_{\mathrm{u}_{n+1}^{(i)}} C$. This term yields the so-called tangent operator of the non-linear problem, calculated at the approximation $\mathrm{u}_{n+1}^{(i)}$. The first term in (2) represents unbalanced forces at the approximation point $\mathrm{u}_{n+1}^{(i)}$. For spatial discretisation the finite element method is applied, [10].

Stability of the solution is verified by the energy criterion given by inequality

$$U_{n+1} - U_n + K_{n+1} - K_n \le \Delta G_{ext} \tag{3}$$

where $U_{n+1} \simeq U(t_{n+1})$ and $U_n \simeq U(t_n)$ are values of the strain energy $U(t)$ numerically calculated at the beginning t_n and at the end t_{n+1} of the time step Δt, respectively, K_{n+1} and K_n are the corresponding values of the kinetic energy $K(t)$, and ΔG_{ext} denotes an increment of the work done by external loads within the time step.

4 Influence of time step size

The twisted T-shaped shell (Fig. 1) exhibits the complex 3D motion comprising large elastic relative deformation and multiple global turns. Two concentrated forces of the ramp function type (from 0 to 10^4 in 1s and back down to 0 in another 1s) are applied at the points (a) and (b). After 2s the shell is free from external loading and moves freely in the space undergoing also large local deformations.

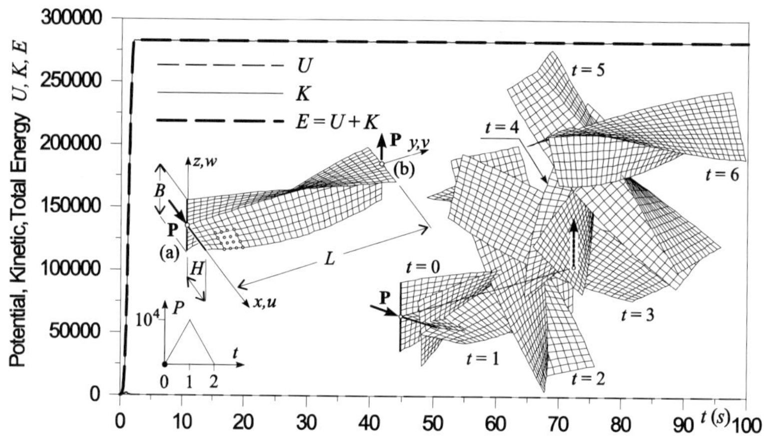

Fig. 1. T-shaped shell structure: $L=50$, $H=10$, $B=14.0112$, $\alpha=90^\circ$, $h_0=0.25$, $E=2\times10^7$, $\nu=0.3$, $\rho_0 h_0=1$; discretisation $(2+2+2)\times 10 = 60e16$ FI (16–nodes elements with full integration); external load function; some configurations in time; energy history

The dynamic analysis is performed for three time steps: $\Delta t = 0.1s$, $\Delta t = 0.025s$, and $\Delta t = 0.005s$. During simulations with the largest time step an instability is noticed after about first 10s of the motion (see Fig. 2). The reduction of the step size allows one to extend duration of the stable simulation provided that the reduction is executed at an appropriate moment. Although the simulation becomes stable again during the analysis with the smaller time steps, visible differences appear in the potential energy values (see Fig. 3).

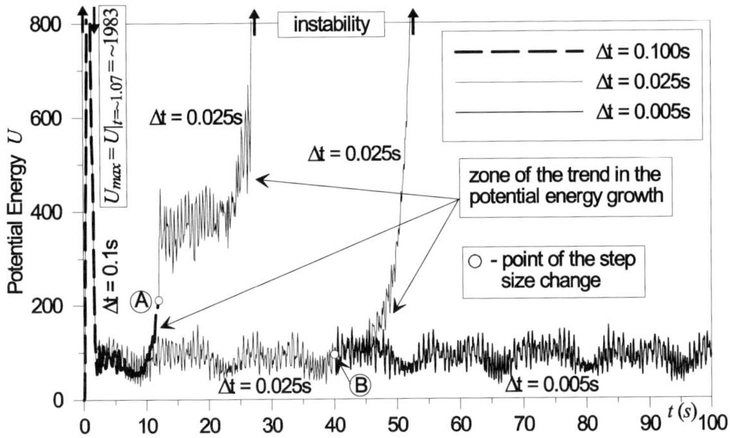

Fig. 2. Potential energy history of twisted T-shaped shell structure

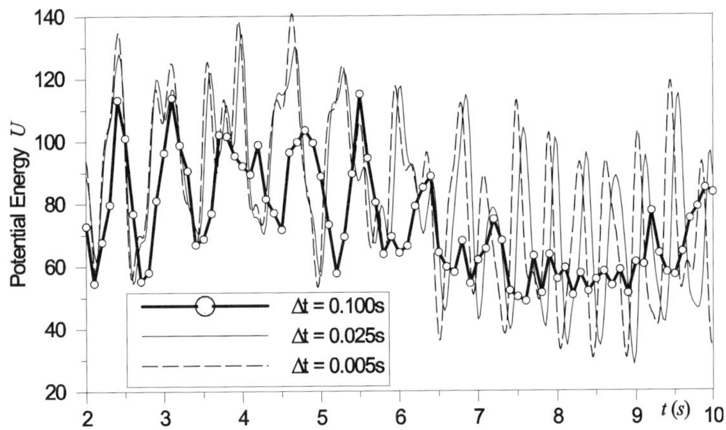

Fig. 3. Magnified potential energy history of twisted T-shaped shell structure for three different time steps Δt

5 Influence of FE mesh refinement

Large overall motion of a flexible cylindrical panel with change of the curvature sign, reinforced by a plate rib (Fig. 4), is analysed. The structure is loaded as in the previous example, but the load function rises here from 0 to 10. The structure, as in the first example, moves freely in the space undergoing multiple turns and large relative deformations.

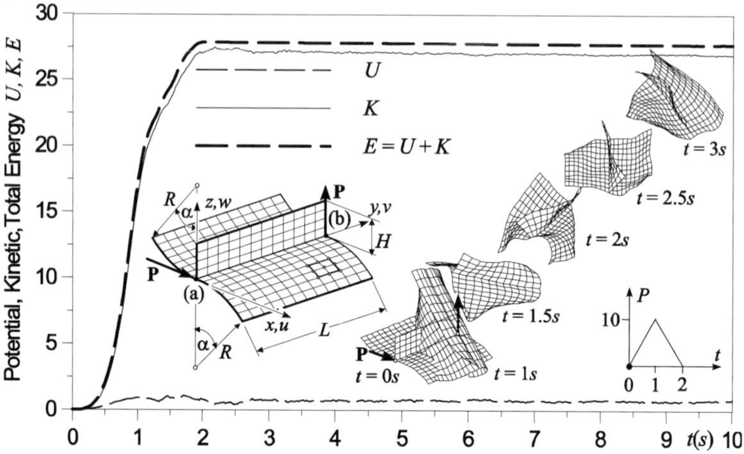

Fig. 4. Wavy cylindrical branched panel: $L=2$, $\alpha=0.4$, $R=1$, $H=0.4$, $h_0=0.01$, $E=10^5$, $\nu=0.25$, $\rho_0 h_0=1$; discretisation; external load function; some configurations in time; energy history

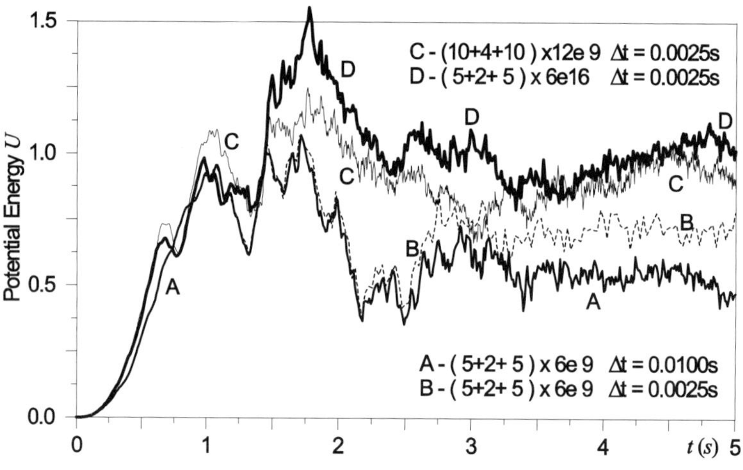

Fig. 5. Variations of potential energy of the cylindrical panel for different meshes, elements, and time steps

The panel is analysed with different discretisations: by $(5+2+5)\times 6=72\,e9\;FI$ (9–nodes elements) with totally 1950 dof, by $(5+2+5)\times 6=72\,e16\;FI$ with totally 4218 dof, and by $(10+4+10)\times 12=288\,e16\;FI$ with totally 7350 dof. The analysis is performed for two different time steps $\Delta t=0.01s$ and $\Delta t=0.0025s$.

The energy histories given in Fig. 5 indicate that there are visible differences in the total energy values, both for different time steps with the same elements and for one time step with different elements. We observed that with the growth of number of dof the numerical problem becomes very fast unstable and requires shortening of the time step size already at the early simulation stage. Moreover, unfortunately, Fig. 5 indicates that by using the same shorter time step $\Delta t = 0.0025s$ different discretisations lead to different plots of the potential energy, and thus to different time changes of stresses in the panel. More information about these examples is included in [7,8].

6 Conclusions

From our experience it follows that the 6-parameter shell theory (with 6 dof in each node of FE mesh) is appropriate to analyse branched shell structures of any shape. The large overall motion of the structures can be analysed without any limits in translation and rotation. When external forces are removed, the convergence of simulated dynamic behaviour is confirmed by two facts: a) the global centre of mass moves along the straight line, and b) the total energy of the system is conserved.

Under considered external loads, as long as they are applied, the total energy initially increases. After removing the loads small oscillations of kinetic and potential energies can still be observed, but the sum of both energy components remains constant. The numerical instability of the solution demonstrates itself after some period of simulation as an explosive increase of the total energy. In these simulations we have noted that the instability is preceded by the rising trend in the potential energy growth (Fig. 2). This observation allows one to choose an appropriate time instant to shorten the size of the time step (point B in Fig. 2) and to proceed with further stable dynamic analysis. However, too late reduction of size of the integration step, already in the rising trend phase of the potential energy growth (point A in Fig. 2), does not assure correct results. By condensing the FE mesh, or using more complex elements, we increase the number of dof, which makes the system more sensitive to the step size (the highest eigenfrequency of the discrete system raises dramatically). As a result, the dynamic analysis with longer time steps is practically impossible without applying special methods. The instability appears here already at a very early stage of numerical simulation.

Significant differences in time changes of potential energy (and thus also in the stresses) obtained for different FE mesh and time step sizes in Fig. 3 and Fig. 5 allow one to question the correctness of the typical results obtained for longer integration steps and coarse mesh, even if the usual stability criteria are satisfied.

Acknowledgement

The author I. Lubowiecka is awarded by Foundation for Polish Science.

References

1. Cardona C, Geradin C (1988) A beam finite element non-linear theory with finite rotations. Int J Num Meth Engng 26: 2403-2438
2. Chróścielewski J (1996) The family of C^0 finite elements in the non-linear sixparameter shell theory (in Polish). Zesz. Nauk. Politechniki Gdańskiej Nr 540, Budownictwo Lądowe LIII: 1-291
3. Chróścielewski J, Makowski J, Pietraszkiewicz W (2002) Non-linear dynamics of flexible shell structures. Computer Assisted Mechanics and Engineering Sciences 9: 341-357
4. Chróścielewski J, Makowski J, Pietraszkiewicz W (2000) Large overall motion of flexible branched shell structures. In: Ambrosio J A C, Kleiber M (eds.) Computational aspects of nonlinear structural systems with large rigid body motion. NATO Advanced Research Workshop, Pułtusk (Poland), July 2-7 2000. IDMEC, Lisboa, pp 201-218
5. Chróścielewski J, Makowski J, Stumpf H (1997) Finite element analysis of smooth, folded and multi-shell structures. Comp Meth Appl Mech Engng 141: 1-46
6. Libai A, Simmonds J G (1998) The nonlinear theory of elastic shells, 2nd ed. Cambridge University Press, Cambridge
7. Lubowiecka I (2001) Integration of nonlinear dynamic equations of motion in structural mechanics. Dynamics of elastic shell structures (in Polish). PhD Dissertation, Gdańsk University of Technology
8. Lubowiecka I, Chróścielewski J (2002) On dynamics of flexible branched shell structures undergoing large overall motion using finite elements. Comp Struct 80: 891-898
9. Makowski J, Pietraszkiewicz W, Stumpf H (1999) Jump conditions in the non-linear theory of thin irregular shells. J Elasticity 54: 1-26
10. Simo JC, Vu-Quoc L (1988) On the dynamics in space of rods undergoing large motions - a geometrically exact approach. Comp Meth Appl Mech Engng 66: 125-161

Dynamic Stiffness Vibration Analysis for Higher Order Plate Models

M. Eisenberger

Faculty of Civil Engineering, Technion, Israel Institute of Technology, Technion City, Israel

1 Introduction

The analysis of plates has been performed over the years mostly using the Kirchhoff plate theory which is an extension of the Bernoulli-Euler beam theory. This theory is based on the assumption that plane sections of the cross section remain plane and perpendicular to the plate mid-surface, and is covered extensively by Leissa [3]. A more refined plate theory is the Mindlin plate theory which relaxes the restriction on the angle of shearing deformations that exist in the simpler theory. In recent years, due to the increase in the use of composite materials the shear deformation theory has been limited in the accuracy of computation for many situations, and although some remedies were devised, locking in the numerical analysis has become a main issue and higher order theories have emerged.

In the higher order theory, the restriction on the warping of the cross section is relaxed, and allows variation in the thickness direction of the plate which is cubic in both directions. The cross section displacements for the 3 models are shown in Figure 1. Then the additional constraints are that the top and bottom surfaces of the plate are traction free. For isotropic materials the resulting displacement field is [4]

$$u(x,y,z,t) = u_0(x,y,t) + z\phi_x(x,y,t) - \frac{4z^3}{3h^2}\left(\phi_x + \frac{\partial w_0}{\partial x}\right) \tag{1}$$

$$v(x,y,z,t) = v_0(x,y,t) + z\phi_y(x,y,t) - \frac{4z^3}{3h^2}\left(\phi_y + \frac{\partial w_0}{\partial y}\right) \tag{2}$$

$$w(x,y,z,t) = w_0(x,y,t) \tag{3}$$

where (u_0, v_0, w_0) are the mid surface displacements in the x, y, z directions, respectively, and ϕ_x, ϕ_y denote slopes of a transverse normal at mid surface. This choice of displacement field satisfies the stress free boundary conditions on the top and bottom faces of the plate.

2 Constitutive equations

Using the displacement field in Eqs. (1-3) the strains are

$$\varepsilon_x = u_{0,x} + z\left[\phi_{x,x} - \frac{4z^2}{3h^2}(\phi_{x,x} + w_{0,xx})\right] \tag{4}$$

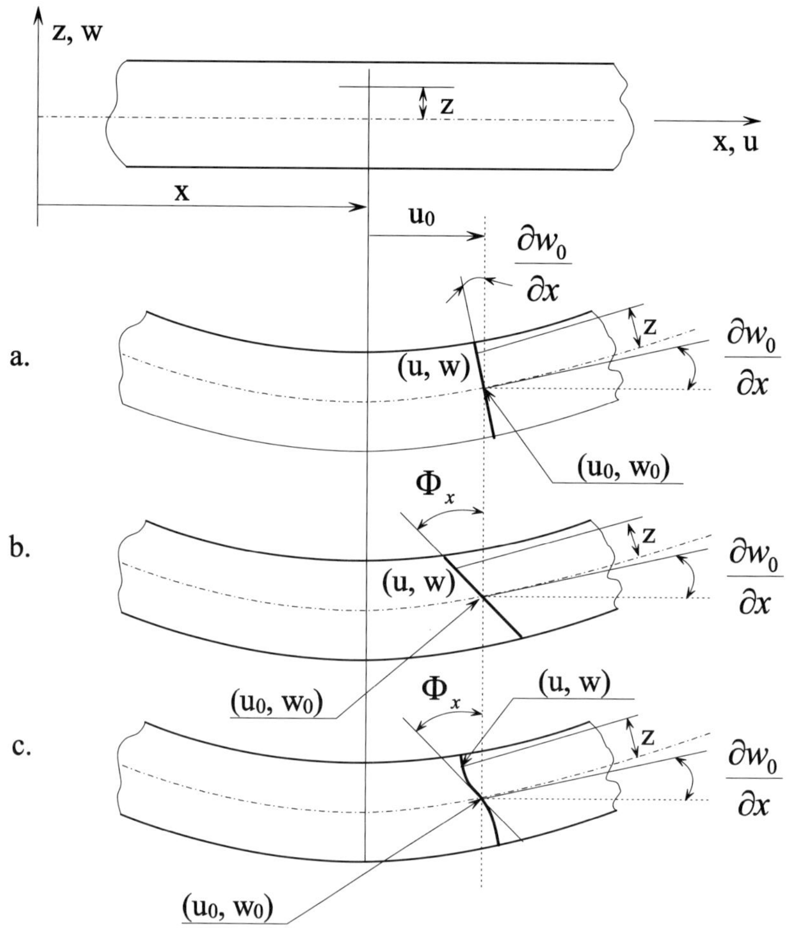

Fig. 1. Cross section displacements in the 3 models: Kirchhoff, Mindlin, and Higher Order Theory

$$\varepsilon_y = v_{0,y} + z\left[\phi_{y,y} - \frac{4z^2}{3h^2}(\phi_{y,y} + w_{0,yy})\right] \tag{5}$$

$$\gamma_{xy} = u_{0,y} + v_{0,x} + z\left[\phi_{x,y} + \phi_{y,x} - \frac{4z^2}{3h^2}(\phi_{x,y} + \phi_{y,x} + 2w_{0,xy})\right] \tag{6}$$

$$\gamma_{yz} = \phi_y + w_{0,y} - \frac{4z^2}{h^2}(\phi_y + w_{0,y}) \tag{7}$$

$$\gamma_{zx} = \phi_x + w_{0,x} - \frac{4z^2}{h^2}(\phi_x + w_{0,x}) \tag{8}$$

For constant thickness plate made of isotropic material the stress strain relationship are given by

$$\begin{bmatrix} \sigma_x \\ \sigma_y \\ \tau_{xy} \end{bmatrix} = \frac{E}{1-\nu^2}\begin{bmatrix} 1 & \nu & 0 \\ \nu & 1 & 0 \\ 0 & 0 & \dfrac{1-\nu}{2} \end{bmatrix}\begin{bmatrix} \varepsilon_x \\ \varepsilon_y \\ \gamma_{xy} \end{bmatrix} \tag{9}$$

$$\begin{bmatrix} \tau_{yz} \\ \tau_{zx} \end{bmatrix} = \begin{bmatrix} G & 0 \\ 0 & G \end{bmatrix}\begin{bmatrix} \gamma_{yz} \\ \gamma_{zx} \end{bmatrix} \tag{10}$$

with E, G, ν and the modulus of elasticity, shear modulus, and Poisson ratio of the material, respectively.

3 Equilibrium equations

The equations of motion are derived using the dynamic version of the principle of virtual displacements. The virtual strain energy and virtual kinetic energy are substituted into the virtual work statement. This is written as

$$\pi = U + T \tag{11}$$

$$U = \int_{-\frac{h}{2}}^{\frac{h}{2}}\int_0^a\int_0^b (\sigma_x\varepsilon_x + \sigma_y\varepsilon_y + \tau_{xy}\gamma_{xy} + \tau_{yz}\gamma_{yz} + \tau_{zx}\gamma_{zx})\, dx\, dy\, dz$$

$$T = \frac{1}{2}\int_{-\frac{h}{2}}^{\frac{h}{2}}\int_0^a\int_0^b \rho\,(\dot{u}^2 + \dot{v}^2 + \dot{w}^2)\, dx\, dy\, dz \tag{12}$$

The variation of the total potential should be equal to zero

$$\delta\pi = \delta U + \delta T =$$
$$\int_{-\frac{h}{2}}^{\frac{h}{2}}\int_0^a\int_0^b (\sigma_x\delta\varepsilon_x + \sigma_y\delta\varepsilon_y + \tau_{xy}\delta\gamma_{xy} + \tau_{yz}\delta\gamma_{yz} + \tau_{zx}\delta\gamma_{zx})\,dx\,dy\,dz +$$
$$+\int_{-\frac{h}{2}}^{\frac{h}{2}}\int_0^a\int_0^b \rho\,(\dot{u}\delta\dot{u} + \dot{v}\delta\dot{v} + \dot{w}\delta\dot{w})\,dx\,dy\,dz = 0 \qquad (13)$$

After integration through the thickness of the plate, the standard procedure is to integrate by parts to relieve the virtual generalized displacements and obtain the Euler-Lagrange equilibrium equations and boundary conditions. Since the stretching deformations are uncoupled from the bending in isotropic plates, only three of the five equations need to be analyzed for the out of plane vibrations of the plate.

4 Solution

The solution in this work is performed using the Extended Kantorovich Method and the order of integration and variation is switched. The unknown functions w_0, ϕ_x, and ϕ_y and can be separated into two functions of x and y, and a function of time. The proposed solution is taken as

$$w_0(x,y,t) = W(x)\,w(y)\sin\omega t \qquad (14)$$

$$\phi_x(x,y,t) = F(x)\,f(y)\sin\omega t \qquad (15)$$

$$\phi_y(x,y,t) = G(x)\,g(y)\sin\omega t \qquad (16)$$

The function F, G, and W are functions of x and are assumed functions. Thus, all its derivatives and integrals can be performed explicitly. The lower case functions f, g, and w are functions of y only, and are the unknowns in this method. Then the integration in the z and x directions can be performed and the potential becomes

$$\pi = \int_0^b (\bar{U} + \bar{T})dy \qquad (17)$$

Substitution of the assumed solution in the potential equation (Eq. 6) and integration through the thickness and along the x direction yields

$$\begin{aligned}\bar{U} = {} & C_1(1-\nu)(F_0F_0f_1f_1 + G_1G_1g_0g_0 + 2F_0G_0f_1g_0 + \\ & 4G_0F_1f_0g_1 + 2F_1F_1f_0f_0 + 2G_0G_0g_1g_1) \\ & + C_2(1-\nu)(2W_1W_1w_1w_1 + W_0W_0w_2w_2 + W_1W_1w_0w_0 + 2W_0W_2w_0w_2) \\ & + C_3(-F_0W_1f_1w_1 - G_1W_1g_0w_1 - \nu G_0W_2g_1w_0 - F_1W_2f_0w_0 \\ & - \nu F_1W_0f_0w_2 - G_0W_0g_1w_2) \\ & + C_4((G_0G_0g_0g_0 + W_0W_0w_1w_1 + 2F_0W_1f_0w_0 + 2G_0W_0g_0w_1 + \\ & F_0F_0f_0f_0 + W_1W_1w_0w_0)(1-\nu^2))\end{aligned} \qquad (18)$$

With the constants from the thickness direction integrations

$$C_1 = \frac{17Eh^3}{630(1-\nu^2)}$$
$$C_2 = \frac{Eh^3}{252(1-\nu^2)}$$
$$C_3 = \frac{8Eh^3}{315(1-\nu^2)} \tag{19}$$
$$C_4 = \frac{4Eh}{5(1-\nu^2)}$$

and the known terms from the integrals in the x direction

$$M_i N_j = \int_0^a \frac{d^i M}{dx^i} \frac{d^j N}{dx^j} dx, \quad i,j = 0,1,2, \quad M,N = F,G,W \tag{20}$$

and the unknowns from the y direction are written as

$$m_i n_j = \frac{d^i m}{dy^i} \frac{d^j n}{dy^j}, \quad i,j = 0,1,2, \quad m,n = f,g,w \tag{21}$$

The variation of the total potential is now reduced to the variation of f, g, and w so that we have

$$\delta(m_i n_j) = \int_0^b \left(\frac{d\delta^i m}{dy^i} \frac{d^j n}{dy^j} + \frac{d^i m}{dy^i} \frac{d\delta^j n}{dy^j} \right) dy, \quad i,j = 0,1,2, \quad m,n = f,g,w \tag{22}$$

Integration by parts of these expressions to relieve the δm and δn yields the following:

$$\begin{aligned} \delta(m_i n_j) &= \int_0^b \left(\frac{d\delta^i m}{dy^i} \frac{d^j n}{dy^j} + \frac{d^i m}{dy^i} \frac{d\delta^j n}{dy^j} \right) dy \\ &= \left. \frac{d\delta^{i-1} m}{dy^{i-1}} \frac{d\delta^j n}{dy^j} \right|_0^b + \left. \frac{d^i m}{dy^i} \frac{d\delta^{j-1} n}{dy^{j-1}} \right|_0^b - \\ &- \int_0^b \left(\frac{d\delta^{i-1} m}{dy^i} \frac{d^{j+1} n}{dy^{j+1}} + \frac{d^{i+1} m}{dy^{i+1}} \frac{d\delta^{j-1} n}{dy^j} \right) dy \end{aligned} \tag{23}$$

This integration is performed once for $i, j = 1$, and twice, recursively, for $i, j = 2$. The same steps are applied to the kinetic energy expression as follows

$$\begin{aligned} \bar{T} &= C_5 \left(F_1 F_1 f_0 f_0 + G_0 G_0 g_1 g_1 \right) - \\ &- C_6 \left(F_1 W_2 f_0 w_0 + G_0 W_0 g_1 w_2 \right) + \\ &+ C_7 \left(W_2 W_2 w_0 w_0 + W_0 W_0 w_2 w_2 \right) + C_8 W_0 W_0 w_0 w_0 \end{aligned} \tag{24}$$

with

$$C_5 = \frac{17\rho\omega^2 h^3}{315} \tag{25}$$

$$C_6 = \frac{8\rho\omega^2 h^3}{315}$$

$$C_7 = \frac{\rho\omega^2 h^3}{252}$$

$$C_8 = \rho\omega^2 h$$

The result of performing these steps on all terms in Eqs. (18) and (24) yields the differential equations of equilibrium by collecting terms with δf, δg, and δw. These equations are:

$$\begin{aligned} &C_1(1-\nu)(2F_0F_0f_2 + 2G_0(F_0+2F_1)g_1 + 4F_1F_1f_0) \\ &\quad - C_3(F_0W_1w_2 + F_1W_2w_0 + \nu F_1W_0w_2) \\ &\quad + C_4(1-\nu^2)(2F_0W_1w_0 + 2F_0F_0f_0) \\ &\quad + 2C_5F_1F_1f_0 - C_6F_1W_2w_0 = 0 \end{aligned} \tag{26}$$

$$\begin{aligned} &C_1(1-\nu)(2G_1G_1g_1 + 2G_0F_0f_2 + 4F_1G_0f_1 + 4G_0G_0g_2) \\ &\quad - C_3(G_0W_1w_1 + G_0W_0w_3 + \nu G_0W_2w_1) \\ &\quad + C_4(1-\nu^2)(2G_0W_0w_1 + 2G_0G_0g_0) \\ &\quad + 2C_5G_0G_0g_1 - C_6G_0W_0w_3 \\ &\quad + 2C_7W_2W_2w_0 + 2C_7W_0W_0w_4 + 2C_8W_0W_0w_0 = 0 \end{aligned} \tag{27}$$

$$\begin{aligned} &C_2(1-\nu)(4W_1W_1w_2 + 2W_0W_0w_4 + 2W_1W_1w_0 + \\ &\quad + 2W_0W_1w_2) - C_3(F_0W_1f_2 + G_1W_1g_1 + \nu G_0W_2g_2 + \\ &\quad + F_1W_2f_0 + \nu F_1W_0f_2 + G_0W_0g_3) + 2C_4(1-\nu^2)(W_0W_0w_2 + \\ &\quad + F_0W_1f_0 + G_0W_0g_1 + W_1W_1w_0) = 0 \end{aligned} \tag{28}$$

This is a set of three ordinary differential equations of order 2, 2, and 4, respectively, with constant coefficients. These are written in compact form as

$$PD = 0 \tag{29}$$

with

$$D^T = \{w'''', w''', w'', w', w, f'', f', f, g''', g'', g', g\} \tag{30}$$

and the terms in matrix $\mathbf{P}$, $P_{i,j}$ are given in the Appendix. These equations are similar to the equations of the higher order beam theory [3] but with more complex coefficients. and the dynamic stiffness method is used for the determination of the exact vibration frequencies.

5 Dynamic stiffness matrix derivation

The dynamic stiffness matrix is a 6 by 6 matrix relating the end forces to the end translations and rotations of the element. In the high order theory a higher order moment, in addition to the shear force and regular bending moment is encountered. The solution procedure for the set of 3 coupled differential equations is by using the exact element method [1]. In this case the degrees of freedom at the two ends of the plate strip are the vertical displacement and two rotations. Then the end forces will be a shear force and two bending moments, one of these due to the higher order assumption in the model. The solutions for *w*, *f*, and *g* are taken as infinite series in one normalized local coordinate. Substitution of these into the differential equations yield a set of 3 recurrence formulas for the w_{i+4}, f_{i+2}, and g_{i+2} in terms of the constants in matrix **P** and the previous 4, 2, and 2, terms in the 3 series, respectively. The stiffnesses are the values of the end forces once unit displacements are applied at each ends of the element and all the other degrees of freedom are kept equal to zero. This procedure is given in [2]. Then the stiffness matrix of the complete structure is assembled, and the external restraints are enforced. The resulting matrix is the stiffness matrix of the all the free degrees of freedom of the plate strip.

The natural frequencies of vibration are the values of frequency the cause the dynamic stiffness matrix of the assembled structure to become singular. This will be denoted the y direction natural frequency.

The solution is now performed again, where the *x* and *y* direction functions in Eq. (14) are switched. The same procedure is applied and a new converged natural frequency is obtained in this direction. This is the *x* direction natural frequency. This completes one cycle of analysis. At this point convergence is checked with the two direction frequencies, and if they differ by more then a preset criteria, another cycle of analysis is performed. The convergence of the procedure was achieved after 4 cycles at most for all combinations of boundary conditions tested.

6 Example

A square plate simply supported on all four edges is analyzed as an example. The thickness is $h=a/10$, where a is the side length. In this case the solution for the displacement *w* is sinusoidal in both the *x* and *y* directions. The natural frequencies from the current solution are compared with results from [4] in Table 1.The CPT results are from the classical plate theory, and FSDT results are from the first order shear deformation theory using Mindlin's formulation. These results are obtained with only one cycle of analysis as the Extended Kantorovich Method converges in the cases with two opposite supports simply supported within one cycle.

Table 1: Four normalized natural frequencies for example plate

m	*n*	CPT	FSDT	HDT
1	1	0.0963	0.0930	0.0931
3	1	0.4816	0.4149	0.4158
2	2	0.3853	0.3406	0.3411
3	3	0.8669	0.6834	0.6862

7 Summary

In this paper the free vibration frequencies of higher order plate model is analyzed. The differential equations of motion are derived using a variational pricple. This results in three coupled partial differential equations for the out of plane vibrations of the plate. Then the equations are reduced to a set of three coupled ordinary differential equations by employing the Kantorovich method. The solution for the resulting strip is found using the exact element method that enables to compute the dynamic stiffness matrix of the strip. Then the natural frequencies are found as the values of frequency that cause the determinant of the structure assembled matrix to become zero. Then, switching the roles of the assumed and calculated solutions in the Kantorovich method another analysis is performed to find the frequencies of the strip in the second direction. This completes a cycle of analysis. The two frequencies are compared to see if they differ by more then a preset value. If needed, another cycle of analysis is performed until satisfactory convergence. High accuracy values of natural frequencies are obtained using this procedure for arbitrary combinations of plate boundary conditions.

References

1. Eisenberger M (2003) Int J Num Meth Engngn, in press
2. Eisenberger M (1990) An Exact Element Method. Int J Num Meth Engngn 30: 363-370
3. Reddy JN (1999) Theory and Analysis of Elastic Plates, Taylor and Francis
4. Leissa AW (1969) Vibration of Plates NASA

Appendix

The terms in the **P** matrix are as follows:

$$P_{1,1} = 0 \tag{A1}$$

$$P_{1,2} = 0 \tag{A2}$$

$$P_{1,3} = -C_3(F_0W_1 - \nu F_1W_0) \tag{A3}$$

$$P_{1,4} = 0 \tag{A4}$$

$$P_{1,5} = 2C_4(1-\nu^2)F_0W_1 - C_3F_1W_2 - C_6F_1W_2 \tag{A5}$$

$$P_{1,6} = 2C_1(1-\nu)F_0F_0 \tag{A6}$$

$$P_{1,7} = 0 \tag{A7}$$

$$P_{1,8} = 4C_1(1-\nu)F_1F_1 + 2C_4(1-\nu^2)F_0F_0 + 2C_5F_1F_1 \tag{A8}$$

$$P_{1,9} = 0 \tag{A9}$$

$$P_{1,10} = 0 \tag{A10}$$

$$P_{1,11} = 2G_0(F_0 + 2F_1) \tag{A11}$$

$$P_{1,12} = 0 \tag{A12}$$

$$P_{2,1} = 2C_7W_0W_0 \tag{A13}$$

$$P_{2,2} = -C_3G_0W_0 - C_6G_0W_0 \tag{A14}$$

$$P_{2,3} = 0 \tag{A15}$$

$$P_{2,4} = -C_3(G_0W_1 + \nu G_0W_2) + 2C_4(1-\nu^2)G_0W_0 \tag{A16}$$

$$P_{2,5} = 2C_7W_2W_2 + 2C_8W_0W_0 \tag{A17}$$

$$P_{2,6} = 2C_1(1-\nu)G_0F_0 \tag{A18}$$

$$P_{2,7} = 4C_1(1-\nu)F_1G_0 \tag{A19}$$

$$P_{2,8} = 0 \tag{A20}$$

$$P_{2,9} = 0 \tag{A21}$$

$$P_{2,10} = 4C_1(1-\nu)G_0G_0 \tag{A22}$$

$$P_{2,11} = 2C_1(1-\nu)G_1G_1 + 2C_5G_0G_0 \tag{A23}$$

$$P_{2,12} = 2C_4(1-\nu^2)G_0G_0 \tag{A24}$$

$$P_{3,1} = 2C_2(1-\nu)W_0W_0 \tag{A25}$$

$$P_{3,2} = 0 \tag{A26}$$

$$P_{3,3} = 2C_2(1-\nu)(2W_1W_1 + W_0W_1) + 2C_4(1-\nu^2)W_0W_0 \tag{A27}$$

$$P_{3,4} = 0 \tag{A28}$$

$$P_{3,5} = 2C_2(1-\nu)W_1W_1 + 2C_4(1-\nu^2)W_1W_1 + C_8W_0W_0 \tag{A29}$$

$$P_{3,6} = -C_3(F_0W_1 + \nu F_1W_0) \tag{A30}$$

$$P_{3,7} = 0 \tag{A31}$$

$$P_{3,8} = -C_3F_1W_2 + 2C_4(1-\nu^2)F_0W_1 \tag{A32}$$

$$P_{3,9} = -C_3G_0W_0 \tag{A33}$$

$$P_{3,10} = -C_3\nu G_0W_2 \tag{A34}$$

$$P_{3,11} = -C_3G_1W_1 + 2C_4(1-\nu^2)G_0W_0 \tag{A35}$$

$$P_{3,12} = 0 \tag{A36}$$

Anisotropic Thermo-Creep-Damage in 3D Thick Plate vs. Reissner's Approach

A. Ganczarski and J. Skrzypek

Institute of Mechanics and Machine Design, Cracow University of Technology, Kraków, Poland

1 Introduction

The objective of this paper is to verify the Reissner approach versus 3D formulation in case of a rotationally-symmetric structure subjected to thermo-mechanical loading under anisotropic creep-damage conditions.

2 Basic equations

2.1 Equations of mechanical state

The general assumptions of rotational symmetry are fulfilled by displacement and temperature fields. The linear strain is decomposed into the elastic, creep and thermal parts $\varepsilon_{ij} = \frac{1}{2}(u_{i,j} + u_{j,i}) = \varepsilon_{ij}^{\mathrm{e}} + \varepsilon_{ij}^{\mathrm{c}} + \alpha\theta\delta_{ij}$, where additionally, the creep strain is incompressible $\varepsilon_{kk}^{\mathrm{c}} = 0$.

2.1.1 Plate as the 3D continuum

The system of displacement equations for the 3D continuum takes the form (cf. Ganczarski and Skrzypek 2001)

$$G\nabla^2 u_i + \frac{G}{1-2\nu} u_{j,ji} = 2G\varepsilon_{ij,j}^{\mathrm{c}} \tag{1}$$

where the stress is defined as follows

$$\sigma_{ij} = 2G\{\varepsilon_{ij} - \varepsilon_{ij}^{\mathrm{c}} + \frac{1}{1-2\nu}[\nu\varepsilon_{kk} - (1+\nu)\alpha\theta]\delta_{ij}\} \tag{2}$$

and $\theta = T - T_{\text{ref}} = \text{const}$.

2.1.2 Reissner's membrane-plate

To describe the Reissner's membrane-plate the simplified von Kármán system of equations is considered (i , j = 1, 2, $i \neq j$ and no sum over j!) (cf. Ganczarski et al. 1997, Ganczarski 2000)

$$\mathsf{B}[\nabla^2 u_i + \frac{1+\nu}{2}(u_{j,i} - u_{i,j})_{,j}] = \frac{h}{2}\frac{\nu}{1-\nu}p_{,i} + N^{\text{c}}_{ij,j} \tag{3}$$

$$\mathsf{D}\nabla^4 w - N_{ij} w_{,ij} = p - \frac{h^2}{10}\frac{2-\nu}{1-\nu}\nabla^2 p + \mathsf{D}\nabla^2(Q^{\text{c}}_{i,i}) - M^{\text{c}}_{ij,ij}$$

where not only shear stress but also the transverse stress are taken into account

$$\sigma_{3i} = -\frac{3Q_i}{2h}(1-4\zeta^2), \quad \sigma_{33} = -\frac{3p}{4}(\tfrac{2}{3} - 2\zeta + \tfrac{8}{3}\zeta^3), \quad \zeta = \frac{x_3}{h} \tag{4}$$

and inelastic and thermal stress resultants take the form (k = 1, 2)

$$N^c_{ij} = \mathsf{B}\int_{-1/2}^{1/2}[(1-\nu)\varepsilon^c_{ij} + \nu\varepsilon^c_{kk})]d\zeta \quad M^c_{ij} = \mathsf{B}h\int_{-1/2}^{1/2}[(1-\nu)\varepsilon^c_{ij} + \nu\varepsilon^c_{kk})]\zeta d\zeta \tag{5}$$

$$Q^c_i = \int_{-1/2}^{1/2}\frac{3\gamma^c_i}{2}(1-4\zeta^2)d\zeta$$

2.2 Constitutive equations of creep and damage

The anisotropic constitutive equations for creep and damage are taken into account in order to describe the damage anisotropy of copper. The Murakami's type creep law (Murakami et al. 1988) is used

$$\dot{\varepsilon}^{\text{c}}_{ij} = \frac{3}{2}[A_1\sigma^{n_1-1}_{\text{eq}}\bar{\alpha}\exp(-\bar{\alpha}t)s_{ij} + A_2\tilde{\sigma}^{n_2-1}_{\text{eq}}\tilde{s}_{ij}] \tag{6}$$

where

$$s_{ij} = \sigma_{ij} - \frac{1}{3}\sigma_{kk}\delta_{ij}, \quad \tilde{s}_{ij} = \tilde{\sigma}_{ij} - \frac{1}{3}\tilde{\sigma}_{kk}\delta_{ij} \tag{7}$$

together with the anisotropic damage evolution equation

$$\dot{D}_{ij} = B\chi^k(\sigma_{ij})(\psi^{-1}_{pq}v^{(1)}_p v^{(1)}_q)^l[(1-\eta)\delta_{ij} + \eta v^{(1)}_i v^{(1)}_j] \tag{8}$$

for the effective stress tensor by Murakami and Ohno 1981

$$\tilde{\sigma}_{ij} = \tfrac{1}{2}(\sigma_{pj}\psi^{-1}_{ip} + \psi^{-1}_{qj}\sigma_{iq}), \quad \psi_{ij} = \delta_{ij} - D_{ij} \tag{9}$$

$v^{(1)}_i$ stands for the unit eigenvector associated with maximum tensile eigenstress.

In order to exclude the inconsistency of original Murakami's damage evolution law (8) in case of rotational-symmetry the following modification of parameter controlling isotropy/anisotropy is postulated (cf. Ganczarski and Skrzypek 2001)

$$\eta = \eta_0 [1 - (1 - \frac{\langle \sigma_3 \rangle}{2\langle \sigma_2 \rangle}) \frac{\langle \sigma_2 \rangle + \langle \sigma_3 \rangle}{\langle \sigma_1 \rangle}] \tag{10}$$

The general form of the Hayhurst's type stress function

$$\chi(\sigma_{ij}) = \xi \sigma_1 + \varsigma \sigma_{\mathrm{eq}} + \frac{1 - \xi - \varsigma}{3} \sigma_{kk} \tag{11}$$

simplifies in case of copper anisotropic damage growth controlled by the maximum tensile eigenstress $(\xi = 1, \varsigma = 0)$.

3 Thermo-mechanical boundary problem

Simple support is assumed for the 3D plate under mechanical load. This condition applied to the structure under constant and homogeneous temperature field essentially constraints thermal transverse expansion at the sidewall so, for the thermal problem a separate boundary conditions are proposed (Fig. 1).

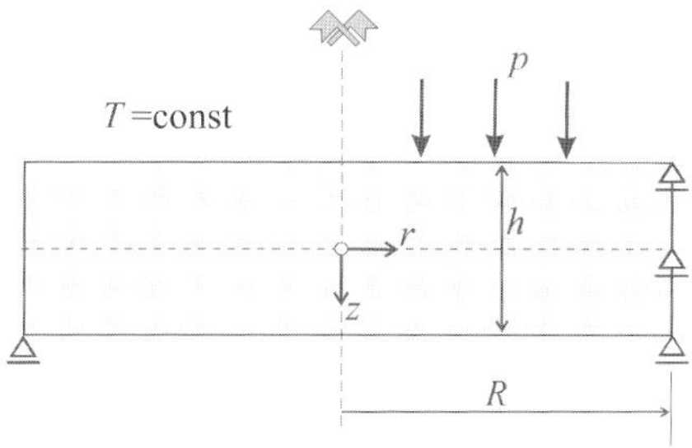

Fig. 1. Support of 3D plate

Finally, two sets of the boundary conditions are used to solve mechanical problem separately for the purely mechanical and the constant temperature field (Table 1), which requires a decomposition of the displacement into purely mechanical and purely thermal components $u_i = u_i^{\mathrm{m}} + u_i^{\mathrm{th}}$.

Table 1. Boundary conditions for 3D plate under combined thermo-mechanical loading

	mechanical loading	thermal loading
upper surface	$\sigma_z = -p \quad \sigma_{rz} = 0$	$\sigma_z = -p \quad \sigma_{rz} = 0$
lower surface	$\sigma_z = 0 \quad \sigma_{rz} = 0$	$\sigma_z = 0 \quad \sigma_{rz} = 0$
symmetry axis	$u_r^{\mathrm{m}} = 0 \quad u_{z,r}^{\mathrm{m}} = 0$	$u_r^{\mathrm{th}} = 0 \quad u_{z,r}^{\mathrm{th}} = 0$
sidewall	$\sigma_z = 0 \quad u_z^{\mathrm{m}} = 0$	$\sigma_z = 0 \quad \sigma_{rz} = 0$
lower corner		$u_z^{\mathrm{th}} = 0$

The boundary conditions of the simply supported Reissner's plate are as follows

$$u(0)=0,\quad w_{,r}(0)=0,\quad N_r(R)=0,\quad M_r(R)=0,\quad w(R)=0 \qquad (12)$$

4 Results

The example presented in this paper is based on the material constants referring to the copper at 523K (cf. Murakami et al. (1988), $T_{\mathrm{ref}} = 0$ °C, $E = 60.24$ GPa, $\sigma_0 = 11.0$ MPa, $\nu = 0.3$, $l = 5.0$, $k = 5.55$, $\xi = 1.0$, $n_1 = 2.66$, $n_2 = 7.10$, $B = 4.46\times10^{-13}\,\mathrm{MPa}^{-k}\,\mathrm{h}^{-k}$, $A_1 = 2.40\times10^{-16}\,\mathrm{MPa}^{-n1}$, $A_2 = 3.00\times10^{-16}\,\mathrm{MPa}^{-n2}\,\mathrm{h}^{-k}$, $\bar{\alpha} = 0.05\,\mathrm{h}^{-k}$, $\alpha = 2.5\times10^{-5}$, characteristic ratio of plate (thickness / radius) and load magnitude are equal to $r/h = 0.7$, $p = 0.1\times\sigma_0$, magnitude of parameter controlling type of anisotropy $\eta_0 = 0.5$.
The maximal principal stress type sensitivity to damage of the copper and purely mechanical load cause the first macrocrack to appear at the center on a bottom plate surface with respect to hoop direction (Fig. 2).

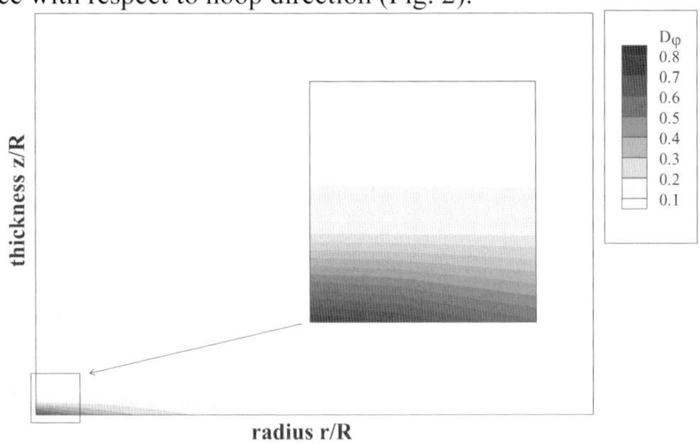

Fig. 2. Zone of hoop damage in 3D plate

Contrary to the Reissner's approach (Ganczarski and Skrzypek 2000), in the 3D formulation the hoop stress is not linear. The other stress components: the axial σ_z and the shear stress σ_{rz} exhibit qualitative and quantitative changes. If the axial stress σ_z may be well approximated by the Reissner cubic parabola (4) during primary and secondary creep in central part of the plate then tertiary creep and specially boundary effect require full 3D analysis (Fig. 3). Similarly, the shear stress σ_{rz} cannot be properly described by the Reissner's parabola (4) even for elastic case because it looses symmetry with respect to middle surface and, what is the most essential, it follows creep-damage process (cf. Okumura and Oguma

1998) (Fig. 4). The shape of supported edge does not obey any of generally applied hypotheses (Fig. 5). Detail analysis reveals essential deformation of the middle surface which no longer coincides the neutral surface.

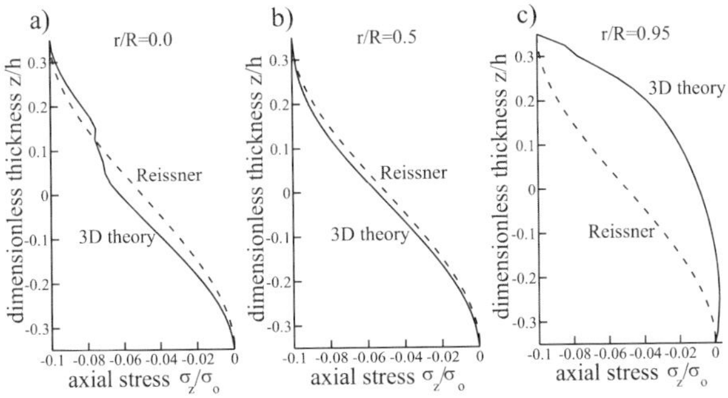

Fig. 3. Comparison of axial stress at the instant of first macrocrack initiation and the Reissner's theory: a) axis of symmetry, b) central part, c) supported edge

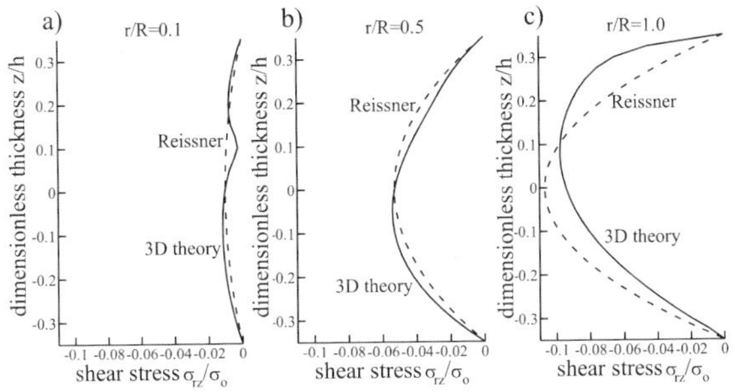

Fig. 4. Comparison of shear stress at the instant of first macrocrack initiation and the Reissner's theory: a) axis of symmetry, b) central part, c) supported edge

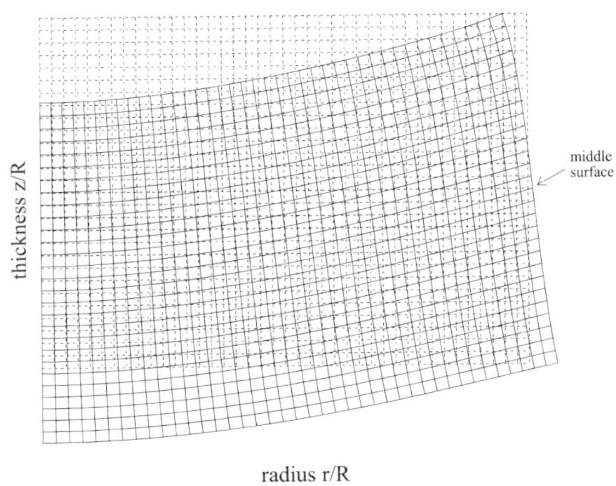

Fig. 5. Deformation of the plate under purely mechanical load (magnification 1250)

5 Conclusions

Applicability of the Reissner's theory is limited to the case of thin-wall structures working in primary and secondary creep conditions (cf. Betten et al. 1989, Naumenko 1996), and conditionally admissible in case of structures of moderate thickness. However, appropriate analysis of thick-wall structures working in tertiary creep-damage conditions requires that any constraints must be ruled out and full 3D formulation of continuum mechanics must be applied.

Acknowlegdement

The grant PB 7 T07A 03819 from the State Committee for Scientific Research KBN Poland is gratefully acknowledged.

References

Betten J, Borrmann M, Butters T (1989) Materialgleichungen zur Beschreibung des primärem Kreichverhaltens innendruckbeanspruchter Zylinderschalen aus isotropem Werkstoff. Ing Arch 67: 99-109

Ganczarski A, Freindl L, Skrzypek J (1997) Orthotropic brittle rupture of Reissner's prestressed plates. Proc. COMPAS V: 1904-1909

Ganczarski A, Skrzypek J (2000) Damage effect on thermo-mechanical fields in a mid-thick plate. J Theor Appl Mech 2, 38: 271-284

Ganczarski A, Skrzypek J (2001) Application of the modified Murakami's anisotropic creep damage model to 3D rotationally-symmetric problem. Tech Mech 21, 4: 251-260

Murakami S, Ohno N (1981) A continuum theory of creep and creep damage. In: Ponter ARS, Hayhurst DR (eds) Creep in Structures. Springer-Verlag pp 442-444

Murakami S (1988) Mechanical modelling of material damage. J Appl Mech 55, 6: 280-286

Murakami S, Kawai M, Rong H (1988) Finite element analysis of creep crack growth by a local approach. Int J Mech Sci 30, 7: 491-502

Naumenko K (1996) Modellierung und Berechnung der Langzeitfestigkeit dünnwandiger Flächentragwerke unter Einbeziehung von Werkstoffkriechen und Schädigung. Otto-von-Guericke Universität Magdeburg, Preprint MBI-96-2

Okumura IA, Oguma Y (1998) Series of solutions for transversely loaded and completely clamped thick rectangular plate based on the three-dimensional theory of elasticity. Arch Appl Mech. 68: 103-121

On a Fast Method for Buckling Load Calculations of Incomplete Ribbed Shells

G. D. Gavrylenko

S P Timoshenko Institute of Mechanics, National Academy of Sciences, Kiev

1 Introduction

A small number of studies (Gavrylenko, 1989,1999) are devoted to the evaluation of the critical loading of incomplete ribbed shells. There are only a few calculations (Singer et al., 1971) for ribbed shells. New approaches are given in Arbocz et al., (2001), and Gavrylenko & Croll, (2001).

The report is devoted to the problem of the stability and load-carrying capacity of imperfect shells in a nonhomogeneous stress-strain state. A finite-difference method was proposed and used for the analysis of critical loads of shells with initial imperfections. The technique takes into account the discreteness in the rib arrangement in two directions and the nonlinear subcritical state of the shells.

2 Procedure for determination of critical loads

An elastic orthotropic thin-walled shell is considered. The shell is stiffened with discretely arranged longitudinal and transverse ribs, which in cylindrical shells are called stringers and frames respectively (Fig. 1). A ribbed shell is a system consisting of a smooth casing and discretely disposed ribs. In actual structures stiffening ribs are usually located on one side of the casing (either inside or outside), i.e., their center of gravity is located with an eccentricity relative to the median surface of the casing. The length of the shell is l, the radius of the median surface is r, and the thickness of the casing is t.

In practice, casing and stiffeners are imperfect and have initial geometrical imperfections in the form of radial deflections. These imperfections are assumed to be given as double trigonometric Fourier series.

It is assumed that the initial deflections are not accompanied by initial stresses. The behavior of the casing in the subcritical state is described by nonlinear differential equations of Mushtari-Donnel-Vlasov type, presented in (Gavrylenko, 1972, 1989; Kornishin, 1964; Arbocz and Babcock, 1969; Donnell, 1934).

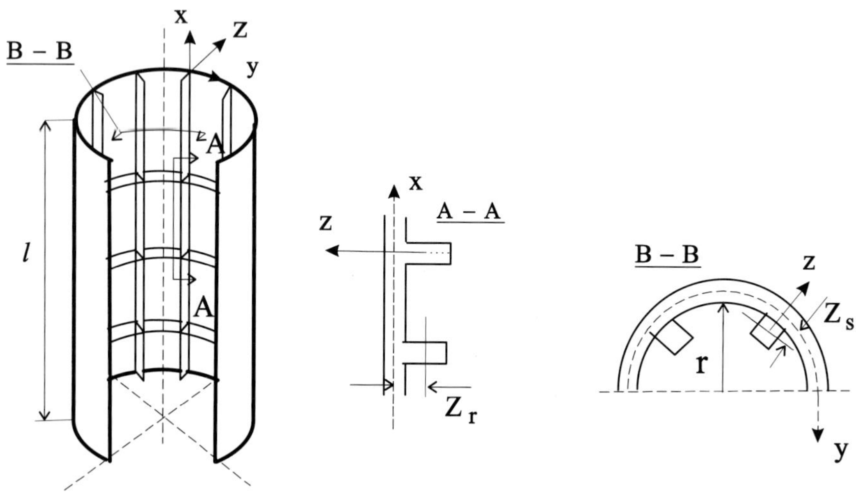

Fig. 1. Notation and convention adopted for geometry

$$\nabla^4\ \overline{\Phi} - \frac{1}{r}\overline{w}_{,xx} + \frac{1}{2}\ L\left(\overline{w}, \overline{w} + 2\overline{w}_H\right) = 0$$

$$\nabla^4\ \overline{w} + \frac{1}{r}\overline{\Phi}_{,xx} - L\left(\overline{\Phi}, \overline{w} + \overline{w}_H\right) + \mathrm{q} = 0 \tag{1}$$

where

$$\nabla^4\ \overline{\Phi} = \mathrm{A}_2\frac{\partial^4\overline{\Phi}}{\partial x^4} + (\mathrm{A}_{12}+\mathrm{A}_{21}+\mathrm{A}_3)\frac{\partial^4\overline{\Phi}}{\partial x^2\partial y^2} + \mathrm{A}_1\frac{\partial^4\overline{\Phi}}{\partial y^4}$$

$$\nabla^4\ \overline{w} = \mathrm{D}_1\frac{\partial^4\overline{w}}{\partial x^4} + (\mathrm{D}_{12}+\mathrm{D}_{21}+4\mathrm{D}_3)\ \frac{\partial^4\overline{w}}{\partial x^2\partial y^2} + \mathrm{D}_2\frac{\partial^4 w}{\partial y^4}$$

$$L\left(\overline{w}, \overline{w}+2\overline{w}_H\right) = \frac{\partial^2\overline{w}}{\partial x^2}\frac{\partial^2\left(\overline{w}+2\overline{w}_H\right)}{\partial y^2} + \frac{\partial^2\overline{w}}{\partial y^2}\frac{\partial^2\left(\overline{w}+2\overline{w}_H\right)}{\partial x^2}$$

$$-2\frac{\partial^2\overline{w}}{\partial y\partial y}\frac{\partial^2\left(\overline{w}+2\overline{w}_H\right)}{\partial x\partial y}$$

$$\mathrm{A}_2 = \frac{1}{E_2 t}, \qquad \mathrm{A}_{21} = \mathrm{A}_{12} = -\nu\mathrm{A}_1, \qquad \mathrm{A}_1 = \frac{1}{E_1 t},\ \mathrm{A}_3 = \frac{1}{Gt}$$

$$\mathrm{D}_\mathrm{i} = \frac{E_i t^3}{12(1-\nu_1\nu_2)}, \qquad i = 1,2 \tag{2}$$

$$D_{12} = \nu_2 D_1, \qquad D_{21} = \nu_1 D_2, \qquad D_3 = \frac{Gt^3}{12}$$

E_1 and E_2 are moduli of elasticity, G is a shear modulus, $\bar{\Phi}$ is a function of forces, $\bar{w}$ is a function of radial deflections, $\bar{w}_H$ is a function of given initial imperfections, q is internal pressure, and the operator $L\left(\bar{\Phi}, \bar{w} + \bar{w}_H\right)$ is similar to $L(\bar{w}, \bar{w} + 2\bar{w}_H)$ by construction.

The system of nonlinear Eqs. (1) is solved using the procedure of successive approximations. To do this, the equations are linearized (Gavrylenko, 1999) using the representations

$$\bar{w} = w_0 + w, \qquad \bar{\Phi} = \Phi_0 + \Phi, \tag{3}$$

where Φ_0 and w_0 are adjusted quantities, and Φ and w are additions or increments.

The system for the additions takes a dimensionless form (the bars with $\bar{w}$ and $\bar{q}$ are omitted)

$$\begin{aligned} &\nabla^4 \varphi - \frac{1}{2} w_{,\alpha\alpha} + \frac{1}{\kappa} L\left(w, w_0 + w_H\right) = \\ &-\left[\nabla^4 \varphi_0 - \frac{1}{2} w_{0,\alpha\alpha} + \frac{1}{2\kappa} L\left(w_0, w_0 + 2w_H\right)\right] \end{aligned} \tag{4}$$

$$\begin{aligned} &\nabla^4 w + \frac{\kappa^2}{2} \varphi_{,\alpha\alpha} - \kappa L\left(w_0 + w_H, \varphi\right) - \kappa L\left(w, \varphi_0\right) = \\ &-\left[\nabla^4 w_0 + \frac{\kappa^2}{2} \varphi_{0,\alpha\alpha} - \kappa L\left(w_0 + w_H, \varphi_0\right) + \frac{\kappa^2}{2} q\right] \end{aligned} \tag{5}$$

where all L (...) are, as earlier, similar to each other in construction,

$$\alpha = x/r, \quad \beta = y/r, \qquad \bar{w} = w\big/\sqrt{D_1 A_2}$$

$$\bar{q} = qr^2 \sqrt{A_2}\big/2\sqrt{D_1}; \qquad \varphi = \Phi \kappa A_2 / 4r^2$$

$$\kappa = 2r\sqrt{1/D_1 A_2}$$

$$\begin{aligned} &\nabla^4 \varphi = \varphi_{,\alpha\alpha\alpha\alpha} + 2s\varphi_{,\alpha\alpha\beta\beta} + d\varphi_{,\beta\beta\beta\beta} \\ &\nabla^4 w = w_{,\alpha\alpha\alpha\alpha} + 2 f_1 w_{,\alpha\alpha\beta\beta} + g w_{,\beta\beta\beta\beta}\,. \end{aligned} \tag{6}$$

Here,

$$d = \frac{E_2}{E_1}; \quad s = \frac{E_2}{E_1}\left(\frac{1}{2GtA_1} - \nu_1\right)$$

$$f_1 = \frac{D_{12} + D_{21} + 4D_3}{2D_1}$$

$$g = D_2 / D_1.$$

The system of Eqs. (4-5) is a generalized linear system of equations. These equations are applicable only to smooth shells. They invite further generalization for analysis of ribbed shells (Gavrylenko, 1989, 1999). Taking into account the forces of interaction between the shell and the ribs, we obtain expressions for jumps in the derivatives of the resolving functions. On further passage to finite differences, we obtain the algebraic system of equations.

The approach suggested is used the nonlinear differential equations Mushtari – Donnell – Vlasov type (Donnell, 1934; Gavrylenko, 1972, 1999). The system of nonlinear equations is solved using the procedure of successive approximation. First of all two used functions may be represented as having two parts one of them is fundamental solution and other is the incremental displacement about this fundamental state. Substituting them in nonlinear equations and neglecting of two order values for incremental displacements may be received the linearized system of equations. In order to solve these equations and to facilitate realizations on a PC, it is suggested the generalized notation. Here, the left sides of the equations form the linearized system of equations of neutral equilibrium, while the right sides constitute the system of nonlinear equations of the subcritical state. The equations are joined in pairs together with their boundary conditions into a single system of symbolic form:

$$L_k^{(1)} = - L_k + f_k \quad (k = 1, 2, 3)$$

$$L_i^{(1)} = - L_i + f_i \quad (i = 1, 2)$$

$$L_s^{(1)} = - L_s + f_s \quad (s = 1, 2, 3). \tag{7}$$

Here, L represents differential operators: $L_k^{(1)}$ and $L_s^{(1)}$ are the boundary conditions (BC) of the linearized problem for the top and the bottom end of the shell; L_k and L_s are the BC of the subcritical state (SS) on the same end; $L_i^{(1)}$ and L_i are the linearized equations of equilibrium and the SS. Here, f_k and f_s are the BC of the SS for the top and the bottom end, f_i are the BC of equations of the SS. The system (9) is used for derivation of the solutions of the initial approximation, nonlinear equations of the SS and determination of the parameters of the critical load. They invite further generalization for analysis of ribbed shells (Gavrylenko, 1999). Taking into account the forces of interaction between the shell and the ribs may be obtained expression for jumps in the derivations of the resolving functions. This method is used only two unknowns in each node of net and is more economical and the most fast among other methods where are used 5 or 6 unknowns in node. It allows prognosticate a carrying capacity of construction taking into account in complex cases the following main factors: laminated structure of material, imperfections of form, presence of discrete stiffened ribs of two directions, inelastic deformations.

3 Numerical examples

The value of $p_{1\min}$ *minimorum* is determined by comparing the different critical loads p_{1min} for various forms of initial deflections. Here, $p_{1\min}$ is the minimum value of the critical load $p_1 = p_{cr} / p_{sh}$, where p_{cr} is the critical load of the imperfect shell, $p_{sh} = 0{,}605\ (E\ t / r)\ F_{sh}$, and F_{sh} is the cross-sectional area of the casing. The influence of deflections is demonstrated in Fig. 2 and Fig. 3 ($r / t = 400$; $l / r = 2.1$; number stringers 32, cross-section of stringers $\angle 6{\times}4{\times}0.5$ mm) for external stringers. The curves $p_1 = f(B_{mn} / t)$ are shown in these figures. Let us consider the stability of compressed ribbed cylindrical shells with initial deflections with allowance for the eccentricity of ribs and the discreteness of their arrangement. Initial imperfections of the following types are presented: regular (R), spreading over the whole surface of the shell, and local (L), covering only a part of the surface and located at the shell center.

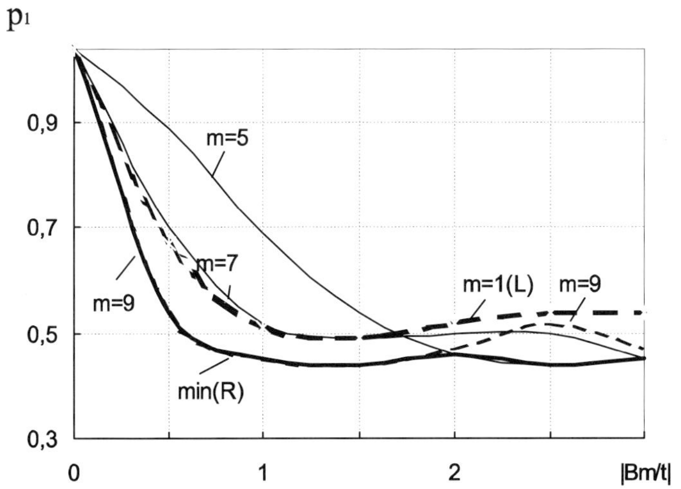

Fig. 2. Dependence of nondimensional parameter p_1 on nondimensional value $|B_m / t|$ for axisymmetric deflection

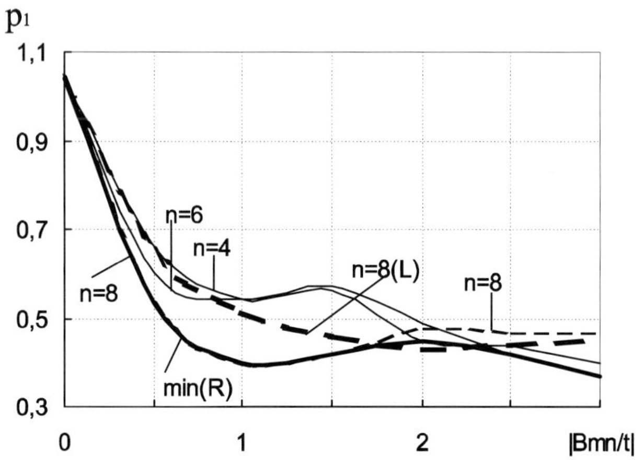

Fig. 3. Dependence of nondimensional parameter p_1 on nondimensional value $|B_{mn}/t|$ for asymmetric deflection

The effect of regular axisymmetric and asymmetric imperfections is shown in each figure for regular and local imperfections.
The dependence of the parameters p_1 on the nondimensional amplitude of the initial deflections is shown:

- for regular axisymmetric deflections $w_H = B_{mn} \sin(m\pi x/l)$ in Fig. 2 for variable m;
- for regular asymmetric deflections $w_H = B_{mn} \sin(m\pi x/l) \cos(ny/r)$ in Fig. 3 for fixed m and variable n;
- for *local axisymmetric deflections* (L) (Fig. 2), located at the center of the shell, for m=1, and they cover l/m_p of the shell length, where m_p is the number of half-waves of a regular axisymmetric deflection for which $p_{1\min}$ is achieved (a **thickened** curve for $p_{1\min}$ determined from the corresponding curves for variable m is presented here for comparison);
- for *local asymmetric deflections* (Fig. 3), which are also located at the shell center (for m= 1) and in the same region of the shell as the previous axisymmetric deflections and given by the formula:
  $w_H = B_{mn} \sin(\pi mx/l) \cos(ny/r)$.

4 Conclusions

- Within the limits of the dimensionless amplitudes considered, we can draw the following conclusions concerning the effect of *regular imperfections:*
  In the analysis of regular axisymmetric imperfections, it is possible to set the number of half-waves m for which $p_{1\min}$ is achieved (**thickened** curves in Fig. 2); in the analysis of regular asymmetric imperfections, it is possible for fixed m to set the number of waves n for which $p_{1\min}$ is achieved (**thickened** curves in Fig. 3).

- The effect of *local imperfections* is as follows:
  The critical loads in the presence of local axisymmetric imperfections (curves in Fig. 2 with index m=1) are usually greater than those for regular axisymmetric imperfections (Fig. 2). The critical loads in the presence of local asymmetric imperfections (curves in Fig. 3 with index m = 1) are usually greater than those for regular asymmetric imperfections (Fig. 3). The behavior of the parameters p_1 as a function of $|B_{mn} / t|$ is conventionally divided into two characteristic stages: the first one is for $|B_{mn} / t| \leq 1$ and the second one for $1 < |B_{mn} / t| \leq 3$.

- We can draw the following conclusions based on the analysis of all the above-mentioned numerical results irrespective of the type of imperfections:
  At the *first stage*, for small amplitudes of initial deflections, the parameters of the critical loads $p_{1\min}$ are sharply decreased (from 1 down to 0.3-0.4) with growth in the dimensionless amplitude of deflections; at the *second stage*, the parameters of the critical loads $p_{1\min}$ reach a minimum or change more smoothly with further growth in the amplitudes of initial deflections ($1 < |B_{mn} / t| \leq 3$); the parameters of the critical loads are virtually stabilized, reaching $p_{1\min}$ (Fig. 2), for some deflections in the same range of amplitudes.

References

Arbocz J, Babcock CD (1969) The effect of general imperfections on the buckling of Cylindrical shells. J Appl Mech Ser E 36, 1: 28-38

Arbocz J, Starnes J, Nemeth M (2001) Buckling predictions of imperfection sensitive shells. In: Euromech colloquium 424, The Netherlands

Donnell LH (1934) A new theory for the buckling of thin cylinders under axial compression and bending. Trans ASME 56: 795-806

Gavrylenko GD (1972) Numerical algorithm of stability analysis of constructively orthotropic shells under arbitrary boundary conditions. Fluid Mechanics and Theory of Elasticity. [in Russian] 14: 131-138

Gavrylenko GD (1989) Stability of ribbed cylindrical shells. In: Inhomogeneous Stress-Strain State. [in Russian], Nauk Dumka, Kiev

Gavrylenko GD (1989) Stability of ribbed cylindrical shells in a Nonuniform stress-strain state. Nauk Dumka, Ukraine

Gavrylenko GD (1999) Stability of ribbed shells of incomplete form. Inst of Mathematics of NAS of Ukraine

Gavrylenko GD, Croll J (2001) Buckling predictions of imperfection sensitive shells. In: Euromech Colloquium 424, The Netherlands, p 34

Kornishin MS (1964) Nonlinear problems of the theory of plates and shallow shells and methods of their solution. [in Russian] Nauka, Moscow

Singer J, Arbocz J, Babcock C (1971) Buckling of imperfect stiffened cylindrical shells under axial compression. AIAA J 9,1: 68-75

Vol'mir AS (1967) The stability of deformed systems. Nauka, Moscow

Adaptive Wavelet-Algorithms for Inelastic Shells

K. Hackl and U. Hoppe

Ruhr–Universität Bochum, Germany

1 Introduction

The development of shell theories has been marked by two competing approaches. There are on the one hand the „degenerated solid formulations", which offer the advantages of great flexibility and easy numerical implementation. On the other hand there are the „stress resultant based formulations" which comprise all classical shell and plate theories. Both approaches have been successfully applied in the analysis of thin elastic shells. However, when considering inelastic shells we are faced with certain difficulties. Because of history-dependent inelastic deformations the stresses become history-dependent themselves and show complicated and a priori unknown behavior across the thickness of the shell. For example, during buckling of a shell structure we typically have plastic loading and elastic unloading in a single cross section. Because of restrictions of the constitutive models used in the „stress resultant based formulations", the resulting theories are not able to capture effects involving simultaneous loading and unloading.

In Hackl [5] it was shown how to derive shell theories from three-dimensional elastoplastic models by expanding across the shell thickness all quantities involved with respect to arbitrary biorthogonal function systems. In this work we will use wavelet–expansions of the stresses across the thickness of the shell which prove to be especially effective in picking up the localized behavior of plastic flow. Special attention is devoted to establish numerical algorithms which are efficient, accurate and stable. Numerical examples for elastoplastic material are presented.

2 Stress resultants

A model of quasi-static inelastic material is used that is based on the concept of *generalized standard media.* We assume an internal stored energy $W = W(\mathbf{E}, \mathbf{K})$, depending on the Green-Lagrange strain tensor $\mathbf{E}$ and a set of internal variables $\mathbf{K}$. In the application we have in mind $\mathbf{K}$ represents the plastic strain, but in general it may account for additional effects, e.g., hardening or damage [6]. The thermo-mechanical dual variables are

$$\mathbf{S} = \partial_{\mathbf{E}} W , \quad \mathbf{Q} = -\partial_{\mathbf{K}} W , \tag{1}$$

where $\mathbf{S}$ denotes the second Piola-Kirchhoff stress tensor and $\mathbf{Q}$ contains the plastic (back-)stress. The flow rule $\dot{\mathbf{K}} = \lambda \partial_{\mathbf{Q}} \Phi$ governs the time evolution of the internal variables. In the context of the generalized standard media it is derived from a dissipation potential by means of the principle of maximum dissipation.

The stationary states of the inelastic problem follow from the equilibrium equations derived by energy minimization with respect to the deformation. In the context of shell theory the three-dimensional domain of integration can be split into a shell-midsurface and a thickness direction, leading to the following variational form of the specific internal energy

$$\begin{aligned} \delta I_{\text{int}} &= \int_{\omega} \int_{-h/2}^{+h/2} \mathbf{S}(\mathbf{E}, \mathbf{K}, \xi) \delta \mathbf{E} j_0 d\xi dA \\ &= \int_{\omega} \sum_{n} \left[\int_{-h/2}^{+h/2} \mathbf{S}(\mathbf{E}, \mathbf{K}, \xi) \xi^{n} j_0 d\xi \, \delta \mathbf{E} \right] dA , \end{aligned} \tag{2}$$

where j_0 constitutes a Jacobian. If the strain is expanded in a power series depending on the coordinate ξ normal to the shell-midsurface, $\mathbf{E} = \sum_{n=0}^{N} \mathbf{E}^{n} \xi^{n}$, the stress resultants read

$$\mathbf{S}^{n} = \int_{-h/2}^{+h/2} j_0 \mathbf{S}(\mathbf{E}, \mathbf{K}, \xi) \xi^{n} d\xi . \tag{3}$$

For nonlinear materials these integrals cannot be evaluated analytically in general and one has to resort to a numerical quadrature scheme:

$$\mathbf{S}^{n} = \sum_{i=1}^{K} \alpha_i j_0(\xi^{i}) \mathbf{S}(\mathbf{E}(\xi^{i}), \mathbf{K}(\xi^{i}), \xi^{i})(\xi^{i})^{n} . \tag{4}$$

Here ξ^{i} denote collocation points and α_i weight factors. The effective evaluation of Eq. (4) constitutes a crucial part of an algorithm for inelastic shells. On the one side it has to be sufficiently accurate, on the other side it is known that employing too many collocation points will render the scheme inefficient or can even cause numerical instabilities.

An appropriate integration scheme should be adaptive: It must provide the minimal number of integration points to compute the quadrature within a desired tolerance. With an increasing number of integration points, the internal variables at newly inserted intergration points must be reconstructable in a consistent way. Finally, it would be desirable if the scheme could also provide an error estimate.

We will discuss now an adaptive integration scheme which is based on wavelet-expansion of the stress distribution across the thickness of the shell.

3 Adaptive wavelet–expansion

A wavelet-expansion [3,4,7] of the stress is given in the form

$$\mathbf{F}(\xi) = \xi^n j_0(\xi)\mathbf{S}(\xi) = \sum_j \mathbf{g}^{0,j}\varphi_j + \sum_{i,j}\mathbf{h}^{i,j}\psi_{i,j}, \tag{5}$$

where φ denotes the scaling-function and ψ the mother wavelet, and φ_j and $\psi_{i,j}$ are obtained by scaling (dilation) and translation

$$\varphi_j = \varphi(2^0\xi - j), \qquad \psi_{i,j} = \psi(2^i\xi - j). \tag{6}$$

Figure 1 shows the corresponding plots of these functions for the so-called Haar-system, which is the simplest possible wavelet basis, and sketches how a piecewise constant function can be decomposed using the two base functions φ and ψ. Note that in the shell-model presented we use more involved wavelets not presented here.

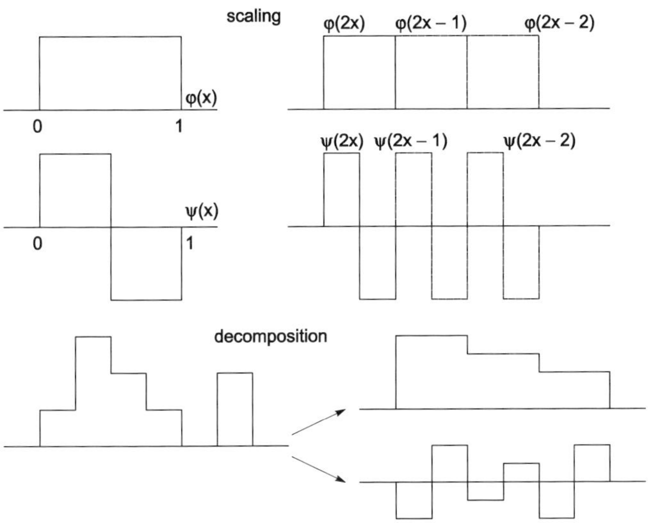

Fig. 1. Haar-function wavelets

The coefficients $\mathbf{h}^{i,j}$ reflect the smoothness of the function $\mathbf{F}(\xi)$. The following table provides an estimate of the order of $\mathbf{h}^{i,j}$ provided that $\tilde{\psi}$ has enough vanishing moments [3,4]:

F discontinuous	$\mathbf{h}^{i,j} \sim 1,$	
F continuous	$\mathbf{h}^{i,j} \sim 1/2^i,$	
F continuously differentiable	$\mathbf{h}^{i,j} \sim 1/4^i,$	(7)
F twice continuously differentiable	$\mathbf{h}^{i,j} \sim 1/8^i,$	

Another feature which is essential for an adaptive algorithm is the ability to reconstruct the values of the internal variables at newly introduced collocation points. In detail, if $\mathbf{K}(\xi_i)$, $i = 1,\ldots,N$, are the previous internal variables, then the new internal variables $\mathbf{K}(\bar{\xi}_i)$, $i = 1,\ldots,M$, must be reconstructed without being able to resort to the (inelastic) material law, whenever the integration order grows, $M > N$. A wavelet-expansion enables reconstruction directly by interpolation of the previous data.

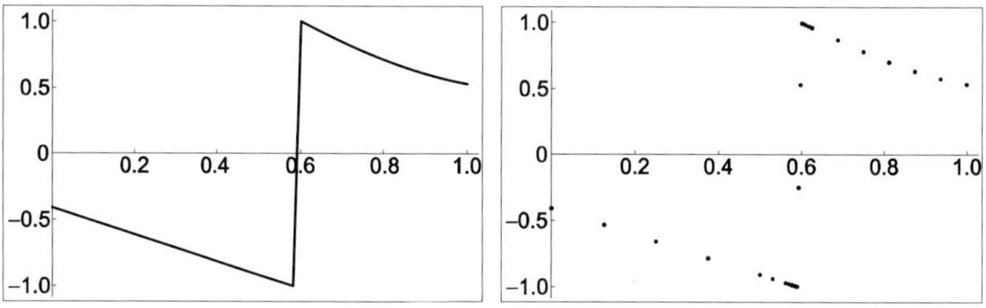

Fig. 2. Almost discontinuous function

Figure 2 shows the location of the quadrature points of a non-smooth function integrated by a wavelet-algorithm. We observe a concentration of the integration points in the vicinity of the discontinuity. The corresponding algorithm (BOX 1) consists of the following steps:

1. For a given scale i the stresses $\mathbf{S}$ are evaluated at the material points $\xi = 2^{-i}$ across the shell thickness. The scaling-function coefficients $\mathbf{g}^{i,j}$ are computed by direct evaluation of the function $\mathbf{F}$.
2. Next the wavelet-function coefficients $\mathbf{h}^{i-1,j}$ are computed by wavelet transformation [3,4].
3. If the magnitude of the wavelet coefficients is lower than a desired tolerance, the resolution of the corresponding scale is accepted and the algorithm is terminated. If the magnitude is above the desired threshold, then the internal variables are reconstructed on the smaller scale $i+1$ and the algorithm is restarted with $i := i+1$.

BOX 1: algorithm

1. $\mathbf{S}\left(\frac{j+c}{2^i}\right) \Rightarrow \mathbf{g}^{i,j} = 2^{i/2}\mathbf{F}\left(\frac{j+c}{2^i}\right)$ given
2. calculate $\mathbf{h}^{i,j} = \frac{1}{2}\sum_k \tilde{b}^{k-2j}\mathbf{g}^{i,k}$
3. $\left\|\mathbf{h}^{i-1,j}\right\| <$ tol? yes $\Rightarrow$ exit

   no $\Rightarrow \mathbf{g}_0^{i+1,k} = \sum_k a^{k-2j}\mathbf{g}_0^{i,j} \quad \Rightarrow \mathbf{K}_0(\bar{\xi}_i), \quad i := i+1$

   return to 1

4 Numerical examples

Scordelis-Lo roof. In order to illustrate some of the points discussed above, we consider a standard example often referred to as "Scordelis-Lo roof". The geometry, displacement boundary conditions and material parameters for this example are given in Figure 3. Due to the symmetry conditions, the analysis can be performed for one quarter of the roof only, restricting deformations to be symmetric along the lines of symmetry.

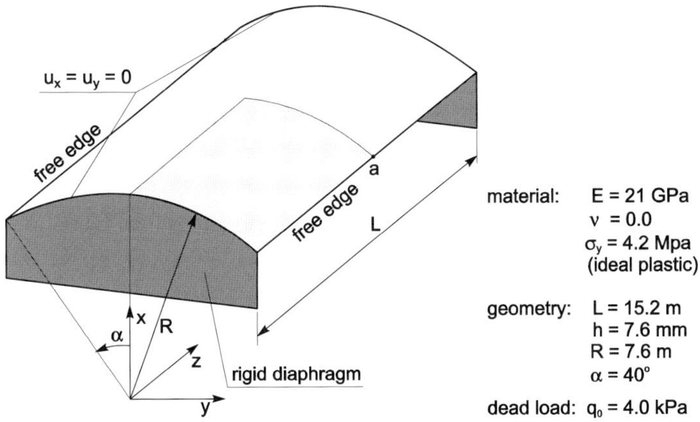

Fig. 3. Scordelis-Lo roof: geometry, boundary conditions and material data

This example has been analyzed by many authors and the respective results are shown in Figure 4. One of the first elastoplastic solutions for this problem has been provided by Peric and Owen [8] and Crisfield and Peng [2], who used the faceted shell formulations based on Morley thin-shell triangular elements which do not account for the transverse shear strains. Subsequently, the problem has been reconsidered by Roehl and Ramm [9], Brank et al. [1], Rouainia and Peric [10] and Schieck et al. [11]. There is a good agreement of the solutions below the maximum load. However, the maximum load and the post-buckling deformation differ significantly.

The most characteristic observation is that some of the obtained solutions predict a sharp drop in the force-displacement path while others do not. We believe that this qualitative difference is related to different modes of failure accompanied by localized deformations and formation of plastic hinges. The type of solution depends strongly on the kind of the numerical integration method for the through-the-thickness integration and the number of integration points. This can be seen obviously from Figure 5, where a transition from 3, 5 to 7 integration points using a Simpson quadrature scheme causes the solution to jump between qualitatively different results.

In the present study, although a kinematically simple shell model is used, the significance of the number of integration points on the result is observed too.

Figure 5 shows the results of the present study for a mesh of 36×36 shell elements with a maximum number of integration points between 5 and 33. We observe that the solution path changes significantly when the number of integration points is increased. For a maximum of 17 and 33 integration points respectively we obtain almost the same result.

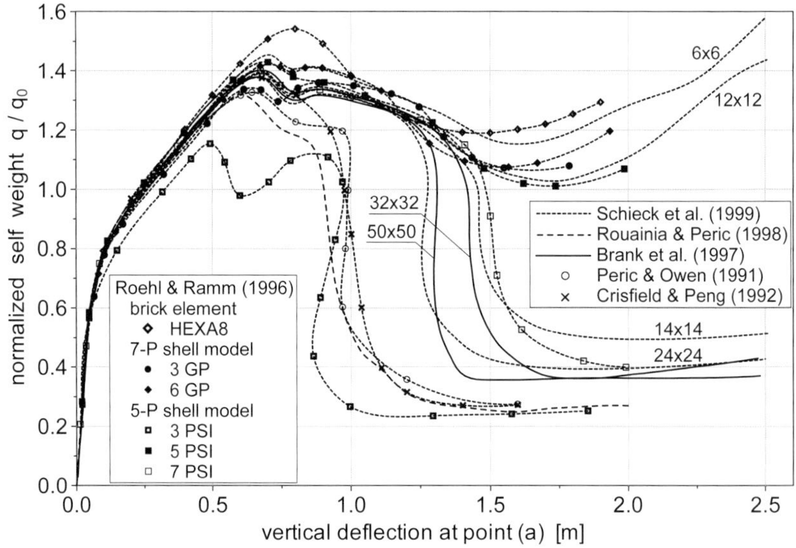

Fig. 4. Scordelis-Lo roof: elastoplastic solutions from the literature

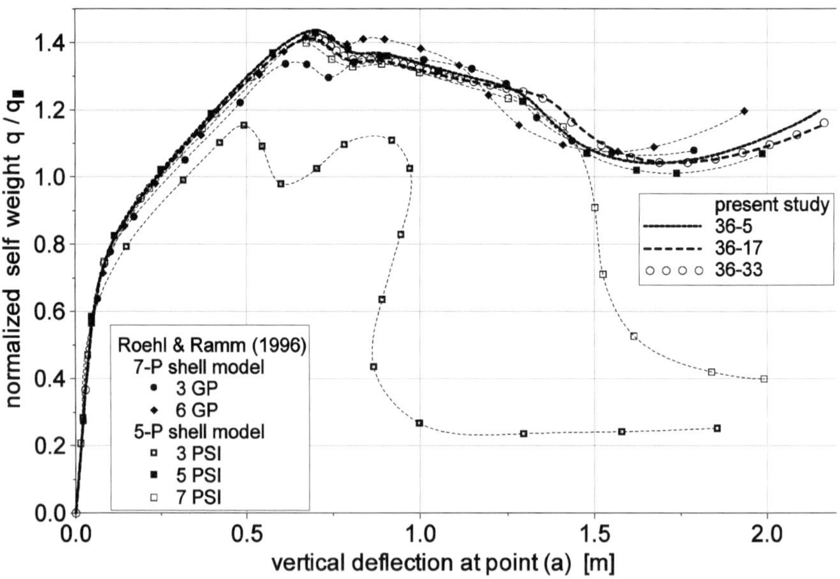

Fig. 5. Scordelis-Lo roof: Roehl and Ramm – present study

Figure 6 shows the number of active integration points for the present example. It can be seen that the highest number of integration points is required at the ridge of the roof, where a localization due to plastic deformation takes place. The number of integration points increases when a cross-section starts to plastify, but also decreases again when the cross-section is fully plastified, because then a nearly uniform distribution of stresses through the thickness is obtained.

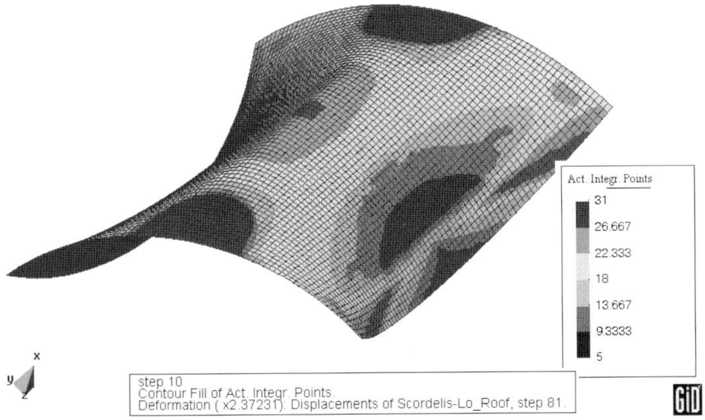

Fig. 6. Scordelis-Lo roof: present study - active integration points

Square plate. A square plate subjected to uniform loading is undergoing large elastoplastic deformations. The geometry and material data are given in Figure 7. At the edges the lower boundary of the plate is fixed in vertical directions whereas the displacements tangential to the plate are free.

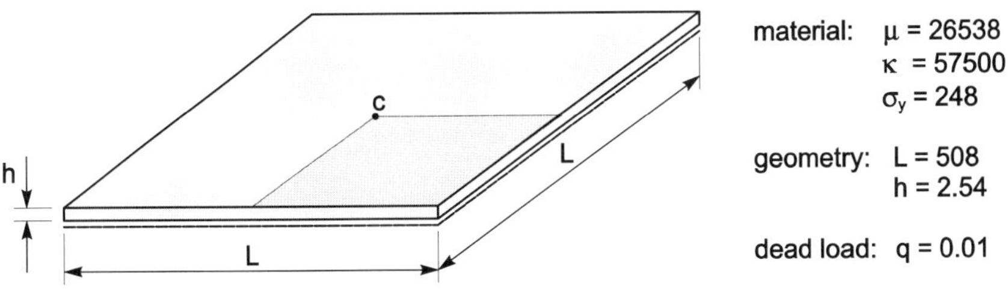

Fig. 7. Square plate: geometry, boundary conditions and material data

Figures 8 shows the number of active integration points used to perform an accurate integration through the thickness. The contour of the number of integration points forms a front that is moving from the edges towards the center of the plate. In the state presented in Figure 8 the edges are already fully plastified, resulting in a low number of integration points in the largely deformed edges.

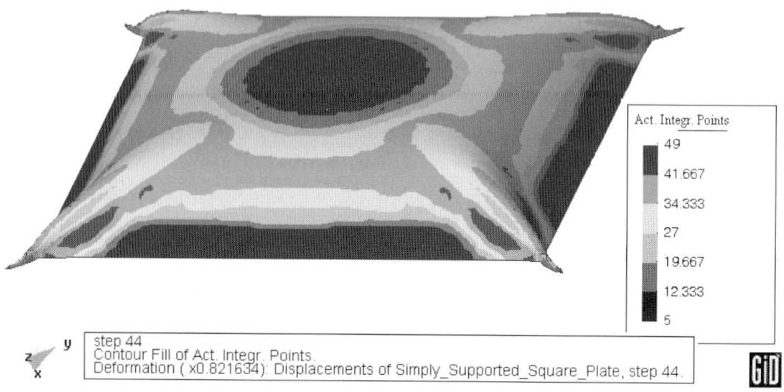

Fig. 8. Square plate: present study - active integration points

References

1. Brank B, Peric D, Damjanic FB (1997) On large deformations of thin elasto-plastic shells: Implementation of a finite rotation model for quadrilateral shell element. Int J Num Meth Engng 40: 689-726
2. Crisfield M, Peng X (1992) Efficient non-linear shell formulations with large rotations and plasticity. In: Owen DRJ et al (eds) Computational Plasticity III: Fundamentals and Applications. Pineridge Press, Swansea pp 1979-1996
3. Dahmen W (1997) Wavelet and multiscale methods for operator equations. Acta Numerica 21: 55-228
4. Daubechies N (1992) Ten Lectures on Wavelets. Society for Industrial and Applied Mathematics, Philadelphia
5. Hackl KA (1997) Framework for nonlinear shells based on generalized stress and strain measures. Int J Solids Structures 34 : 1609–1632
6. Hackl KA (1997) Generalized standard media and variational principles of classical and finite strain elastoplasticity. J Mech Phys Solids 45: 667-688
7. Hackl KA (1996) Wavelet based elastoplastic beam model. ZAMM 76, Suppl. 1: 175–178
8. Peric D, Owen DRJ (1991) The morley thin shell finite element for large deformations problems: simplicity versus sophistication. In: Bicanic N (ed.), Proc Int Conf on Nonlinear Engineering Computations, Pineridge Press, Swansea, pp 121-142
9. Roehl D, Ramm E (1996) Large elasto-plastic finite element analysis of solids and shells with the enhanced assumed strain concept. Int J Solids Struct 33: 3215-3237
10. Rouainia M, Peric D (1998) A computational model for elasto-viscoplastic solids at finite strain with reference to thin shell applications. Int J Numer Meth Engng 42: 289-311
11. Schieck B, Smolenski WM , Stumpf H (1999) A shell finite element for large strain elastoplasticity with anisotropies – Part II: Constitutive equations and numerical applications. Int J Solids Structures 36: 5425-5451

On the Effect of Transverse Core Compressibility on the Buckling of Sandwich Shells

J. Hohe [1] and L. Librescu [2]

[1] Fraunhofer Institut für Werkstoffmechanik, Freiburg, Germany
[2] Department of Engineering Science and Mechanics
Virginia Polytechnic Institute and State University, Blacksburg, USA

1 Introduction

Structural sandwich panels are important elements in lightweight construction, where plates and shells with high bending stiffness and low specific weight are required. The classical structural sandwich panel is a layered medium consisting of three principal layers: Two high-density face sheets which carry the in-plane and bending loads and a low-density core which transmits the transverse normal and shear loads and keeps the face sheets at their desired distance (see Fig 1). Fiber reinforced laminae are common face sheet materials whereas the core is typically made from a low density cellular or foamed material.

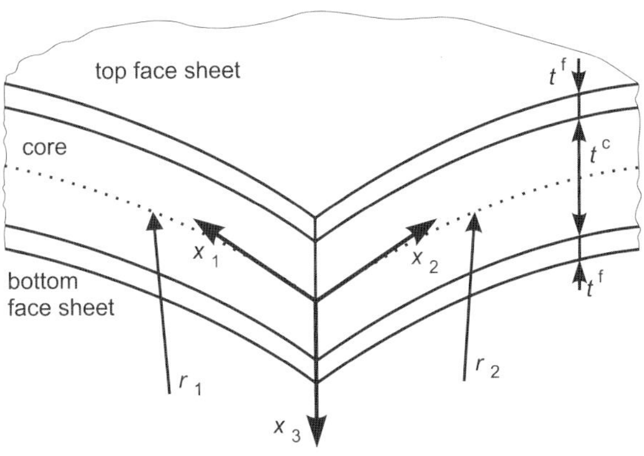

Fig. 1. Structural sandwich panel

In the classical sandwich theories it is assumed that the core is transversely incompressible (see e.g. Allen [1]) whereas in the recent years, increasing interest is directed to refined models which eliminate this assumption. In this context, it was shown by Frostig [4] that the transverse flexibility of the core is an important feature since it enables the development of the face wrinkling instability which might affect the overall instability mode due to nonlinear coupling effects. For this reason, the theoretical critical buckling load determined by the standard models (see e.g. Vinson [6]) might be overestimated. Refined theories for flat sandwich panels that account for transverse normal strains in the core have recently been published by Barut et al. [2], Pai and Palazotto [5] as well as Frostig [4].

The present study is concerned with a comprehensive shell model for structural sandwich panels with transversely compressible core. Geometric nonlinearities are included in the v. Kármán sense. The model accounts for initial geometric imperfections as well as for inertia effects. In its basic form, the model is not restricted to any specific type of constitutive behavior. The model is applied to the buckling and postbuckling analysis of sandwich panels with square projection. The mathematical problem is solved by means of an extended Galerkin procedure.

2 Enhanced sandwich shell theory

2.1 Kinematic relations

The present contribution is concerned with a geometrically nonlinear multilayer model for sandwich shells. For the face sheets, the standard Kirchhoff-Love hypothesis is adopted. Thus, the displacements of the top and bottom face sheets are approximated by

$$v_1^t = u_1^a + u_1^d - \left(x_3 + \frac{t_c + t_f}{2} \right) u_{3,1}^a - \left(x_3 + \frac{t_c + t_f}{2} \right) u_{3,1}^d \tag{1}$$

$$v_2^t = u_2^a + u_2^d - \left(x_3 + \frac{t_c + t_f}{2} \right) u_{3,2}^a - \left(x_3 + \frac{t_c + t_f}{2} \right) u_{3,2}^d \tag{2}$$

$$v_3^t = u_3^a + u_3^d \tag{3}$$

and

$$v_1^b = u_1^a - u_1^d - \left(x_3 - \frac{t_c + t_f}{2} \right) u_{3,1}^a + \left(x_3 - \frac{t_c + t_f}{2} \right) u_{3,1}^d \tag{4}$$

$$v_2^b = u_2^a - u_2^d - \left(x_3 - \frac{t_c + t_f}{2} \right) u_{3,2}^a + \left(x_3 - \frac{t_c + t_f}{2} \right) u_{3,2}^d \tag{5}$$

$$v_3^b = u_3^a - u_3^d \tag{6}$$

where the vectors $\boldsymbol{u}^a(x_1, x_2)$ and $\boldsymbol{u}^d(x_1, x_2)$ are the average and the half difference of the tangential displacements of the face sheet mid-surfaces respectively. For the core displacements, a first and second order power-series expansion is employed. The displacement field for the core is approximated by

$$v_1^c = u_1^a - \frac{t_f}{2} u_{3,1}^d - \frac{2}{t_c} u_1^d x_3 + \frac{t_f}{t_c} u_{3,1}^a x_3 + \frac{4}{t_c^2} \Phi_1 \left(x_3^2 - \frac{t_c^2}{4} \right) \tag{7}$$

$$v_2^c = u_2^a - \frac{t_f}{2} u_{3,2}^d - \frac{2}{t_c} u_2^d x_3 + \frac{t_f}{t_c} u_{3,2}^a x_3 + \frac{4}{t_c^2} \Phi_2 \left(x_3^2 - \frac{t_c^2}{4} \right) \tag{8}$$

$$v_3^c = u_3^a - \frac{2}{t_c} u_3^d x_3 \tag{9}$$

where $\Phi_i(x_1, x_2)$ are additional displacement functions associated with the warping of the core. The transverse core compressibility is included via the x_3-dependent term in Eq. (9).

In addition to the load dependent displacements according to Eqs. (1) to (9), small initial geometric imperfections

$$\overset{0}{v}{}_3^t = \overset{0}{u}{}_3^a + \overset{0}{u}{}_3^d, \qquad \overset{0}{v}{}_3^b = \overset{0}{u}{}_3^a - \overset{0}{u}{}_3^d, \qquad \overset{0}{v}{}_3^c = \overset{0}{u}{}_3^a - \frac{2}{t_c} \overset{0}{u}{}_1^d x_3 \tag{10 - 12}$$

with respect to the transverse direction are considered. Since independent imperfections are assumed for the top and bottom face sheets, the transverse imperfection according to Eq. (12) depends on the thickness x_3- direction.

The geometrical nonlinearity is included in the v. Kármán sense. Furthermore, the principles of shallow shell theory are adopted. Thus, the components of the Green-Lagrange strain tensor are given by:

$$\gamma_{11} = v_{1,1} - \frac{1}{r_1} v_3 + \frac{1}{2}(v_{3,1})^2 + v_{3,1} \overset{0}{v}_{3,1} \tag{13}$$

$$\gamma_{22} = v_{2,2} - \frac{1}{r_2} v_3 + \frac{1}{2}(v_{3,2})^2 + v_{3,2} \overset{0}{v}_{3,2} \tag{14}$$

$$\gamma_{33} = v_{3,3} + \frac{1}{2}(v_{3,3})^2 + v_{3,3} \overset{0}{v}_{3,3} \tag{15}$$

$$\gamma_{23} = \frac{1}{2}(v_{3,2} + v_{2,3}) + \frac{1}{2} v_{3,2} v_{3,3} + \frac{1}{2} v_{3,2} \overset{0}{v}_{3,3} + \frac{1}{2} \overset{0}{v}_{3,2} v_{3,3} \tag{16}$$

$$\gamma_{13} = \frac{1}{2}(v_{3,1} + v_{1,3}) + \frac{1}{2} v_{3,1} v_{3,3} + \frac{1}{2} v_{3,1} \overset{0}{v}_{3,3} + \frac{1}{2} \overset{0}{v}_{3,1} v_{3,3} \tag{17}$$

$$\gamma_{12} = \frac{1}{2}(v_{2,1} + v_{1,2}) + \frac{1}{2} v_{3,1} v_{3,2} + \frac{1}{2} v_{3,1} \overset{0}{v}_{3,2} + \frac{1}{2} \overset{0}{v}_{3,1} v_{3,2} \tag{18}$$

Under the assumption that no antisymmetric distribution of the transverse shear stresses and strains are permitted for the core layer, the warping functions Φ_i can be eliminated using Eqs. (16) and (17).

2.2 Equations of motion

A consistent set of equations of motion and boundary conditions are obtained in a natural manner by means of Hamilton's principle

$$\int_{t_0}^{t_1} \left(\delta U - \delta W - \delta T \right) dt = 0 \tag{20}$$

where δU, δW and δT are the variations of the strain energy, the work done by the external loads and the kinetic energy respectively. The variations of the energy components are expressed in terms of the variations $\delta \boldsymbol{u}^{\mathrm{a}}$ and $\delta \boldsymbol{u}^{\mathrm{d}}$ of the displacement functions, the components τ_{ij} of the second Piola-Kirchhoff stress tensor and the external loads. Subsequently, the stress components τ_{ij} and the integrals with respect to the transverse direction are replaced by introduction of the generalized stress resultants and stress couples:

$$\{N^t_{\alpha\beta}, M^t_{\alpha\beta}\} = \int_{-t^f - t^c/2}^{-t^c/2} \tau_{\alpha\beta}\{1, (x_3 + \frac{1}{2}(t^c + t^f))\} dx_3 \tag{21}$$

$$\{N^b_{\alpha\beta}, M^b_{\alpha\beta}\} = \int_{t^c/2}^{t^f + t^c/2} \tau_{\alpha\beta}\{1, (x_3 - \frac{1}{2}(t^c + t^f))\} dx_3 \tag{22}$$

$$\{N^c_{i3}, M^c_{i3}\} = \int_{-t^c/2}^{t^c/2} \tau_{i3}\{1, x_3)\} dx_3 \tag{23}$$

In accordance with the definition of the alternative displacement vectors $\delta \boldsymbol{u}^{\mathrm{a}}$ and $\delta \boldsymbol{u}^{\mathrm{d}}$, alternative stress resultants and stress couples according to

$$\{N^a_{\alpha\beta}, M^a_{\alpha\beta}\} = \frac{1}{2}\{(N^t_{\alpha\beta} + N^b_{\alpha\beta}), (M^t_{\alpha\beta} + M^b_{\alpha\beta})\} \tag{24}$$

$$\{N^d_{\alpha\beta}, M^d_{\alpha\beta}\} = \frac{1}{2}\{(N^t_{\alpha\beta} - N^b_{\alpha\beta}), (M^t_{\alpha\beta} - M^b_{\alpha\beta})\} \tag{25}$$

are introduced for the face sheets. Substituting the generalized stress resultants and stress couples into the variational expression, neglecting all tangential and rotatory inertia terms, integrating by parts wherever possible and collecting the coefficients corresponding to the different virtual displacements yields the equations of motion

$$0 = N^a_{11,1} + N^a_{12,2}, \qquad 0 = N^a_{12,1} + N^a_{22,2} \tag{26, 27}$$

$$0 = N^d_{11,1} + N^d_{12,2} + \frac{1}{t^c} N^c_{13}, \qquad 0 = N^d_{12,1} + N^d_{22,2} + \frac{1}{t^c} N^c_{23} \tag{28, 29}$$

$$\begin{aligned}
0 = {} & (u^a_{3,11} + \overset{0}{u}{}^a_{3,11} + \frac{1}{r_1}) N^a_{11} + 2(u^a_{3,12} + \overset{0}{u}{}^a_{3,12}) N^a_{12} + \\
& + (u^a_{3,22} + \overset{0}{u}{}^a_{3,22} + \frac{1}{r_2}) N^a_{22} + M^a_{11,11} + 2M^a_{12,12} + M^a_{22,22} + \\
& + (u^d_{3,11} + \overset{0}{u}{}^d_{3,11}) N^d_{11} + 2(u^d_{3,12} + \overset{0}{u}{}^d_{3,12}) N^d_{12} + (u^d_{3,22} + \overset{0}{u}{}^d_{3,22}) N^d_{22} + \\
& + \frac{1}{t^c}(\frac{t^c + t^f}{2} - u^d_3 - \overset{0}{u}{}^d_3)(N^c_{13,1} + N^c_{23,2}) - \frac{2}{t^c}(u^d_{3,1} + \overset{0}{u}{}^d_{3,1}) N^c_{13} - \\
& - \frac{2}{t^c}(u^d_{3,2} + \overset{0}{u}{}^d_{3,2}) N^c_{23} + \hat{q}^a_3 - m^f \ddot{u}^a_3
\end{aligned} \tag{30}$$

$$
\begin{aligned}
0 = &(u^a_{3,11} + \overset{0}{u}{}^a_{3,11} + \frac{1}{r_1})N^d_{11} + 2(u^a_{3,12} + \overset{0}{u}{}^a_{3,12})N^d_{12} + \\
&+(u^a_{3,22} + \overset{0}{u}{}^a_{3,22} + \frac{1}{r_2})N^d_{22} + M^d_{11,11} + 2M^d_{12,12} + M^d_{22,22} + \\
&+(u^d_{3,11} + \overset{0}{u}{}^d_{3,11})N^a_{11} + 2(u^d_{3,12} + \overset{0}{u}{}^d_{3,12})N^a_{12} + (u^d_{3,22} + \overset{0}{u}{}^d_{3,22})N^a_{22} + \\
&+\frac{2}{t^c}(\frac{t^c}{2} - u^d_3 - \overset{0}{u}{}^d_3)N^c_{33} + \hat{q}^d_3 - m^f \ddot{u}^d_3
\end{aligned}
\tag{31}
$$

and the corresponding nonlinear boundary conditions (see Hohe and Librescu [3] for details). Since no specific constitutive equation has been considered so far, the equations of motion (26) to (31) hold irrespectively of the material behavior.

3 Analytical solution technique

The general sandwich shell theory developed in Sect 2 will subsequently be applied to a postbuckling analysis of curved sandwich panels with rectangular projection. The panels are assumed to be simply supported along all four edges. Under this assumption

$$u^a_3 = w^a_{mn} \sin(\frac{m\pi}{l_1}x_1)\sin(\frac{n\pi}{l_2}x_2) \tag{32}$$

$$u^d_3 = w^d_{pq} \sin(\frac{p\pi}{l_1}x_1)\sin(\frac{q\pi}{l_2}x_2) \tag{33}$$

is an appropriate form for the transverse displacements. In Eqs. (32) and (33), l_1 and l_2 are the panel edge lengths. The modal amplitudes w remain unknown at this stage. The initial geometric imperfections are assumed in the same form

$$\overset{0}{u}{}^a_3 = \overset{0}{w}{}^a_{mn} \sin(\frac{m\pi}{l_1}x_1)\sin(\frac{n\pi}{l_2}x_2) \tag{34}$$

$$\overset{0}{u}{}^d_3 = \overset{0}{w}{}^d_{pq} \sin(\frac{p\pi}{l_1}x_1)\sin(\frac{q\pi}{l_2}x_2) \tag{35}$$

with the same numbers m, n, p and q of modal half waves.

The material behavior for both, the core and the face sheets is assumed to be linear elastic. Under this assumption, a consistent solution for the in-plane displacement functions can be obtained by substituting the constitutive equations for orthotropic elasticity together with Eqs. (32) to (35) into the first four equations of motion (26) to (29). Details can be found in a recently published paper by the present authors (Hohe and Librescu [3]). Substituting this solution for the displacement functions into Hamilton's principle (20) and collecting the coefficients for the unknown variations of the modal amplitudes of the transverse displacements according to Eqs. (32) and (33) yields to coupled nonlinear equations for the unknown transverse modal amplitudes. The system is solved numerically by means of Newton's method.

4 Numerical examples

4.1 Validation

As a first example for validation of the present sandwich shell theory, the buckling loads of a square sandwich plate with $l_1 = l_2 = 500$ mm are analyzed. Aluminum face sheets with a thickness of $t^f = 1$ mm are considered. The core is assumed to consist of an isotropic low-density material with 1% of the stiffness of the face sheet material. The sandwich plate is loaded by a tangential edge load N_{11} whereas the edges perpendicular to the loading direction are freely movable with respect to the x_1-x_2-plane.

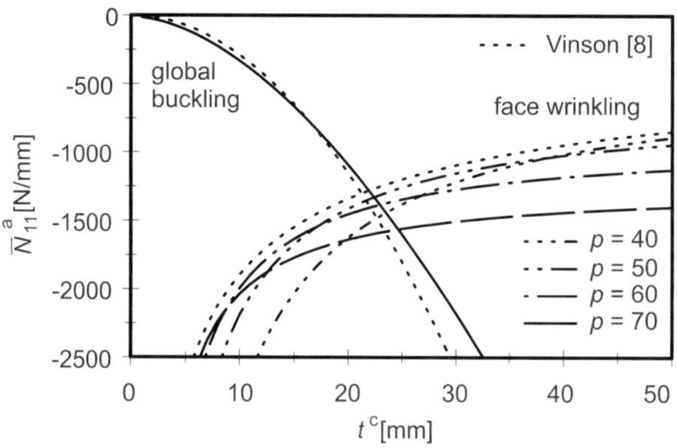

Fig. 2. Global and local buckling loads for sandwich plate loaded in tangential compression

In Fig 2, the compressive buckling loads for both, the global and the local (face wrinkling) buckling modes are presented in dependence on the core thickness t^c. Dotted lines indicate the respective buckling loads obtained by the approximate formulae given in the book by Vinson [6]. The results based on the present sandwich theory are found in a good agreement with the results based on Vinson's [6] formulae. Since these approximate formulae are well validated experimentally, the present sandwich shell model can also be regarded as being validated experimentally. In this context, care has to be taken with respect to the buckling mode of the face wrinkling instability, since the buckling load obtained by the present sandwich model strongly depends on the number p of modal half waves (see Fig 2).

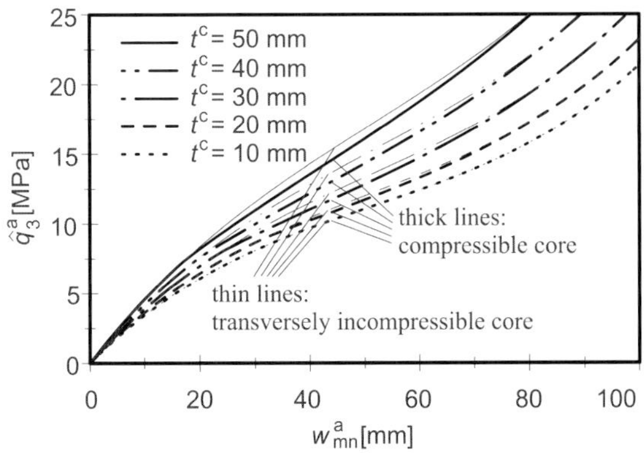

Fig. 3. Overall load-deflection curves for spherical sandwich cap under transverse pressure

4.2 Spherical sandwich cap under transverse load

As a second example, the mechanical response of a spherical sandwich cap with square projection is considered. The overall geometry is defined by the edge lengths $l_1 = l_2 = l = 500$ mm and the radii of curvature $r_1 = r_2 = r = 500$ mm. The thickness and the material properties of the core and the face sheets are the same as in the previous example. All external edges are assumed to be immovable with respect to the tangential directions. The cap is loaded by a unilateral transverse pressure applied to the top face sheet.

In Fig 3, the load-deflection behavior of the sandwich cap is presented for five different levels of the core thickness t^c. A comparison with a classical analysis under the assumption of a transversely incompressible core shows distinct effects in the intermediate range of the load-deflection curves. In this range, a face wrinkling instability develops due to the high resulting in-plane edge load in this range. If the development of the face wrinkling instability is suppressed by the assumption of transverse incompressibility of the core, the overall stiffness of the sandwich structure is overestimated by up to 10%. In case of the resulting compressive edge load (see Fig 3), a more severe effect is observed, especially for thick cores. Here, the development of the face wrinkling instability results in a distinctively weaker structural response compared to the conventional model.

5 Conclusion

The present study is concerned with a comprehensive sandwich shell theory accounting for non-classical effects such as the transverse compressibility of the core and initial geometric imperfections. In addition, geometrical nonlinearities and inertia effects are included. Consistent equations of motion and boundary conditions are obtained by means of Hamilton's principle.

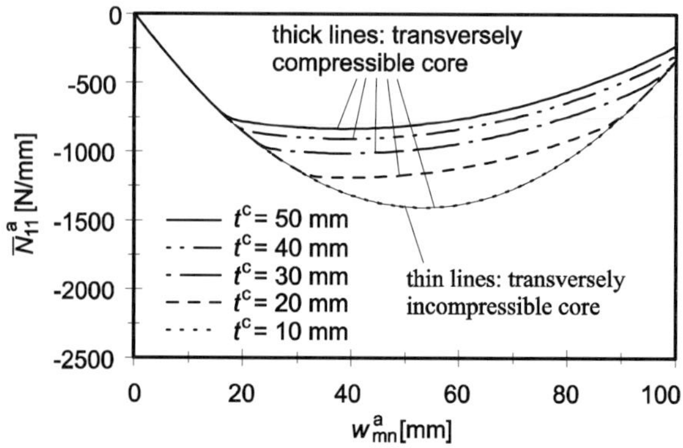

Fig. 4. Resulting tangential edge load for spherical sandwich cap under transverse pressure

Owing to its general formulation, the present model enables the efficient high-precision analysis of sandwich structures including all major effects. Especially the development of the local face wrinkling instability is observed to be an important feature in the structural response of sandwich shells which can have distinct effects even on the global structural response. Therefore, the local instability and thus the transverse core compressibility have to be considered within the model for the global structural analysis.

Acknowledgement

The financial support of the Office of Naval Research under Grant no. N 00014-02-1-0594, and the interest and encouragement of the Grant Monitor, Dr YDS Rajapakse, are gratefully acknowledged.

References

1. Allen HG (1969) Analysis and design of structural sandwich panels. Pergamon Press, Oxford
2. Barut A, Madenci E, Heinrich J, Tessler A (2001) Analysis of thick sandwich construction by a {3, 2}-order theory. Int J Solids Struct 38: 6063-6077
3. Hohe J, Librescu L (2003) A nonlinear theory for doubly curved anisotropic sandwich shells with transversely compressible core. Int J Solids Struct 40: 1059-1088
4. Frostig Y (2003) Classical and high-order computational models in the analysis of modern sandwich panels. Compos B 34: 83-100
5. Pai PF , Palazotto AN (2001) A higher-order sandwich plate theory accounting for 3-D stresses. Int J Solids Struct 38: 5045-5063
6. Vinson JR (1999) The behavior of sandwich structures of isotropic and composite materials. Technomic Publishing, Lancaster, PA

Drape Simulation of Flexible Structures

Ł. Jański and V. Ulbricht

Institute of Solid State Mechanics, Dresden University of Technology, Germany

1 Introduction

Drape is a special deformation mode of textile surfaces. It occurs when a textile surface is supported only partially and the gravitational force is the dominating external load [5]. These conditions are fulfilled in the cases of e.g. tablecloths, curtains and most of all clothes (garments). The freely hanging parts of the mentioned textile surfaces usually form smooth folds. The form of the folds depends not only on the geometry of a particular piece of material, but also on the material properties of the chosen cloth.

In case of a designer or tailor who wants to prepare optimal clothes for a customer, the form of folds in drape is very important. Certain forms can assure certain esthetic aims. The nowadays standard behaviour of a designer is to assume some form of folds in case of some human shape and then choose a material. The next step is to test the assumption in an experimental way. This behaviour is expensive and slow. The usage of some CAD system, similar as in mechanical engineering, would be helpful to save the costs and to hasten the process.

Drape is a mechanical problem. A textile surface deforms because of an action of the gravitational field. As usual in mechanics, the deformation character depends on the geometry of a structure, boundary conditions, material and load. In view of the structural mechanics, the special geometry, boundary conditions, material and load characterizing drape cause this deformation mode to be classified as buckling or post-buckling behaviour. The very small bending stiffness of textile surfaces, compared with their tensile and in plane shear stiffness, ensures drape to be the bending dominated deformation mode.

We investigate the drape of textile surfaces from the mechanical point of view and try to give the background for a construction of a CAD system. This system can be used by designers and tailors in the future.

The present publication describes a numerical model which can be applied for drape simulations. The efficiency of the model is proved on the basis of two examples. Thereby the drape behaviour of the square tablecloth supported on the square table with rounded corners and of the round tablecloth supported on the round table is simulated. The last system coincides with the drape experiment in [3]. The publication contains the proof that the ambiguous behaviour of a textile specimen in the drape experiment can be explained by the highly non-linear response of the system.

2 Model description

The basis for the mechanical model forms the two-dimensional one-director Cosserat continuum embedded in the three-dimensional space [8]. Let the manifold

$$\mathsf{C} = \{(\mathbf{R},\mathbf{D}) \mid (\mathbf{R},\mathbf{D}) : \mathsf{A} \subset \mathbb{R}^2 \to \mathbb{R}^3 \times \mathsf{S}^2\} \tag{1}$$

be the reference configuration. The set S^2 is a sphere of the following structure

$$\mathsf{S}^2 = \{\mathbf{D} \in \mathbb{R}^3 \mid \|\mathbf{D}\| = 1\} \tag{2}$$

Let the set

$$\mathsf{P} = \{\mathbf{R} \mid \mathbf{R} : \mathsf{A} \subset \mathbb{R}^2 \to \mathbb{R}^3\} \tag{3}$$

describe the surface of the Cosserat continuum in the reference configuration C. The actual configuration has the form of the manifold

$$\mathsf{L} = \{(\mathbf{r},\mathbf{d}) \mid (\mathbf{r},\mathbf{d}) : \mathsf{A} \subset \mathbb{R}^2 \to \mathbb{R}^3 \times \mathsf{S}^2\} \tag{4}$$

The mapping

$$\phi : \mathsf{C} \to \mathsf{L} \tag{5}$$

is the deformation. Two displacement fields

$$\mathbf{u} = \mathbf{r} - \mathbf{R} \qquad \text{and} \qquad \Delta\mathbf{d} = \mathbf{d} - \mathbf{D} \tag{6}$$

can be defined to characterize the change of the configuration. Let the elements of the set A be identified with the coordinate pairs (Θ^1, Θ^2) and let $f,_\alpha$ state for a partial derivative of some function f with respect to the coordinates Θ^α. The deformation measures

$$\mathbf{e} = e_{\alpha\beta} \mathbf{A}^\alpha \otimes \mathbf{A}^\beta \tag{7}$$

$$\kappa = \kappa_{\alpha\beta} \mathbf{A}^\alpha \otimes \mathbf{A}^\beta \tag{8}$$

$$\gamma = \gamma_\alpha \mathbf{A}^\alpha \tag{9}$$

are introduced to describe the deformation. The vectors $\mathbf{A}^\alpha$ form the contravariant basis of the surface P. The coordinates of the tensors **e**, $\boldsymbol{\kappa}$ and of the vector $\boldsymbol{\gamma}$ can be expressed as functions of the displacement fields **u** and $\boldsymbol{\Delta}\mathbf{d}$ in the form

$$2e_{\alpha\beta} = \mathbf{A}_\alpha \cdot \mathbf{u},_\beta + \mathbf{u},_\alpha \cdot \mathbf{A}_\beta + \mathbf{u},_\alpha \cdot \mathbf{u},_\beta \tag{10}$$

$$\kappa_{\alpha\beta} = \mathbf{D},_{\beta} \cdot \mathbf{u},_{\alpha} + \Delta\mathbf{d},_{\beta} \cdot \mathbf{A}_{\alpha} + \Delta\mathbf{d},_{\beta} \cdot \mathbf{u},_{\alpha} \tag{11}$$

$$\gamma_{\alpha} = \mathbf{D} \cdot \mathbf{u},_{\alpha} + \Delta\mathbf{d} \cdot \mathbf{A}_{\alpha} + \Delta\mathbf{d} \cdot \mathbf{u},_{\alpha} . \tag{12}$$

The deformation measure $\boldsymbol{\gamma}$ serves for the formulation of the constraint

$$\gamma = \mathbf{0} \tag{13}$$

The existence of a strain energy density function Σ per unit area of the surface P is assumed in the first approximation of a textile surface behaviour. In this case, the force and the couple tensors can be obtained from the equations

$$\mathbf{N} = \frac{\partial \Sigma(\mathbf{e}, \underline{\kappa})}{\partial \mathbf{e}} \qquad \mathbf{M} = \frac{\partial \Sigma(\mathbf{e}, \underline{\kappa})}{\partial \underline{\kappa}} . \tag{14}$$

In the formula (14), $\underline{\boldsymbol{\kappa}}$ means the symmetric part of the tensor $\boldsymbol{\kappa}$. Let Σ be an orthotropic function. Let $E_{\alpha\beta}$, $\underline{K}_{\alpha\beta}$, $N_{\alpha\beta}$ and $M_{\alpha\beta}$ represent the coordinates of the tensors **e**, $\underline{\boldsymbol{\kappa}}$, **N** and **M** in some from Θ^{α} independent cartesian coordinate system $\mathbf{e}_{\alpha}$. Considering the boundary conditions of the Kawabata Evaluation System for Fabrics [7] and assuming no material coupling between the membrane and bending response of a textile surface, the following constitutive equations

$$\begin{bmatrix} N_{11} \\ N_{22} \\ N_{12} \end{bmatrix} = \begin{bmatrix} D_1 & \nu_1 D_1 & 0 \\ \nu_2 D_2 & D_2 & 0 \\ 0 & 0 & G \end{bmatrix} \begin{bmatrix} E_{11} \\ E_{22} \\ E_{12} \end{bmatrix} \tag{15}$$

$$\begin{bmatrix} M_{11} \\ M_{22} \\ M_{12} \end{bmatrix} = \begin{bmatrix} B_1 & \mu_1 B_1 & 0 \\ \mu_2 B_2 & B_2 & 0 \\ 0 & 0 & \Omega \end{bmatrix} \begin{bmatrix} \underline{K}_{11} \\ \underline{K}_{22} \\ \underline{K}_{12} \end{bmatrix} \tag{16}$$

can be formulated. The additional restrictive expressions

$$\nu_1 D_1 = \nu_2 D_2 \qquad \text{and} \qquad \mu_1 B_1 = \mu_2 B_2 \tag{17}$$

must be fulfilled.

To be able to write the equilibrium condition and to impose the Eq. (13), the total potential in the form

$$\Pi = \int_{\mathsf{P}} \left[\frac{1}{2} \left(\mathbf{N} : \mathbf{e} + \mathbf{M} : \underline{\kappa} \right) + m_G \mathbf{e}_3 \cdot \mathbf{u} + \frac{p}{2} \gamma^2 \right] dA \tag{18}$$

is constructed. In the formula (18), m_G, $\mathbf{e}_3$ and p indicate the surface force describing the weight of a textile material, the action direction of the surface force and the penalty factor [9] respectively. The necessary and sufficient condition for the stationary value of the total potential, equivalent to the equilibrium condition, is expressed by the equation

$$\delta\Pi = 0\,. \tag{19}$$

To solve the Eq. (19) with respect to the displacements (6), the finite element method [1] is used . To avoid locking effects, the MITC finite elements [2] are implemented. Because the algebraic system of equations, obtained after the finite element discretisation, is nonlinear, the Newton method [6] in connection with the arc-length method [4] is applied as a solution strategy.

3 Simulation results

The first set of the material parameters in the table 1 corresponds to the simulation of the square probe and the second one to the simulation of the round probe.

Table 1. Material parameters used in the simulation

Set no.	$D_1=D_2$ [N/m]	$\nu_1=\nu_2$ [1]	G [N/m·rad]	$B_1=B_2$ [N·m]	$\mu_1=\mu_2$ [1]	Ω [N·m]	m_G [N/m²]
1	547	0.1	41.0	$7.30\cdot10^{-6}$	0.1	$3.28\cdot10^{-6}$	2.06
2	950	0.1	23.8	$7.13\cdot10^{-6}$	0.1	$3.21\cdot10^{-6}$	2.75

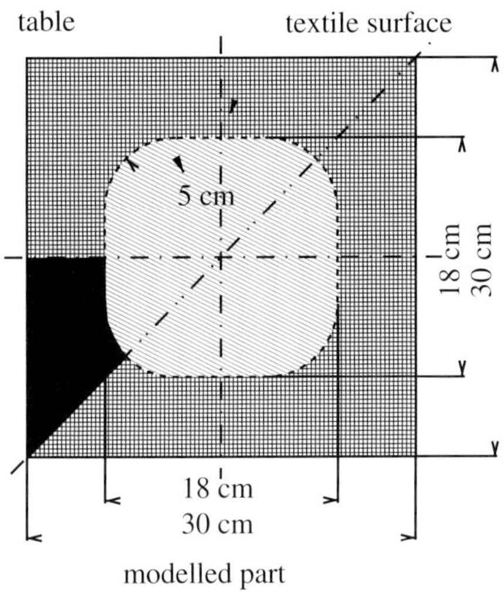

Fig. 1. The geometrical model of the square tablecloth supported on the square table with rounded corners

Figures 1 and 2 show the geometrical and the finite element models of the square tablecloth supported on the square table with rounded corners respectively. The surface force acts in the Z direction.

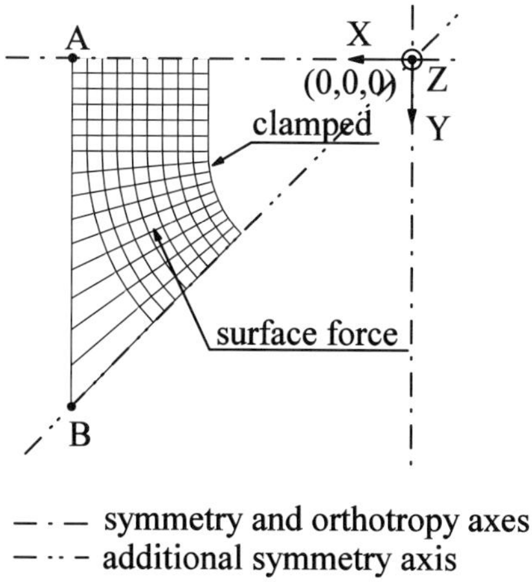

Fig. 2. The finite element model of the square tablecloth supported on the square table with rounded corners

The two curves, describing the relationship between the surface force and the displacement of the two points A and B depicted in the figure 2, are presented in the figure 3. The dashed horizontal line represents the surface load level $m_G = 2,06\, N/m^2$ The quasi-static response of the investigated system shows

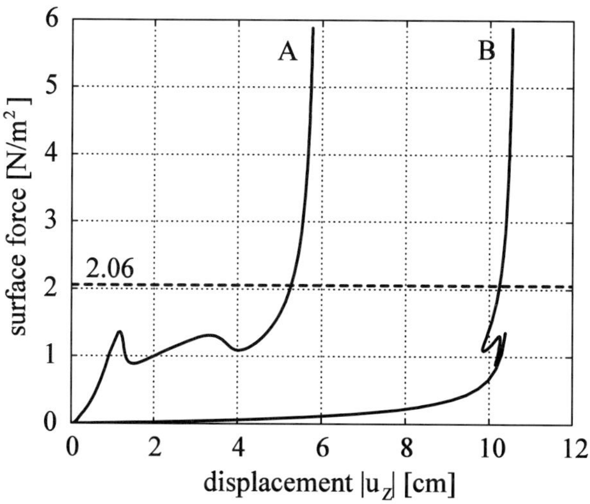

Fig. 3. Surface force versus displacement curves of the two points A and B

an essential non-linearity. The behaviour of the tablecloth does not reveal any ambiguity for the surface force $m_G = 2,06\,N/m^2$.

The figure 4 presents the actual configuration of the tablecloth for the surface force $m_G = 2,06\,N/m^2$.

Fig. 4. The actual configuration for the surface force $m_G = 2,06\,N/m^2$

Figures 5 and 6 show the geometrical and the finite element models of the round tablecloth supported on the round table respectively. The surface force acts in the Z direction.

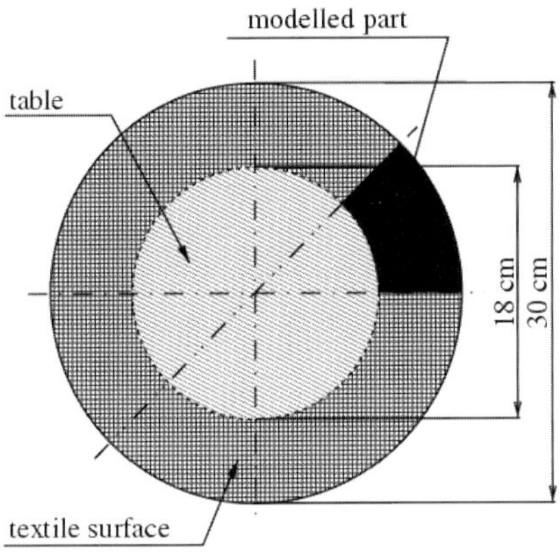

Fig. 5. The geometrical model of the round tablecloth supported on the round table

The two curves, describing the relationship between the surface force and the displacement of the two points A and B depicted in the figure 6, are presented in the figure 7. The dashed horizontal line represents the surface load level $m_G = 2,75\,N/m^2$. The quasi-static response of the investigated system shows an essential non-linearity. The behaviour of the tablecloth has for the surface force $m_G = 2,75\,N/m^2$ a non-unique character.

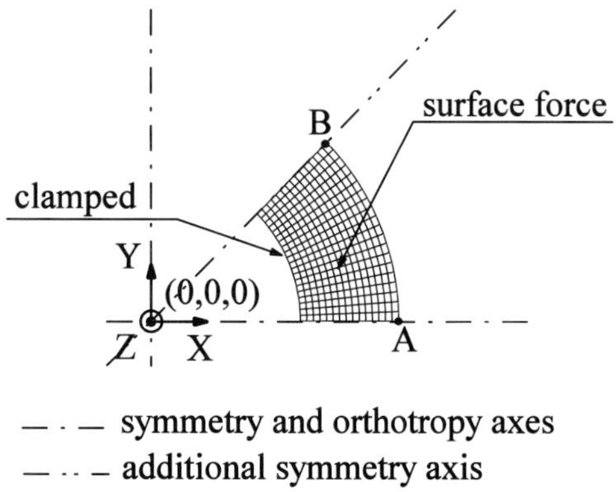

Fig. 6. The finite element model of the round tablecloth supported on the round table

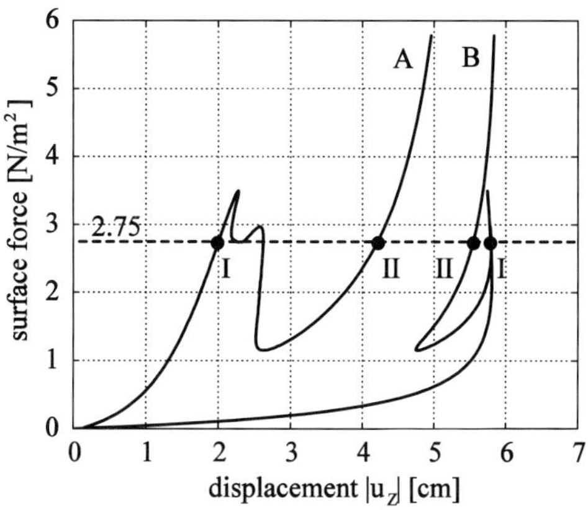

Fig. 7. Surface force versus displacement curves of the two points A and B

The figure 8 presents two possible (under the consideration of the assumed symmetry) actual configurations I and II of the tablecloth for the surface force $m_G = 2,75\, N / m^2$. The two configurations have different numbers of folds 4 and 12 respectively.

Fig. 8. Two possible actual configurations I (left) and II (right) for the surface force $m_G = 2,75\, N / m^2$.

4 Conclusions

We have described a mechanical model based on the two-dimensional one-director Cosserat continuum which can be used for the prediction of the drape behaviour of textile surfaces. On the basis of the two simulations of the tablecloth drape, we have proven the efficiency of the described model. The simulations have shown that the quasi-static responses of both modelled systems reveal an essential non-linearity . The surface force versus displacement curves exhibit in both cases load ranges where the drape behaviour of the tablecloth becomes non-unique. Because the round tablecloth supported on the round table coincides with the drape experiment configuration, the second of the presented simulations proves the thesis that the ambiguous behaviour of a textile probe in the drape experiment can be explained by the highly nonlinear response of the system.

References

1. Bathe KJ (1996) Finite element procedures. Prentice-Hall International
2. Bucalem ML, Bathe KJ (1993) Higher–order MITC general shell elements. Int J Num Meth Eng 36: 3729-3754
3. Chu CC, Cummings CL, Teixeira NA (1950) Mechanics of elastic performance of textile materials Part V: A study of the factors affecting the drape of fabrics – The development of a drape meter. Text Res Journ 20: 539-548
4. Crisfield MA (1997) Non-linear finite element analysis of solids and structures - Essentials. John Wiley & Sons
5. Cusick GE (1965) 46-The dependence of fabric drape on bending and shear stiffness. J Text Inst 56: T596-T606
6. Engeln-Müllges G, Reutter F (1996) Numerik-Algorithmen. VDI Verlag
7. Kawabata S, Niwa M (1989) Fabric performance in clothing and clothing manufacture. Journ Text Inst 80: 19-50
8. Naghdi PM (1972) The theory of shells and plates. In: Flügge S, Truesdell C (eds) Handbuch der Physik-Festkörpermechanik II, Band VI/2. Springer-Verlag
9. Reddy JN (1984) Energy and variational methods in applied mechanics. John Wiley & Sons

Universal Dynamic Theory of Shells

J. Kaplunov

Department of Mathematics, The University of Manchester, UK

1 Introduction

Two fundamental restrictions are usually characteristic of approximate shell theories including the classical Kirchhoff-Love theory. The first of them imposes limits on wavelengths and time scales in question. In fact, conventional shell models tackle only motions that vary slowly enough in time and space. Another important restriction relates to boundary conditions on shell faces. 2D shell theories always assume traction free faces or prescribed surface tractions. More general boundary conditions, e.g. those corresponding to fixed faces, are not considered. These may be useful when modeling thin coating and interfaces.

In this publication we attempt to remove the aforementioned limitations. An asymptotic approach is adapted which is traditional for thin structures (e.g. see [3,4,7]). Along with 'classical' low-frequency long-wave motions we also take into consideration high-frequency ones. The latter may demonstrate both long-wave and short-wave behavior and are not only restricted to shells with traction free faces.

The formulae (6)-(9) given below specify asymptotic behaviors of displacements and stresses for various ranges of wavelengths and time scales. They represent the basic concept underlying universal dynamic treatment. In particular, these formulae initiate rather straightforward derivation of numerous shortened equations. A brief discussion of most important dynamic models is presented in the paper. It is empathized that long-wave and short-wave approximations allow matching. We write out estimations for relevant overlap regions and conditions that determine ranges of validity of various approximations including higher-order ones.

The paper makes use of asymptotic developments reported in the monograph [16]. We also cite below key journal publications on the subject.

2 General background

Consider transient motion of a thin shell of arbitrary shape. 3D dynamic equations in linear isotropic elasticity may be schematically presented as

$$\sum_{j=1}^{3} L_{ij}\sigma_{ij} + \rho \frac{\partial^2 u_i}{\partial t^2} = 0, \quad i = 1,2,3...., \quad \text{with}\ \sigma_{ij} = \sum_{k=1}^{3} l_{ijk} u_k \tag{1, 2}$$

where $\sigma_{ij}(\alpha_1, \alpha_2, \alpha_3)$ are stresses, $u_i(\alpha_1, \alpha_2, \alpha_3)$ are displacements, L_{ij} and l_{ijk} are symbolic notations for differential operators in static elasticity, ρ is mass density, t is time.

For the sake of definiteness we concentrate only on two types of boundary conditions on the shell faces $\alpha_3 = \pm h$ (h is the shell half-thickness). They are ($i = 1,2,3$)

$$\sigma_{3i}(\alpha_1, \alpha_2, \pm h) = q_i^{\pm}, \qquad \text{or}\ u_i(\alpha_1, \alpha_2, \pm h) = v_i^{\pm} \tag{3, 4}$$

where $q_i^{\pm}$ and $v_i^{\pm}$ denote prescribed surface tractions and displacements, respectively.

It should be mentioned that traditional shell theories usually operate only with the static boundary conditions on faces presented by the formula (3). The kinematical boundary conditions (4) are, as a rule, out of the scope of these theories.

As usual we introduce the small geometric parameter $\eta = h / R$, where R is a typical curvature radius. In contrast to many of the considerations on the subject, we also specify explicitly two other problem parameters characterizing the time scale T and wavelength l. According to Gol'denveiser's methodology [1-3] the latter may be expressed in terms of the basic geometric parameter as

$$l = h\eta^{q-1}, \quad T = c^{-1}h\eta^{a-1}, \tag{5}$$

where q and a are the so-called 'variability and dynamicity indices', respectively, and c denotes a typical wave speed, e.g. $c = c_k$, $k = 1,2$ (c_1 is dilation wave speed, c_2 is distortion wave speed).

All widespread shell theories usually deal only with low-frequency phenomena ($a < 1$ or $T >> h/c$), which are necessarily long-wave ones ($q < 1$ or $l >> h$). Below we take into consideration high-frequency phenomena ($a = 1$ or $T \sim h/c$) as well. These may involve both long-wave and short-wave components; for the latter $q = 1$ or $l \sim h$. We also mention that long-wave motions may often be divided into tangential ($u_1, u_2 >> u_3$) and transverse ($u_3 >> u_1, u_2$) ones.

3 Universal asymptotic behaviors

In the case of long-wave motions of a shell with traction free faces ($q_i^{\pm} = 0$ in the formula (3) asymptotic behaviors of original displacements and stresses are given by:

for low-frequency tangential motions ($q = 1$)

$$\sigma_{ii} = E\eta^{1-q}(\sigma_{ii}^0 + \eta\sigma_{ii}^1)$$

$$\sigma_{ij} = E\eta^{1-q}(\sigma_{ij}^0 + \eta\sigma_{ij}^1)$$

$$\sigma_{i3} = E\eta^{4-4q}(\sigma_{i3}^0 + \eta^{2q-1}\sigma_{i3}^1)$$

$$\sigma_{33} = E\eta^{3-3q}(\sigma_{33}^0 + \eta\sigma_{33}^1) \tag{6}$$

$$u_i = h(u_i^0 + \eta u_i^1)$$

$$u_3 = h\eta^{1-q}(u_3^0 + \eta^{2q-1}u_3^1)$$

for low-frequency transverse motions ($q = (a+1)/2$)

$$\sigma_{ii} = E\eta^{2-2q}(\eta^{2q-1}\sigma_{ii}^0 + \sigma_{ii}^1)$$

$$\sigma_{ij} = E\eta^{2-2q}(\eta^{2q-1}\sigma_{ij}^0 + \sigma_{ij}^1)$$

$$\sigma_{i3} = E\eta^{3-3q}(\eta\sigma_{i3}^0 + \sigma_{i3}^1)$$

$$\sigma_{33} = E\eta^{4-4q}(\eta^{2q-1}\sigma_{33}^0 + \sigma_{33}^1) \tag{7}$$

$$u_i = h\eta^{1-q}(\eta^{2q-1}u_i^0 + u_i^1)$$

$$u_3 = h(\eta u_3^0 + u_3^1)$$

for high-frequency tangential motions ($h^2 c^{-2}\dfrac{\partial^2 u_i}{\partial t^2} + \Lambda_{sh}^2 u_i \sim \eta^{2-2q}u_i$)

$$\sigma_{kk} = E\eta^{1-q}(\sigma_{kk}^0 + \eta\sigma_{kk}^1)$$

$$\sigma_{ij} = E\eta^{1-q}(\sigma_{ij}^0 + \eta\sigma_{ij}^1) \tag{8}$$

$$\sigma_{i3} = E(\sigma_{i3}^0 + \eta\sigma_{i3}^1)$$

$$u_i = h(u_i^0 + \eta u_i^1)$$

$$u_3 = h\eta^{1-q}(u_3^0 + \eta u_3^1)$$

for high-frequency transverse motions ($h^2 c^{-2}\dfrac{\partial^2 u_3}{\partial t^2} + \Lambda_{st}^2 u_3 \sim \eta^{2-2q}u_3$)

$$\sigma_{kk} = E(\sigma_{kk}^{0} + \eta\sigma_{kk}^{1})$$

$$\sigma_{ij} = E\eta^{2-2q}(\sigma_{ij}^{0} + \eta\sigma_{ij}^{1}) \tag{9}$$

$$\sigma_{i3} = E\eta^{1-q}(\sigma_{i3}^{0} + \eta\sigma_{i3}^{1})$$

$$u_i = h\eta^{1-q}(u_i^{0} + \eta u_i^{1})$$

$$u_3 = h(u_3^{0} + \eta u_3^{1})$$

In the formulae above $i \neq j = 1,2$, $k = 1,2,3$, E is Young's modulus, Λ_{sh} and Λ_{st} are thickness shear and stretch resonance frequencies, respectively. The quantities with the superscripts '0' and '1' are assumed to be of the same asymptotic order. In addition, in the case of low-frequency asymptotic behaviors (6) and (7) the quantities marked with '0'correspond to symmetric motions with respect to the mid-surface $\alpha_3 = 0$, while those marked with '1' are associated with anti-symmetric motions. At the same time these quantities may specify both symmetric and anti-symmetric motions in the case of high-frequency asymptotic behaviors (8) and (9); in doing so, the quantities with the superscript '0' are symmetric if those with the superscript '1' are anti-symmetric and *vice versa.*

A shell with fixed faces ($v_i^{\pm} = 0$ in the formula (4)) does not possess low-frequency vibration modes, whereas the asymptotic settings (8) and (9) for high-frequency long-wave motions are still valid near the thickness resonance frequencies Λ_{sh} and Λ_{st} specific of fixed faces (see [8] for more detail). When $q = 1$ the formulae (6)-(9) transform to those for short-wave motions. Thus they express the most general asymptotic concept in shell dynamics.

4 Leading-order approximations

Direct asymptotic integration of the original 3D equations in elasticity based on the asymptotic assumptions (6)-(9) including the limiting case $q = 1$ results in various approximate models.

If the variation of the surface tractions $q_i^{\pm}$ in time and space agrees with the conditions $q = a$ and $q = (a+1)/2$ underlying the asymptotic behaviors (6) and (7), then a superposition of the leading-order tangential and transverse low-frequency approximations corresponds to a slightly modified version of the classical Kirchhoff-Love theory. The only refinements consist of taking into account the normal compression and the thickness variation. The former leads to an improved formula for the normal stress resultants T_i. It becomes

$$T_i = \frac{2Eh}{1-\nu^2}\left(\varepsilon_i + \nu\varepsilon_j\right) + \boxed{\frac{\nu h}{1-\nu}\left(q_3^+ + q_3^-\right)} \qquad i \neq j = 1,2 \tag{10}$$

where ν is Poisson's ratio, ε_i denote strains.

Another improvement assumes a two-term approximation along the thickness for the 3D transverse displacement $u_3(\alpha_1, \alpha_2, \alpha_3)$. It is

$$u_3 = w + \boxed{\frac{\alpha_3}{h}\frac{\nu}{2E}\left(T_1 + T_2\right)} \tag{11}$$

where $w(\alpha_1, \alpha_2)$ is the transverse displacement of the mid-surface.

It is interesting that the aforementioned refinements, associated with the framed terms in the formulae (10) and (11), occur as asymptotically secondary correctors in the general static theory of shells formulated by A.L. Gol'denveiser [5]. In dynamics, however, these enter in asymptotically leading equations governing tangential motions [11]. In addition, the phenomenon of the normal compression and the thickness variation is of particular interest for advanced low-frequency models of fluid-structure interaction (e.g. see [1]).

It is also worth mentioning that a two-parameter asymptotic analysis (the variation and dynamicity indexes q and a are assumed to be independent) of forced vibrations may require a more substantial revision of the classical theory. In particular, for plate bending, transverse shear deformation, rotation inertia and etc. sometimes have to be introduced into the leading-order approximation [12].

Direct asymptotic integration based on the high-frequency long-wave settings (8) and (9) also leads to approximate 2D models. In particular, the leading-order equation of transverse motions under surface tractions or displacements (see the original 3D boundary conditions (3) and (4)) can be written in a unified form as

$$Q\Delta_2 u_3^a + Q_R u_3^a + h^2 c^{-2}\frac{\partial^2 u_3^a}{\partial t^2} + \Lambda_{st} u_3^a = F \tag{12}$$

where $u_3^a(\alpha_1, \alpha_2)$ is 2D slowly varying transverse amplitude, Δ_2 denotes 2D Laplacian specified on the mid-surface, Q is a constant coefficient that is strongly dependent on the mode number and Poisson's ratio, Q_R is a variable coefficient involving curvature effect, F is integral external load calculated in terms of the surface tractions $q_i^\pm$ or the surface displacements $v_i^\pm$; all the quantities entering an Eq. (12) are specified in the book [16].

In the case of tangential high-frequency long-wave motions asymptotic integration based on the formula (8) results in a vector analogue of the scalar Eq. (12). The former operates with the slowly varying amplitudes of tangential displacements and is valid near thickness shear resonance frequencies.

The 2D asymptotic theory of high-frequency long-wave vibrations has many interesting applications including those related to radiation and scattering by thin walled structures (see also [10]). The leading-order short-wave approximation usually does not allow reduction to a 2D form. The only advantage of the latter, in comparison with 3D elasticity, is that the variable coefficients in the linear differential operators L_{ij} and l_{ijk} (see the Eqs. (1) and (2)) may be fixed on the mid-surface $\alpha_3 = 0$. However, for some important specific problems, e.g. harmonic vibrations of a shell of revolution with a relatively small wave number in the circumferential direction, the leading-order short-wave approximation may be transformed to rather elegant 2D equations that represent natural generalizations of plane and anti-plane problems in elasticity for shell meridian cross-sections (e.g. see [6,9]).

5 Domains of validity and matching

The truncation error $O\left(\eta^{2-2q}\right)$, characteristic of all leading-order long-wave models following the asymptotic behaviours (6)-(9), results in the error $O\left(\eta^{2-3q}\right)$ when calculating propagating vibration modes. In the latter case the range of validity for the aforementioned models becomes (e.g. see [3] for further detail)

$$q < 2/3 \tag{13}$$

i.e. it does not cover the entire long-wave domain defined by the inequality $q < 1$.

Asymptotic analysis of higher-order plate and shell theories [3,4] shows that the range of validity of the Kirchhoff-Love theory in dynamics can be easily extended to the domain $q < 3/4$ by a proper correction of original inertial terms. The result is

$$\frac{\partial^2 w}{\partial t^2} \to \left[1 + h^2 \frac{7\nu - 17}{15(1-\nu)} \Delta_2\right] \frac{\partial^2 w}{\partial t^2} \tag{14}$$

and

$$\frac{\partial^2 \mathbf{u}_\tau}{\partial t^2} \to \left[1 + h^2 \frac{\nu^2}{3(1-\nu^2)} \operatorname{grad}_2 \operatorname{div}_2\right] \frac{\partial^2 \mathbf{u}_\tau}{\partial t^2} \tag{15}$$

where $\mathbf{u}_\tau(\alpha_1, \alpha_2)$ denotes the tangential displacement vector of the mid-surface; all 2D differential operators are specified in [16].

The theory with the refined inertial terms (14)-(15) possesses many important features. In particular, the substitution (15) transforms the 2D extensional wave front in the Kirchhoff-Love theory and the theory of generalized plane stress (its velocity is $c_3 = \left[E/\rho(1-\nu^2)\right]^{1/2}$) to a quasi-front (see [7,16] and many others).

A further refinement of inertial terms leads to asymptotic series in the right hand sides of the formulae (14) and (15). In particular, for harmonic flexural vibrations of a plate a correction of the circular frequency ω takes the form

$$\omega^2 \to \omega^2 \sum_{k=0}^{n} A_k(\nu) \left(\frac{\omega h}{c} \right)^k \qquad n = 0,1,2,3,... \tag{16}$$

where $A_0 = 1$ and the expressions of A_1, A_2 and A_3 are given in [16].

It is particularly important that the ranges of validity for long-wave approximations overlap those for short-wave ones; in doing so, the adapted matching principle states

$$\lim_{l/R \to 0} \left\{ \begin{array}{c} \text{long-wave} \\ \text{approximations} \end{array} \right\} = \lim_{l/h \to \infty} \left\{ \begin{array}{c} \text{short-wave} \\ \text{approximations} \end{array} \right\} \tag{17}$$

It exploits the fact that in the short-wave limit ($l/R \to 0$) the effect of curvature becomes, in a sense, negligible for long-wave approximations, whereas in the long-wave limit ($l/h \to \infty$) short-wave models may be reduced to a 2D form.

For all the motions except the low-frequency transverse (bending) one, overlap regions, corresponding to leading-order approximations, are given by

$$0 < q < 2/3 \tag{18}$$

For the bending motion, matching assumes taking into account some extra terms expressing the effect of curvature (see [16] for detail). The sought for estimation becomes

$$1/2 \le q < 2/3 \tag{19}$$

Numerical analysis of overlap regions is presented in the paper [2] devoted to a total synthesis of the 'exact' dispersion curves for a cylindrical shell on the basis of long-wave and short-wave approximations.

6 Concluding remarks

The consideration above is not necessarily restricted to a linear isotropic continuum. The developed approach also demonstrates a high efficiency for pre-stressed structures including, for example, the above mentioned phenomenon of a quasi-front [13-15].

At the same time further considerations dealing, say, with effects of non-linearity and non-homogeneity may require novel ideas and techniques. Thus the proposed methodology is universal only in sense that it is useful for overcoming two fundamental limitations discussed in the paper.

Acknowledgement

The research was partly supported by EPSRC (GR/R53692/01) and the Russian Foundation of Basic Research (02-01-00843).

References

1. Belov AV, Kaplunov JD, Nolde EV (1999) A refined asymptotic model of fluid-structure interaction in scattering by elastic shells. Flow Turb and Comb 61: 255-267
2. Beresin VL, Kossovich LYu, Kaplunov JD (1995) The synthesis of the dispersion curves for a cylindrical shell on the basis of approximate theories. Int J Sound Vibration 186: 37-53
3. Gol'denveizer AL (1976) Theory of elastic thin shells (in Russian), 2nd Edn. – Nauka, Moscow
4. Gol'denveizer AL (1966) The principles of reducing three-dimensional problems of elasticity to two-dimensional problems of the theory of plates and shells. In: Görtler H (ed) Proc. 11th Int Congr Appl Mech, Springer, Berlin pp 306-311
5. Gol'denveizer AL (1961) Theory of elastic thin shells. Pergamon Press, Oxford
6. Gol'denveizer AL, Kaplunov JD (1988) Dynamic boundary layer in problems of vibration of shells (in Russian). Izv Akad Nauk SSSR, Ser. Mekhanika Tverdogo Tela 23 No.4, pp 152-162 (English Transl Mechanics of Solids 23 (1988))
7. Gol'denveizer AL, Kaplunov JD, Nolde EV (1993) On Timoshenko-Reissner type theories of plates and shells. Int J Solids Structures. 30: 675-694
8. Kaplunov JD (1995) Long-wave vibrations of a thin body with fixed faces. QJMAM 48: 311-327
9. Kaplunov JD (1990) Integration of the dynamic boundary layer equations (in Russian). Izv Akad Nauk SSSR, Ser Mekhanika Tverdogo Tela 25, No.1: 148-160. (English Transl Mechanics of Solids 25 (1990))
10. Kaplunov JD, Markushevich DG (1993) Plane vibrations and radiation of an elastic layer lying on a liquid half-space. Wave Motion 17: 199-211
11. Kaplunov JD, Nolde EV (1996) On the effect of transverse compression in shell dynamics (in Russian). Prik Matem Mechan 60: 644-650 (English Transl PMM Russia 60 (1996))
12. Kaplunov JD, Nolde EV (1992) Two-parametric asymptotic analysis of the dynamic equations of elasticity in the case of plate bending (in Russian). Prik Matem Mekhan 56: 750-755 (English Transl PMM Russia 56 (1992))
13. Kaplunov JD, Nolde EV, Rogerson GA (2002) An asymptotically consistent model for long-wave high-frequency motion of a pre-stressed elastic plate. Mech Math Solids 7: 581-606
14. Kaplunov JD, Nolde EV, Rogerson GA (2002) Short-wave motion of a pre-stressed plate. IMA J. Appl Math 67: 383-399
15. Kaplunov JD, Nolde EV, Rogerson GA (2000) A low frequency model for dynamic motion in pre-stressed incompressible elastic structures. Proc R Soc Lond A 456: 2589-2610
16. Kaplunov JD, Kossovich LYu, Nolde EV (1998) Dynamics of thin walled elastic bodies. Academic Press, New-York

On Consistent Second-Order Plate Theories

R. Kienzler

Universität Bremen, Fachbereich Produktionstechnik,
Fachgebiet Technische Mechanik - Strukturmechanik,
Bremen, Germany

1 Introduction

In his pioneering work, Kirchhoff [4] established the so-called classical plate theory. With a set of (partly self-contradictory) a priori assumptions and the concepts of ersatz-shear forces as well as corner forces, Kirchhoff's theory delivers accurate results for thin plates undergoing small deformations. In a series of papers, Reissner [10-12] extended the theory in considering effects of shear deformations and stresses acting transversely to the plate mid-surface. The refinement of plate theories in various directions is a subject of still ongoing research. Some of the latest achievements are collected in a special issue of ZAMP [14] dedicated to Eric Reissner.

In this paper, we apply a rather classical approach. Starting from the equations of the linear theory of elasticity, the displacements are developed in power series with respect to the thickness coordinate ζ. The kinematical equations are satisfied by comparing coefficients of equal order in ζ. The strain-energy density and the density of the potential of the external forces are integrated with respect to the thickness coordinate and the principle of virtual work, finally, leads to a system of differential equations of Navier-Lamé type with a set of proper boundary conditions. In order to obtain a consistent plate (or shell) theory it is important that all governing equations are approximated uniformly (Naghdi [8], Koiter and Simmonds [5], Lo et al. [7], Krätzig [6]; Kienzler [2]). In a previous paper (Kienzler [3]), the uniform-approximation argument, which is proposed for the *derivation* of the plate equations, was extended to the requirement that it must be applied also to the *solution* of the respective differential equations. In this contribution, we employ a different solution strategy and provide some additional results.

2 Governing equations

The development is given in a framework as simple as possible: linear elasticity, homogeneous and isotropic material, constant plate thickness and Cartesian coordinates. An extension of the theory, however, towards a more general setting is easily possible. Let us consider a rectangular plate embedded in a Cartesian coordinate system as depictured in Fig. 1. The plane $x^3 = 0$ is designated as the plate mid-surface.

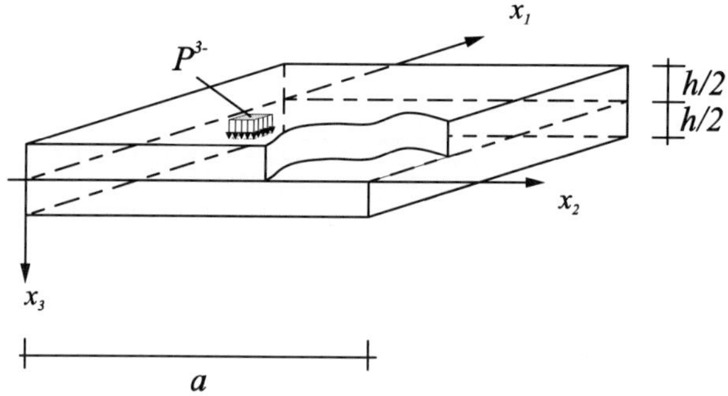

Fig. 1. Rectangular plate

The constant plate thickness h is adopted as the characteristic length in thickness direction and the plate extension a as the characteristic in-plane length with the condition $h/a << 1$. The external loads have components in x_3 - direction only and may be either applied by surface tractions (P^{3+}, P^{3-}) or by body forces $(f^3 = f)$. We introduce dimensionless coordinates as

$$\xi^\alpha = \frac{x^\alpha}{a}, \quad \zeta = \frac{x^3}{a}, \quad \alpha = 1, 2 \tag{1}$$

and partial derivations as

$$\frac{\partial()}{\partial x^1} = \frac{1}{a}\frac{\partial()}{\partial \xi^1} = \frac{1}{a}(),_1 = \frac{1}{a}()',$$

$$\frac{\partial()}{\partial x^2} = \frac{1}{a}\frac{\partial()}{\partial \xi^2} = \frac{1}{a}(),_2 = \frac{1}{a}()^\bullet. \tag{2}$$

The notation $(),_1$ and $(),_2$ or $()'$ and $()^\bullet$ will be used simultaneously.
The displacements $u_i (i = 1, 2, 3)$ are expanded in thickness direction in a power series

$$u_i(\xi^\alpha, \zeta) = a\sum_{l=0}^{\infty} {}^l u_i(\xi^\alpha)\zeta^l \ . \tag{3}$$

The index on the upper left-hand side of the dimensionless displacement quantities ${}^l u_i$ indicates the counting order of the series expansion.

In order to obtain a tractable set of equations, the infinite series has to be truncated at some specific value. Due to the thickness-integration procedure and the use of dimensionless quantities, (1)-(3), terms multiplied with different orders of $(c^2)^n (n = 0, 1, 2, ...)$ occur, with the small plate parameter

$$c^2 = \frac{h^2}{12a^2} << 1. \tag{4}$$

If we restrict ourselves to a consistent second-order $(n = 2)$ approximation, all terms up to the order c^4 are considered and higher-order terms are neglected. It turns out [3] that it is sufficient, firstly, to restrict the series expansion to $l \leq 5$ in (3) and to consider, secondly, even numbers $(l = 0, 2, 4)$ for u_3 and odd numbers $(l = 1, 3, 5)$ for u_α, since the plate problem and the disc problem (plane stress) are decoupled in this setting. We choose, therefore, the following ansatz

$$u_\alpha = a(\psi_\alpha \zeta + \Psi_\alpha \zeta^3 + \mathscr{H}_\alpha \zeta^5) \ , \tag{5}$$

$$u_3 = a(w - \omega\zeta^2 + \Omega\zeta^4) \ .$$

The derivation of the set of differential equations along the line indicated in the introduction may be found in Preusser [9] and Kienzler [3], and will not be repeated here.

With the abbreviations

$$\varphi = \psi_1{}' + \psi_2{}^\bullet, \quad \Phi = \Psi_1{}' + \Psi_2{}^\bullet, \quad \mathscr{P} = \mathscr{H}_1{}' + \mathscr{H}_2{}^\bullet;$$

$$\psi = \psi_2{}' - \psi_1{}^\bullet, \quad \Psi = \Psi_2{}' - \Psi_1{}^\bullet, \quad \mathscr{H} = \mathscr{H}_2{}' - \mathscr{H}_1{}^\bullet; \tag{6}$$

$$\Delta() = ()'' + ()^{\bullet\bullet},$$

shear modulus $G = E/(2(1+\nu))$, Young's modulus E, Poisson's ratio ν, we arrive at

$$\Delta w+\varphi+c^2\left[\Delta\omega+3\Phi\right]+\frac{9}{5}c^4\left[\Delta\Omega+5\mathscr{P}\right] \qquad =-\frac{{}^0Pa}{Gh}, \tag{7a}$$

$$\begin{aligned} &w' +\psi_1+c^2\left[-2\frac{1-\nu}{1-2\nu}\varphi' +\psi^\bullet+\frac{1-6\nu}{1-2\nu}\omega' +3\Psi_1\right]+ \\ &+\frac{9}{5}c^4\left[-2\frac{1-\nu}{1-2\nu}\Phi' +\Psi^\bullet+\frac{1-10\nu}{1-2\nu}\Omega' +5\mathscr{H}_1\right] \quad =0, \end{aligned} \tag{7b}$$

$$\begin{aligned} &w^\bullet +\psi_2+c^2\left[-2\frac{1-\nu}{1-2\nu}\varphi^\bullet -\psi' +\frac{1-6\nu}{1-2\nu}\omega^\bullet +3\Psi_2\right]+ \\ &+\frac{9}{5}c^4\left[-2\frac{1-\nu}{1-2\nu}\Phi^\bullet -\Psi' +\frac{1-10\nu}{1-2\nu}\Omega^\bullet +5\mathscr{H}_2\right] \quad =0, \end{aligned} \tag{7c}$$

$$c^2\left[\Delta w+\frac{1-6\nu}{1-2\nu}\varphi-8\frac{1-\nu}{1-2\nu}\omega\right]+\frac{9}{5}c^4\left[\Delta\omega+\frac{3-10\nu}{1-2\nu}\Phi-16\frac{1-\nu}{1-2\nu}\Omega\right]=-c^2\frac{{}^2Pa}{Gh}, \tag{7d}$$

$$c^2\left[3w' +3\psi_1\right]+\frac{9}{5}c^4\left[-2\frac{1-\nu}{1-2\nu}\varphi' +\psi^\bullet+\frac{3-10\nu}{1-2\nu}\omega' +9\Psi_1\right]=0, \tag{7e}$$

$$c^2\left[3w^\bullet +3\psi_2\right]+\frac{9}{5}c^4\left[-2\frac{1-\nu}{1-2\nu}\varphi^\bullet -\psi' +\frac{3-10\nu}{1-2\nu}\omega^\bullet +9\Psi_2\right]=0, \tag{7f}$$

$$\frac{9}{5}c^4\left[\Delta w+\frac{1-10\nu}{1-2\nu}\varphi-16\frac{1-\nu}{1-2\nu}\omega\right] \qquad =0, \tag{7g}$$

$$\frac{9}{5}c^4\left[5w' +5\psi_1\right] \qquad =0, \tag{7h}$$

$$\frac{9}{5}c^4\left[5w^\bullet +5\psi_2\right] \qquad =0. \tag{7i}$$

The load resultants ${}^0P$ and ${}^2P$ are given as

$$\begin{aligned} {}^0P &= h\left({}^0f+{}^2fc^2\right)+\left(P^{3+}+P^{3-}\right), \\ {}^2P &= h\,{}^0f+3\left(P^{3+}+P^{3-}\right), \end{aligned} \tag{8}$$

$$f = {}^0f + {}^2f\zeta^2 + 0\left(\zeta^4\right).$$

By differentiation, addition and subtraction of equations, the system (7) may be easily decoupled (cf. Preusser, [9]) into one system of differential equations (6 x 6) for the 6 unknowns $w, \varphi, \Phi, \mathscr{P}, \omega, \Omega$

$$\Delta w + \varphi + 3c^2\Phi + 9c^4\mathscr{P} - \frac{9}{5}c^4\Delta\Omega + c^2\left(\Delta\omega + \frac{18}{5}c^2\Delta\Omega\right) = -\frac{{}^0Pa}{Gh}, \tag{9a}$$

$$\begin{aligned}&\Delta w + \varphi + 3c^2\Phi + 9c^4\mathscr{P} - \frac{9}{5}c^4\Delta\Omega + \\ &+\frac{1-6v}{1-2v}c^2\left(\Delta\omega + \frac{18}{5}c^2\Delta\Omega\right) - 2\frac{1-v}{1-2v}c^2\left(\Delta\varphi + \frac{9}{5}c^2\Delta\Phi\right) = 0,\end{aligned} \tag{9b}$$

$$\begin{aligned}&c^2\Delta w + \frac{9}{5}c^4(\Delta\omega + 2\Phi) + \frac{1-6v}{1-2v}c^2\left(\varphi + \frac{9}{5}c^2\Phi\right) - \\ &\qquad -8\frac{1-v}{1-2v}c^2\left(\omega + \frac{18}{5}c^2\Omega\right) = -c^2\frac{{}^2Pa}{Gh},\end{aligned} \tag{9c}$$

$$c^2\left(3\Delta w + 3\varphi\right) + \frac{9}{5}c^4\left(-2\frac{1-v}{1-2v}\Delta\varphi + \frac{3-10v}{1-2v}\Delta\omega + 9\Phi\right) = 0, \tag{9d}$$

$$\frac{9}{5}c^4\left(\Delta w + \frac{1-10v}{1-2v}\varphi - 16\frac{1-v}{1-2v}\omega\right) = 0, \tag{9e}$$

$$\frac{9}{5}c^4\left(5\Delta w + 5\varphi\right) = 0, \tag{9f}$$

and one system of differential equations (3 x 3) for the three unknowns $\psi, \Psi, \mathscr{H}$

$$\psi + 3c^2\Psi + 9c^4\mathscr{H} - c^2\Delta\psi - \frac{9}{5}c^4\Delta\Psi = 0, \tag{10a}$$

$$3c^2\psi + \frac{81}{5}c^4\Psi - \frac{9}{5}c^4\Delta\psi = 0, \tag{10b}$$

$$9c^4\psi = 0. \tag{10c}$$

3 Second-order plate equations

Instead of solving the systems of differential Eqs. (9) and (10) right away, we neglect the terms of order c^6 during the elimination of the unknowns, since terms of this order of magnitude have been neglected already during the process of establishing the equations. On the other hand, for the calculation of stress resultants and boundary conditions, the knowledge of each individual quantity is not necessary rather than certain linear combinations as will be shown in the following.

Starting with the system (9), we keep in mind that the transverse deflection w is a smooth function and that - due to the dimensionless representation - $w, \Delta w$, etc. are of equal order of magnitude. Preceding in strictly a similar way as in [3] we find

$$c^4\varphi = -c^4\Delta w + 0(c^6)\,, \tag{11a}$$

$$c^4\omega = +c^4\frac{1}{2}\frac{\nu}{1-\nu}\Delta w + 0(c^6)\,, \tag{11b}$$

$$c^4\Phi = +c^4\frac{1}{6}\frac{2-\nu}{1-\nu}\Delta\Delta w + 0(c^6)\,, \tag{11c}$$

$$c^2\varphi = -c^2\Delta w - c^4\frac{3}{1-\nu}\Delta\Delta w + 0(c^6)\,, \tag{11d}$$

$$\begin{aligned} c^2\left(\omega + \frac{18}{5}c^2\Omega\right) = {} & c^2\frac{1}{2}\frac{\nu}{1-\nu}\Delta w + c^4\frac{3}{20}\frac{-1+10\nu+\nu^2}{(1-\nu)^2}\Delta\Delta w + \\ & + c^2\frac{1-2\nu}{8(1-\nu)}\frac{{}^2Pa}{Gh} + 0(c^6)\,, \end{aligned} \tag{11e}$$

$$\begin{aligned} \varphi + 3c^2\Phi + 9c^4\mathscr{P} - \frac{9}{5}c^4\Delta\Omega = {} & -\Delta w - c^2\frac{1}{2}\frac{4+\nu}{1-\nu}\Delta\Delta w - \\ & - c^4\frac{3}{20}\frac{31+18\nu+\nu^2}{(1-\nu)^2}\Delta\Delta\Delta w - c^2\frac{1-6\nu}{8(1-\nu)}\frac{\Delta^2 Pa}{Gh} + 0(c^6)\,. \end{aligned} \tag{11f}$$

With (11), Eqs. (9. b-f) are satisfied identically and (9. a) yields in a first instance

$$\frac{K}{a^2}\left(\Delta\Delta w + \frac{3}{5}c^2\frac{4+\nu}{1-\nu}\Delta\Delta\Delta w\right) = a\left({}^0P + \frac{1}{2}c^2\frac{\nu}{1-\nu}\Delta^2 P\right) \tag{12}$$

with the plate stiffness K given by

$$K = \frac{Eh^3}{12(1-\nu^2)}. \tag{13}$$

If the terms multiplied by c^2 in (12) are neglected, the differential equation of the consistent first-order approximation, i. e., the classical Kirchhoff equation is recovered. Since the factor K/a^2 involves with (13) the factor c^2, multiplication of (12) with c^2 delivers

$$\frac{c^2 K}{a^2}\Delta\Delta w = ac^{2\,0}P + 0(c^6), \tag{14}$$

which is used to transform (12) into its final form

$$\frac{K}{a^2}\Delta\Delta w = a\left[{}^0P + \frac{c^2}{10(1-\nu)}\left(5\nu\Delta\,{}^2P - 6(4+\nu)\Delta\,{}^0P\right)\right]. \tag{15}$$

In order to compare (15) with plate equations of other authors we suppose that the external forces are applied only by surface tractions.

With (8) we have

$${}^0P = P^{3+} + P^{3-} \equiv P, \qquad {}^2P = 3P, \tag{16}$$

and (15) leads to

$$\frac{K}{a^2}\Delta\Delta w = a\left(1 - \frac{3(8-3\nu)}{10(1-\nu)}c^2\Delta\right)P. \tag{17}$$

Eq. (17) coincides with that given by Brod [1]. In Kienzler [3], it has been shown that Preusser's [9] approach delivers the same results for all $l \geq 3$, if terms of the order c^6 are consequently neglected. It has been shown further that (17) is identical to the plate equation given by Reissner [10-12] once his definition of the transverse displacement w_R is transformed to our notation by

$$w_R = w + \frac{3}{10}c^2\frac{\nu}{1-\nu}\Delta w. \tag{18}$$

Zhilin [13] advanced a similar plate equation

$$\frac{K}{a^2}\Delta\Delta w_S = a\left[1 - \frac{2}{1-\nu}c^2\left(\frac{1}{\Gamma} - \frac{\nu}{2}\right)\Delta\right]P. \tag{19}$$

With

$$w_S = \frac{1}{h}\int_{-\frac{h}{2}}^{+\frac{h}{2}} \frac{u_3}{a}\, dx^3 = w + c^2\omega + 0(c^4) = w + c^2 \frac{1}{2}\frac{\nu}{1-\nu}\Delta w + 0(c^4) \tag{20}$$

and Zhilin's shear correction factor Γ

$$\Gamma = \frac{5}{6-\nu}, \tag{21}$$

Eq. (19) transforms exactly to (17).

It follows, therefore, that the plate equations of all authors mentioned above are equivalent within the uniform second-order approach.

Turning now our attention to the system of differential Eqs. (10), we have to take into account that the quantity $\psi = \psi_2^{\prime} - \psi_1^{\bullet}$ describe edge effects of plates. They decay rapidly with increasing distance from the boundary. Thus ψ and $\Delta\psi$ are not of the same order of magnitude. Experience rather shows that

$$\Delta\psi >> \psi\,. \tag{22}$$

Although it follows from (10.c) that

$$c^4\psi = 0(c^6)\,, \tag{23}$$

the term $c^4\Delta\psi$ may not be neglected in (10. b). Otherwise, the elimination process works through as above and delivers

$$\begin{aligned} c^4\Psi &= -\frac{1}{9}c^2\psi + 0(c^6)\,, \\ \psi + c^2\Psi + 9c^4\mathscr{H} &= +\frac{8}{15}c^2\psi + 0(c^6) \end{aligned} \tag{24}$$

with the remaining differential equation for ψ as

$$c^2\left(\psi - \frac{3}{2}c^2\Delta\psi\right) = 0\,. \tag{25}$$

4 Stress resultants and boundary conditions

Equations (15) and (25) are the two differential equations of the consistent second-order plate theory to be solved. Once w and ψ are known, the stress resultants can be expressed in terms of w and ψ. From the general form given in Kienzler [3], (Eqs. (30)-(32)), we obtain the following relations for the bending and twisting moments $M_{\alpha\beta}$ and transverse shear forces Q_α by using (11), (13), and (24) and the abbreviations

$$\kappa = \frac{3}{10}\frac{8+\nu}{1-\nu}, \quad \varepsilon = \frac{1}{10}\frac{\nu}{1-\nu}, \quad \tilde{P} = 5\,{}^2P - 3\,{}^0P\,; \tag{26}$$

$$M_{11} = -\frac{K}{a}\left[\left(1+c^2\kappa\Delta\right)\left(w'' + \nu w^{\bullet\bullet}\right) + c^2\frac{6}{5}(1-\nu)\psi'^{\bullet}\right] + c^2a^2\varepsilon\tilde{P},$$

$$M_{22} = -\frac{K}{a}\left[\left(1+c^2\kappa\Delta\right)\left(w^{\bullet\bullet} + \nu w''\right) - c^2\frac{6}{5}(1-\nu)\psi'^{\bullet}\right] + c^2a^2\varepsilon\tilde{P},$$

$$M_{12} = -\frac{K}{a}(1-\nu)\left[\left(1+c^2\kappa\Delta\right)w'^{\bullet} - c^2\frac{3}{5}\left(\psi'' - \psi^{\bullet\bullet}\right)\right],$$

$$\begin{aligned} Q_1 &= -\frac{K}{a^2}\left[\left(1+c^2\kappa\Delta\right)\Delta w' + \frac{2}{5}(1-\nu)\psi^{\bullet}\right] + c^2a\varepsilon\tilde{P}' \\ &= -\frac{K}{a^2}\left[\Delta w' + \frac{2}{5}(1-\nu)\psi^{\bullet}\right] + \frac{c^2a}{10(1-\nu)}\left(5\nu\,{}^2P' - 6(4+\nu)\,{}^0P'\right), \end{aligned} \tag{27}$$

$$\begin{aligned} Q_2 &= -\frac{K}{a^2}\left[\left(1+c^2\kappa\Delta\right)\Delta w^{\bullet} - \frac{2}{5}(1-\nu)\psi'\right] + c^2a\varepsilon\tilde{P}^{\bullet} \\ &= -\frac{K}{a^2}\left[\Delta w^{\bullet} - \frac{2}{5}(1-\nu)\psi'\right] + \frac{c^2a}{10(1-\nu)}\left(5\nu\,{}^2P^{\bullet} - 6(4+\nu)\,{}^0P^{\bullet}\right). \end{aligned}$$

Employing Kronecker's tensor of unity, $\delta_{\alpha\beta}$ $(\delta_{11} = \delta_{22} = 1, \delta_{12} = \delta_{21} = 0)$, and Levi-Cevita's permutation tensor, $\varepsilon_{\alpha\beta}(\varepsilon_{11} = \varepsilon_{22} = 0, \varepsilon_{12} = -\varepsilon_{21} = 1)$, Eqs. (27) read in index notation

$$\begin{aligned} M_{\alpha\beta} = -\frac{K}{a}\Big[&(1+\kappa c^2\Delta)\left((1-\nu)w,_{\alpha\beta} + \nu\delta_{\alpha\beta}w,_{\gamma\gamma}\right) + \\ &+\frac{3}{5}(1-\nu)(\varepsilon_{\alpha\gamma}\psi,_{\beta\gamma} + \varepsilon_{\beta\gamma}\psi,_{\alpha\gamma})\Big] + \delta_{\alpha\beta}c^2a^2\varepsilon\tilde{P}, \end{aligned} \tag{28}$$

$$Q_\alpha = -\frac{K}{a^2}\left[(1+\kappa c^2\Delta)\Delta w,_\alpha + \frac{2}{5}(1-\nu)\varepsilon_{\alpha\beta}\psi,_\beta\right] + c^2a\varepsilon\tilde{P},_\alpha\,.$$

The formulae (76)-(78) given in Kienzler [3] are incomplete since ψ has not been considered there. Due to (25), the bending-equilibrium equations

$$\frac{1}{a} M_{\alpha\beta,\alpha} - Q_\beta = 0 \tag{29}$$

are satisfied and the force-equilibrium equation

$$\frac{1}{a} Q_{\alpha,\alpha} + {}^0P = 0 \tag{30}$$

leads to (15).

Neglecting ψ and all terms multiplied by c^2 in (27) the stress resultants of the Kirchhoff-plate theory are recovered. Higher-order stress resultants can also be calculated (cf. Kienzler, [3]) but are of minor importance since, firstly, the equilibrium conditions, in which they are involved, are satisfied identically and, secondly, they are not needed for the boundary conditions.

The boundary conditions given in Kienzler [3] have to be corrected to

$$\begin{aligned} M^*_{nn} &= M_{nn} \quad \text{or} \quad \psi^*_n = -(1 + c^2\kappa\Delta)w' - \frac{6}{5}c^2\psi^\bullet , \\ M^*_{nn} &= M_{nt} \quad \text{or} \quad \psi^*_t = -(1 + c^2\kappa\Delta)w^\bullet + \frac{6}{5}c^2\psi' , \\ Q^*_n &= Q_n \quad \text{or} \quad w^* = (1 + 3c^2\varepsilon\Delta)w. \end{aligned} \tag{31}$$

The indices n and t indicate the normal and tangential direction along the boundary, respectively.

5 Discussion

With the differential equations (15) and (25), the boundary conditions (31) and the equations for the relevant stress resultants (28), we have established a consistent second-order plate theory without introducing any *a priori* assumptions. Hereby, we used the displacement representation (5) involving nine parameters: $w, \omega, \Omega, \psi_\alpha, \Psi_\alpha$ and $\mathcal{H}_\alpha$, or, alternatively w, ω, Ω, φ, ψ, Φ, Ψ, $\mathcal{P}$ and $\mathcal{H}$. From (11e, f) and (24b) it turns out that the quantities Ω, $\mathcal{P}$ and $\mathcal{H}$ or Ω and $\mathcal{H}_\alpha$ are not specified by the calculation. They can be chosen arbitrarily, without changing the governing equations , i.e., a cubic representation in ζ with the six parameters $w, \omega, \varphi, \psi, \Phi$ and Ψ (or w, ω, ψ_α and Ψ_α) would be sufficient. In addition, from (11c) and (24a) we have

$$c^2\Phi = c^2 \frac{1}{6}\frac{2-\nu}{1-\nu} \Delta\Delta w + c^4\hat{\Phi} , \tag{32}$$

$$c^2\Psi = -c^2 \frac{1}{9}\psi + c^4\hat{\Psi} ,$$

and it follows also that no specific restriction is put on the terms multiplied by c^4, i.e., $\hat{\Phi}$ and $\hat{\Psi}$. Instead of neglecting the terms Ω, $\mathcal{P}$, $\mathcal{H}$, $\hat{\Phi}$ and $\hat{\Psi}$ one can use them to satisfy *a posteriori* the boundary conditions for the tractions on the plate faces $x_3 = \pm\, h/2$.

The uniform-approximation technique has been already applied to shell theories in [2]. The zeroth-order approximation results in the membrane theory, whereas the first-order approximation delivers a consistent bending theory of shells. Warping of the cross sections and change-of-thickness effects must be taken into account. Shear-deformation effects are of second order. Due to the complexity of the equations, caused mainly by the curvature of the shell's middle surface, the second-order equations for an individual shell geometry should be established using the benefits of computer devices allowing for symbolic manipulations.

The derivation of the equations of the second-order plate theory in curvilinear coordinates does not encounter any difficulties. Also the treatment of temperature effects, anisotropic elastic materials, laminated structures and materials with a non-linear characteristic is straightforward. The coupling between disc and plate terms, however, has to be observed.

6 Concluding remarks

The uniform-approximation technique leads to self-consistent plate theories of any desired degree of accuracy without the necessity of introducing any *a priori* assumptions or neglections. The second-order theory discussed in the paper is governed by two uncoupled differential equations, one of fourth order in the transverse displacement w and one of second order in ψ, the quantity describing the influence of shear deformation. Along the boundary, three boundary conditions (31) have to be satisfied to guarantee a unique solution. Once w and ψ are known, the stress resultants - bending and twisting moments, transverse shear forces - can be evaluated (28).

It has been shown that the theories of other authors, e. g., Reissner and Zhilin, have different appearances, but are equivalent in terms of a uniform second-order approximation. A more detailed discussion and the development of the first-order approximation can be found in Kienzler [3].

As a result of the present investigation, it can be said that it is of no use to extend the series expansion term by term and to analyse its impact on the theory. The uniform-expansion technique delivers identical results for a desired order of approximation as long as *enough* terms are considered.

References

1. Brod K (1972) Herleitung der Plattengleichungen der klassischen Elastizitätstheorie durch systematische Entwicklung nach einem Dickenparameter. Bericht des Max-Planck-Instituts für Strömungsforschung, Göttingen
2. Kienzler R (1982) Eine Erweiterung der klassischen Schalentheorie; der Einfluss von Dickenverzerrungen und Querschnittverwölbungen. Ing Arch 52: 311-322. Extended version in: Dissertation, Technische Hochschule Darmstadt (1980)
3. Kienzler R (2002) On consistent plate theories. Arch Appl Mech 72: 229-247
4. Kirchhoff GR (1850) Über das Gleichgewicht und die Bewegung einer elastischen Scheibe. C J Angew Math 40: 51-88
5. Koiter WT, Simmonds JG (1973) Foundations in shell theories. In: Becker E, Mikhailov GK (eds) Theoretical and applied mechanics, Proc 13th Int Congr Theor Appl Mech, pp 150-176
6. Krätzig WB (1989) On the structure of consistent shell theories. In: Koiter WT, Mikhailov GK (eds) Proc 3rd IUIAM Symp on Shell Theory. North-Holland, Amsterdam pp 353-368
7. Lo KH, Christensen RM, Wu EM (1977) A higher-order theory of plate deformation. Part 1: Homogeneous plates. ASME J Appl Mech 44: 663-668
8. Naghdi PM (1972) The theory of plates and shells. In: Flügge S (ed) Handbuch der Physik. Springer, Berlin VI, A2: 425-646
9. Preusser G (1984) Eine systematische Herleitung von verbesserten Plattengleichungen. Ing Arch 54: 51-61. Extended version in: Dissertation, Technische Hochschule Darmstadt (1982)
10. Reissner E (1944) On the theory of bending of elastic plates. J Math Phys 23: 184-191
11. Reissner E (1945) The effect of transverse shear deformation on the bending of elastic plates. ASME J Appl Mech 12: A69-A77
12. Reissner E (1947) On bending of elastic plates. Quart Appl Math 5: 55-68
13. Steigmann D, Kaplunov JD (eds) (2000) Z Angew Math Phys (ZAMP). Special issue dedicated to Prof Eric Reissner 51: 507-649
14. Zhilin PA (1992) On the Poisson and Kirchhoff plate theories from the point of view of the modern plate theory (in Russian) Izvestia Akedemii Nauk Rossii, Mekhanika tverdogo tela, Moscow, 3: 48-64

2-D Model of Stressed Plate Derived From the 3-D Elasticity Theory Model

A.G. Kolpakov and I.G. Sheremet

Novosibirsk, Russia

1 Introduction

The classical method of deriving 2-D models of plate with initial stresses is based on the Kirchhoff-Love kinematics hypothesis [7] and it is suitable for uniform plates made of a homogeneous material. If the plate is non-uniform or made of inhomogeneous material, the classical method cannot be applied. The investigators developed the homogenization method [1,2,8] suitable to derive 2-D models from 3-D models and applied it to plates without initial stresses.

Considering plates with initial stresses, we meet new problems. The first one is the problem of the original model. The linear model of the theory of elasticity does not describes bodies with initial stresses. For example, it does not describes the lose of stability. It is the general opinion that the 3-D non-linear model of elasticity will describes all such effects. Unfortunately, the 3-D non-linear model is very complex and difficult for analysis. This paper demonstrates that the models of plates with initial stresses can be derived from the relatively simple 3-D linearized model of elastic bodies with initial stresses [9].

In this paper, 2-D plate models will be derived from 3-D linearized models of elastic bodies with initial stresses using two-scale expansion technique. Applying this technique to a 3-D model of an elastic body with initial stresses, we will meet a new situation (which has no analogy in the classical theory) related with the influence of initial stresses. From the mathematical point of view this situation is connected with the loosing of some symmetries in the governing equations as compared with the governing equations of the body without initial stresses. Considering this situation, we will find that it is not a formal mathematical problem, but this loses of symmetries is the reason which "transform" the initial stresses of the original 3-D model into terms of the 2-D models.

Using the 3-D model as the original model, we consider both the uniform structures made of homogeneous materials and the non-uniform structures made of inhomogeneous materials. As a result, the method developed in this paper can be applied to both the non-uniform and the inhomogeneous plates widely used in modern constructions.

2 Formulation of the problem

We will examine 3-D body of periodic structure obtained by repeating a certain small periodicity cell εY over the Ox_1x_2-plane (Fig.1). Here ε is the periodicity cell characteristic dimension, which is assumed to be small (that is formalized in the form $\varepsilon \to 0$). For $\varepsilon \to 0$ the 3-D plate "tightens" to the domain D at the Ox_1x_2-plane - a 2-D plate. The 3-D plate is subjected to the initial stresses σ^*_{ij}. These stresses can influence the limit (as $\varepsilon \to 0$) 2-D model. Our aim is to derive 2-D plate model incorporating the initial stresses.

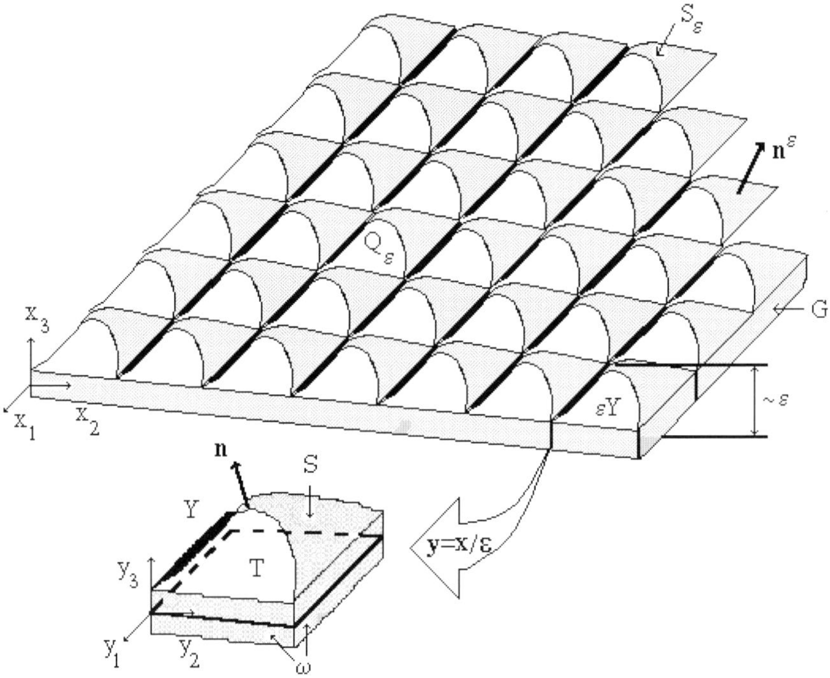

Abb. 1. The plate of periodic structure and those periodicity cell in the "fast" variables **y**=**x**/ε.

The starting point of our investigation is the exact 3-D formulation of an elasticity problem for a body with initial stresses [9]. The equilibrium equations of the plate as a 3-D elastic body with initial stresses can be written in the form

$$\partial/\partial x_j(A_{ijmn}(x_1,x_2,\mathbf{x}/\varepsilon)\partial u^\varepsilon_m/\partial x_n)=f^\varepsilon_i(x_1,x_2,\mathbf{x}/\varepsilon) \text{ in } Q_\varepsilon \tag{2.1}$$

$$A_{ijmn}(x_1,x_2,\mathbf{x}/\varepsilon)\partial u^\varepsilon_m/\partial x_n n^\varepsilon_j=g_i(x_1,x_2,\mathbf{x}/\varepsilon) \text{ on } S_\varepsilon, \ \mathbf{u}^\varepsilon(\mathbf{x})=0 \text{ on } S_u. \tag{2.2}$$

Q_ε is 3-D region occupied by the plate; S_ε is the lateral surface of the plate; $\mathbf{n}^\varepsilon$ is a normal to S_ε; the plate fastened to the surface S_u (Fig.1); $\mathbf{u}^\varepsilon$ are the displacements; $\delta_{ii}=1$, $\delta_{im}=0$ if $i\neq m$; $\mathbf{f}^\varepsilon=(f_1,f_2,\varepsilon^{-1}f_3)$, $\mathbf{g}$ are the mass and surface forces; A_{ijmn} are combinations of the tensor of elasticity constants $\varepsilon^{-3}a_{ijmn}$ of body with no initial stresses and the initial stresses σ^*_{ij}, which, in the case under consideration, are taken in the form

$$A_{ijmn}(x_1,x_2,\mathbf{x}/\varepsilon)=\varepsilon^{-3}a_{ijmn}(\mathbf{x}/\varepsilon)+\varepsilon^{-1}b_{ijmn}(x_1,x_2,\mathbf{x}/\varepsilon)\,, \tag{2.3}$$

where

$$b_{ijmn}(x_1,x_2,\mathbf{x}/\varepsilon)=\sigma^*_{jn}(x_1,x_2,\mathbf{x}/\varepsilon)\delta_{im}. \tag{2.4}$$

Denote by T the projection of the periodicity cell Y in the "fast" variables $\mathbf{y}=\mathbf{x}/\varepsilon$ on the Oy_1y_2-plane (Fig.1) and Π(T). It is assumed that the functions $a_{ijmn}(\mathbf{y})$ and $\sigma^*_{jn}(x_1,x_2,\mathbf{y})$ are periodic in y_1, y_2 with the periodicity cell T.

The stresses σ_{ij} determined by the formula $\sigma_{ij}=A_{ijmn}(x_1,x_2,\mathbf{y})\partial u^\varepsilon_m/\partial x_n$ are called the additional stresses [9]. For $A_{ijmn}(x_1,x_2,\mathbf{y})$ determined by (2.3), the relationship between the additional stresses σ_{ij} and displacements $\mathbf{u}^\varepsilon$ takes the form:

$$\sigma_{ij}=(\varepsilon^{-3}a_{ijmn}(\mathbf{y})+\varepsilon^{-1}b_{ijmn}(x_1,x_2,\mathbf{y}))\partial u^\varepsilon_m/\partial x_n\ . \tag{2.5}$$

The formula (2.5) can be considered as a local constitutive equation of body with initial stresses [5].

In contrast to the elastic constants a_{ijmn}, which are *symmetric* in i and j [6] the coefficients A_{ijmn} (2.3) are *not symmetric* in i and j. This non-symmetry is related with the asymmetry of the coefficient b_{ijmn} (2.4), incorporating the initial stresses, and plays the determined role in the analysis of structures with initial stresses.

3 Asymptotic expansions and the equations of equilibrium

We will study global deformations of the plate, in particular, buckling in global forms, as $\varepsilon\to 0$. To do this, we use the following asymptotic expansions for the displacements $\mathbf{u}^\varepsilon$ and stresses σ_{ij}:

$$\mathbf{u}^\varepsilon=\mathbf{u}^{(0)}(x_1,x_2)+\varepsilon\mathbf{u}^{(1)}(x_1,x_2,\mathbf{y})+\ldots=\mathbf{u}^{(0)}(x_1,x_2)+\sum_{k=1}^{\infty}\varepsilon^k\mathbf{u}^{(k)}(x_1,x_2,\mathbf{y}) \tag{3.1}$$

$$\sigma_{ij}=\sum_{p=-3}^{\infty}\varepsilon^p\sigma^{(p)}_{ij}(x_1,x_2,\mathbf{y})\ . \tag{3.2}$$

Here x_1,x_2 are a "slow" variable in the plane of the plate D, $\mathbf{y}=\mathbf{x}/\varepsilon$ are the "fast" variables. It is assumed that $\mathbf{u}^{(k)}$, $\sigma^{(p)}_{ij}$ are periodic in y_1, y_2 with the periodicity cell T.

Analysis of the problem (2.1)-(2.4) breaks down into two stages. The first entails obtaining the equations of equilibrium for the plate considered as a 2-D structure. This stage is not involved with local governing equations (in the case under consideration, with the Eq. (2.5)), and it is the same stage for any governing Eqs. [3].

The resultant stresses introduced by the formula $N^{(p)}_{ij}=\langle\sigma^{(p)}_{ij}\rangle$ and the resultant moments introduced by the formula $M^{(p)}_{\alpha\beta}=-\langle\sigma^{(p)}_{\alpha\beta}y_3\rangle$ (i,j=1,2,3; α,β=2,3; p=−3,−2,...) satisfy the following 2-D equilibrium Eqs. [2]:

$$N^{(p)}_{\alpha\beta,\beta x}=0 \ (p=-3,-2) \tag{3.3}$$

$$N^{(-1)}_{3\beta,\beta x}=\langle f_3\rangle+\langle g_3\rangle_S \tag{3.4}$$

$$M^{(-2)}_{\alpha\beta,\beta x}-N^{(-1)}_{\beta 3}=0 \ . \tag{3.5}$$

Here and below the Latin indices take the values 1,2,3 and the Greek indices take the values 1,2; $,\alpha x$ means $\partial/\partial x_\alpha$ and $,3y$ means $\partial/\partial y_3$;

$$\langle\rangle=(\text{meas}T)^{-1}\int_Y \mathbf{dy} \text{ and } \langle\rangle_S=(\text{meas}T)^{-1}\int_S \mathbf{dy}$$

are the average value over the periodicity cell Y and those lateral surface S in the "fast" variables $\mathbf{y}$.

The Eqs. (3.3)-(3.5) look like the classical equations of equilibrium [7], but $N^{(-1)}_{3\beta}\neq N^{(-1)}_{\beta 3}$!

The following equations are satisfied for the $\sigma^{(p)}_{ij}$ [2]:

$$\sigma^{(p)}_{ij,jy}+\sigma^{(p-1)}_{i\beta,\beta x}=0 \text{ in } Y,\ \sigma^{(p)}_{ij}n_j=0 \text{ on } S\ (p=-3,-2,..,),\ \sigma^{(-4)}_{ij}=0 \tag{3.6}$$

4 The constitutive equations of plate with initial stresses

The second stage of analysis of the problem consists of obtaining the governing equations for the plate as 2-D structure and the excluding of unknown quantities from the equilibrium equations. In contrast to the first stage, this stage does involve local constitutive equations and it is the main stage in our investigation.

With the use of two-scale expansion, the differential operators are presented in the form of sum of operators in $\{x_i\}$ and in $\{y_i\}$ [1,8]. For the function $f(x_1,x_2,\mathbf{y})$ this representation takes the form [3]

$$\partial f/\partial x_3=\varepsilon^{-1}f_{,3y},\ \partial f/\partial x_\alpha=f_{,\alpha x}+\varepsilon^{-1}f_{,\alpha y}\quad (\alpha=1,2)\ . \tag{4.1}$$

Substituting (3.1),(3.2) into the local governing Eq. (2.5), with allowance for (4.1) and equating the terms with identical powers of ε in (4.2), we obtain

$$\sigma^{(p)}_{ij}=a_{ijm\beta}(\mathbf{y})u^{(p+3)}_{m,\beta x}+a_{ijmn}(\mathbf{y})u^{(p+4)}_{m,ny}\quad (p=-3,-2), \tag{4.2}$$

$$\sigma^{(-1)}_{ij}=a_{ijm\beta}(\mathbf{y})u^{(2)}_{m,\beta x}+a_{ijmn}(\mathbf{y})u^{(3)}_{m,ny}+b_{ijmn}u^{(1)}_{m,ny}\ . \tag{4.3}$$

Let us consider the problem (3.6, p=-3),(4.2, p=-3),(4.5,k=1), which can be written as

$$(a_{ijmn}(\mathbf{y})u^{(1)}_{m,ny}+a_{ijm\beta}(\mathbf{y})u^{(0)}_{m,\beta x}(x_1,x_2)_{,jy}=0 \text{ in } Y,$$

$$(a_{ijmn}(\mathbf{y})u^{(1)}_{m,ny}+a_{ijm\beta}(\mathbf{y})u^{(0)}_{m,\beta x}(x_1,x_2)n_j=0 \text{ on } S,$$

$\mathbf{u}^{(1)}$ is periodic in y_1, y_2 with the periodicity cell T.

Solution of this problem can be found in the form

$$\mathbf{u}^{(1)}=-y_3u^{(0)}_{3,\alpha x}(x_1,x_2)\mathbf{e}_\alpha+\mathbf{V}(x_1,x_2)$$
$$u^{(0)}_{\alpha}=0\ (\alpha=1,2) \tag{4.4}$$

$\{\mathbf{e}_i\}$ are basis vectors of the co-ordinate system; $\mathbf{V}(x_1,x_2)$ is a function of the argument x_1,x_2.
Substitution of (4.4) into (4.2) gives the following equations

$$\sigma^{(-3)}_{ij}=0, \tag{4.5}$$

$$\sigma^{(-2)}_{ij}=a_{ijmn}(\mathbf{y})u^{(2)}_{m,ny}-a_{ij\alpha\beta}(\mathbf{y})y_3u^{(0)}_{3,\alpha x\beta x}(x_1,x_2)+a_{ij\alpha\beta}(\mathbf{y})V_{\alpha,\beta x}(x_1,x_2)\ .$$

Let us examine the problem (3.6, p=-2),(4.5, p=-2),(4.5,k=2), which can be written as

$$(a_{ijmn}(\mathbf{y})u^{(2)}_{m,ny}-a_{ij\alpha\beta}(\mathbf{y})y_3u^{(0)}_{3,\alpha x\beta x}(x_1,x_2)+a_{ij\alpha\beta}(\mathbf{y})V_{\alpha,\beta x}(x_1,x_2))_{,jy}=0 \text{ in } Y$$
$$(a_{ijmn}(\mathbf{y})u^{(2)}_{m,ny}-a_{ij\alpha\beta}(\mathbf{y})y_3u^{(0)}_{3,\alpha x\beta x}(x_1,x_2)+a_{ij\alpha\beta}(\mathbf{y})V_{\alpha,\beta x}(x_1,x_2))n_j=0 \text{ on } S \tag{4.6}$$

$\mathbf{u}^{(2)}$ is periodic in y_1, y_2 with the periodicity cell T.

In order to solve problems of this kind, the so-called basic cell problems are introduced

$$(a_{ijmn}(\mathbf{y})\mathbf{X}^{\nu\alpha\beta}_{m,ny}+(-1)^{\nu}y_3^{\nu}a_{ij\alpha\beta}(\mathbf{y}))_{,jy}=0 \quad \text{in } Y$$

$$(a_{ijmn}(\mathbf{y})\mathbf{X}^{\nu\alpha\beta}_{m,ny}+(-1)^{\nu}y_3^{\nu}a_{ij\alpha\beta}(\mathbf{y}))n_j=0 \quad \text{on } S$$

$\mathbf{X}^{\nu\alpha\beta}(\mathbf{y})$ is periodic in y_1, y_2 with the periodicity cell T (ν=0,1).

The solution of the problem (4.6) can be expressed as follows

$$\mathbf{u}^{(2)}=-\mathbf{X}^{1\alpha\beta}(\mathbf{y})u^{(0)}_{3,\alpha x\beta x}(x_1,x_2)-y_3V_{3,\beta x}(x_1,x_2)+\mathbf{X}^{0\alpha\beta}(\mathbf{y})V_{\alpha,\beta x}(x_1,x_2)\ . \tag{4.7}$$

Substituting (4.7) into (4.5), we have

$$\begin{gathered}\sigma^{(-3)}_{ij}=(a_{ijmn}(\mathbf{y})\mathbf{X}^{0\alpha\beta}_{m,ny}+a_{ij\alpha\beta}(\mathbf{y}))V_{\alpha,\beta x}(x_1,x_2)+\\ +(a_{ijmn}(\mathbf{y})\mathbf{X}^{1\alpha\beta}_{m,ny}(\mathbf{y})-y_3a_{ij\alpha\beta}(\mathbf{y}))u^{(0)}_{3,\alpha x\beta x}(x_1,x_2)\ .\end{gathered} \tag{4.8}$$

4.1 Governing equations for the in-plane forces and moments

Averaging (4.8) with ij=γδ (γ,δ=1,2) over the periodicity cell Y, we obtain

$$N^{(-2)}_{\gamma\delta}=A^{0}_{\gamma\delta\alpha\beta}V_{\alpha,\beta x}+A^{1}_{\gamma\delta\alpha\beta}u^{(0)}_{3,\alpha x\beta x}\ . \tag{4.9}$$

Multiplying (4.8) with ij=γδ (γ,δ=1,2) by $-y_3$ and averaging over the periodicity cell Y, we obtain the following equation

$$M^{(-2)}_{\gamma\delta}=A^{1}_{\gamma\delta\alpha\beta}V_{\alpha,\beta x}+A^{2}_{\gamma\delta\alpha\beta}u^{(0)}_{3,\alpha x\beta x}\ . \tag{4.10}$$

The coefficients in the Eqs. (4.9),(4.10) are

$$\begin{gathered}A^{\mu+\nu}_{\gamma\delta\alpha\beta}=\langle((-1)^{\nu}a_{\gamma\delta\alpha\beta}(\mathbf{y})y_3+a_{\gamma\delta mn}(\mathbf{y})X^{\nu\alpha\beta}_{m,ny}(\mathbf{y}))(-1)^{\mu}y_3\rangle\\ (\nu,\mu=0,1;\ \alpha,\beta,\gamma,\delta=1,2)\ .\end{gathered} \tag{4.11}$$

The obtained Eqs. (4.9),(4.10) are governing equations of the plate considered as a 2-D structure. The coefficients (4.11) are the stiffnesses of the plate. The derived governing equations determine the resultant in-plane forces and moments but not the shear forces.

4.2 A non-symmetry of the shear forces in the plate with initial stresses and the governing equations for shear forces

For the case of the absence of the initial stresses we have $N^{(-1)}_{ij}=N^{(-1)}_{ji}$ and can exclude the quantities $N^{(-1)}_{\beta3}$, $N^{(-1)}_{3\beta}$ (called the shear forces) from (3.5),(3.6) using this symmetry with respect the indices i and j, only. This procedure of exclusion is used both in the classical theory [7] and asymptotic theory [2] for plate with no initial stresses. In our case $N^{(-1)}_{ij}$ *does not have the symmetry* with respect the indices i and j. Then we need to examine $N^{(-1)}_{ij}$ in the details, to obtain some additional information on the shear forces, which will allow us exclude the shear forces in another manner.

Let us insert $\mathbf{u}^{(1)}$ into (4.3) in accordance with (4.4). Then we obtain

$$\sigma^{(-1)}_{ij}=a_{ijm\beta}(\mathbf{y})u^{(2)}_{m,\beta x}+a_{ijmn}(\mathbf{y})u^{(3)}_{m,ny}-b_{ij\beta 3}(x_1,x_2,\mathbf{y})u^{(0)}_{3,\beta x}(x_1,x_2)\ . \qquad (4.11)$$

The first and second terms in the right-hand side in (4.11) *are symmetric* with respect i and j by virtue of symmetry of the elastic constants. Then the following relation can be derived from (4.11)

$$\sigma^{(-1)}_{ij}-\sigma^{(-1)}_{ji}=b_{ij\beta 3}(x_1,x_2,\mathbf{y})u^{(0)}_{3,\beta x}(x_1,x_2)-b_{ji\beta 3}u^{(0)}_{3,\beta x}(x_1,x_2).$$

Averaging this equality over the periodicity cell Y, we obtain

$$N^{(-1)}_{ij}-N^{(-1)}_{ji}=<b_{ij\beta 3}(x_1,x_2,\mathbf{y})>u^{(0)}_{3,\beta x}(x_1,x_2)-<b_{ji\beta 3}(x_1,x_2,\mathbf{y})>u^{(0)}_{3,\beta x}(x_1,x_2) \qquad (4.12)$$

For i=3 and j=γ one obtains from (4.12) and the definition of b_{ijmn} (2.4):

$$N^{(-1)}_{3\gamma}-N^{(-1)}_{\gamma 3}=K_\gamma\ , \qquad (4.13)$$

where $K_\gamma(x_1,x_2)$ is defined by the formula

$$\begin{aligned}K_\gamma&=<b_{\gamma 3\beta 3}(x_1,x_2,\mathbf{y})>u^{(0)}_{3,\beta x}(x_1,x_2)-<b_{3\gamma\beta 3}(x_1,x_2,\mathbf{y})>u^{(0)}_{3,\beta x}(x_1,x_2)=\\&=\delta_{\gamma\beta}<\sigma^*_{33}(x_1,x_2,\mathbf{y})>u^{(0)}_{3,\beta x}(x_1,x_2)-<\sigma^*_{\gamma\beta}(x_1,x_2,\mathbf{y})>u^{(0)}_{3,\beta x}(x_1,x_2)=\\&=N^*_{33}u^{(0)}_{3,\gamma x}(x_1,x_2)-N^*_{\gamma\beta}u^{(0)}_{3,\beta x}(x_1,x_2)\ .\end{aligned} \qquad (4.14)$$

The formula (4.14) relates the asymmetrical part K_γ of the shear forces with the deformation characteristics of 2-D plate.

Proposition 1

Let the initial stresses σ^*_{ij} satisfy the equilibrium equations:

$$\begin{aligned}&\partial\sigma^*_{ij}/\partial x_j=\varepsilon^a F_i \text{ in } Q_\varepsilon\\&\sigma^*_{ij}n_j=g_i \text{ on } S_\varepsilon\end{aligned} \qquad (4.15)$$

If a<-1 and **g**=0 then $<b_{ijmn}Z_{,ny}>=0$ for any $Z(\mathbf{y})\in\Pi(T)$.

Proposition 2

Under the conditions of proposition 1, $<\sigma^*_{i3}>=0$.

5 The limit 2-D problem

Now, we can exclude the quantities $N^{(-1)}_{3\beta}$, $N^{(-1)}_{\beta 3}$ from the equilibrium Eqs. (3.4),(3.5) and obtain a closed 2-D model of the plate. Differentiating (3.5) and using (3.4),(4.13),(4.14) we obtain

$$(A^{1}_{\gamma\delta\alpha\beta}V_{\alpha,\beta x}+A^{2}_{\gamma\delta\alpha\beta}u^{(0)}_{3,\alpha x\beta x})_{,\gamma x\delta x}=K_{\delta,\delta x}=[N^*_{\delta\beta}\,u^{(0)}_{3,\beta x}]_{,\delta x} \tag{5.1}$$

Let us write the limit model for the special case, when the initial stresses are proportional to a parameter: $\sigma^*_{ij}=\lambda\sigma^{0}_{ij}$ (λ is the parameter) and satisfy the conditions of proposition 1. Then the limit model has the following form:

$$(A^{o}_{\gamma\delta\alpha\beta}V_{\alpha,\beta x}+A^{1}_{\gamma\delta\alpha\beta}u^{(0)}_{3,\alpha x\beta x})_{,\delta x}=0 \tag{5.2}$$

$$(A^{1}_{\gamma\delta\alpha\beta}V_{\alpha,\beta x}+A^{2}_{\gamma\delta\alpha\beta}u^{(0)}_{3,\alpha x\beta x})_{,\gamma x\delta x}=\lambda[N^*_{\delta\beta}\,u^{(0)}_{3,\beta x}]_{,\delta x}\ . \tag{5.3}$$

The boundary conditions are

$$V_{\alpha}(x_1,x_2)=u^{(0)}_{3}(x_1,x_2)=\partial u^{(0)}_{3}/\partial x_j(x_1,x_2)n_j=0 \text{ on } \partial D=0\ . \tag{5.4}$$

∂D means the boundary of the domain D.

6 The initial stresses corresponding to the bending moments

Consider the case of plate subjected to the initial stresses of the order of ε^{-2}. The starting point of the investigation is the elasticity problem (2.1),(2.2) with the coefficients

$$A_{ijmn}(x_1,x_2,\mathbf{x}/\varepsilon)=\varepsilon^{-3}a_{ijmn}(\mathbf{x}/\varepsilon)+\varepsilon^{-2}b_{ijmn}(x_1,x_2,\mathbf{x}/\varepsilon)\ , \tag{6.1}$$

where

$$b_{ijmn}(x_1,x_2,\mathbf{x}/\varepsilon)=\sigma^*_{jn}(x_1,x_2,\mathbf{x}/\varepsilon)\delta_{im}\ . \tag{6.2}$$

We assume that the initial stresses σ^*_{jn} satisfy the following condition

$$<\sigma^*_{\alpha\beta}>=0\ (\alpha,\beta=1,2)\ . \tag{6.3}$$

We will comment briefly on the initial stresses under consideration. The term $\varepsilon^{-2}b_{ijmn}=\varepsilon^{-2}\sigma^*_{jn}\delta_{im}$ corresponds to the bending moments which are independent of ε. In fact, the moments are computed by means of integration of the stresses multiplied by y_3 over the thickness of the plate. The integral of such kind has the characteristic value equal to ε^2. The condition (6.3) means that the in-plane forces, corresponding to the initial stresses under consideration, are equal to zero. The case of non-zero in-plane forces (corresponding to the initial stresses of the order of ε^{-1}) was considered above.

For $A_{ijmn}(x_1,x_2,\mathbf{y})$ determined by (6.1), the relationship between additional stresses σ_{ij} and displacements $\mathbf{u}^\varepsilon$ takes the form:

$$\sigma_{ij} = (\varepsilon^{-3} a_{ijmn}(\mathbf{y}) + \varepsilon^{-2} b_{ijmn}(x_1,x_2,\mathbf{y}))\partial u^\varepsilon_m/\partial x_n \ . \tag{6.4}$$

In order to analyse the problem we use the asymptotic expansions (3.1),(3.2). As above, analysis of the problem (2.1),(2.2),(6.1)-(6.4) breaks down into two stages. The first stage (obtaining the equations of equilibrium for the plate considered as a 2-D structure) is not involved with local governing equations (see Section 2). Thus the equilibrium Eqs. (3.3)-(3.6) remain valid in the case under consideration.

The second stage of analysis of the problem consists of obtaining the governing equations for the plate as 2-D structure and the excluding of unknown quantities from the equilibrium equations. Carrying out asymptotic analysis of the problem, we find that all the moments of the initial stresses in the limit 2-D equations vanish (see [4] for details). It means that the moments of initial stresses do not influence 2-D model of plate with initial stresses.

7 Conclusions

- 2-D models of plates with initial stresses can be derived from the linearized 3-D elasticity problem with initial stresses using the two-scale expansion method.
- In 2-D plate models the terms incorporating the initial stresses arise as a result of the loosing of symmetries in the constitutive equations of the elastic body with initial stresses.
- The method developed can be applied to both the classical uniform plates made of homogeneous material and the non-uniform e.g., net, ribbed, etc. or made of composite materials.

References

1. Bakhvalov NS, Panasenko GP (1989) Homogenization: Averaging processes in periodic media. Kluwer, Dordrecht
2. Caillerie D (1984) Thin elastic and periodic plates. Math Meth Appl Sci 6: 159-191
3. Kalamkarov AL, Kolpakov AG (1997) Analysis, design and optimization of composite structures. John Wiley & Sons, Chichester, New York
4. Kolpakov AG (2001) On the calculation of rigidity characteristics of the stressed constructions. Int J Solids and Structures 38(15): 2469-2485

5. Kolpakov AG (1992) On the dependence of velocity of elastic waves in composite media on initial stresses. Computers & Structures 44: 97-101
6. Timoshenko S, Goodier JN (1970) Theory of elasticity. McGraw-Hill, New York
7. Timoshenko S, Woinowsky-Kricger K (1959) Theory of plates and shells. McGraw-Hill, New York
8. Sanchez-Palencia E (1980) Non-homogeneous media and vibration theory. Lect Notes in Physics 127, Springer, New York
9. Washizu K (1982) Variational methods in elasticity and plasticity. Pregamon, Oxford, New York

Thickness Flexible Theory With Generalized Core Warping for Plane Sandwich Structures

A. Kühhorn and M. Golze

Institute for Structural Mechanics and Vibration of Vehicles, Brandenburg University of Technology, Cottbus, Germany

1 Introduction

An advanced design of sandwich structures does not only require the knowledge of the global stress- and deformation behavior, but also the knowledge of the local effects, such as load singularities and the loss of stability (Fig. 1) caused by the short wave wrinkling of one (bending) or both (pressure) sandwich skins.

Based on the nonlinear theory for sandwich shells with seven kinematic degrees of freedom, introduced from Kühhorn and Schoop [2,3,5], an improved theory for plane sandwich shells with eight degrees of freedom (position vector $\boldsymbol{r}$, director $\boldsymbol{d}$, and intensity factors α_{z1}, α_{z2} for including core flexibility) will be presented, enabling a much better representation of the sandwich core behavior.

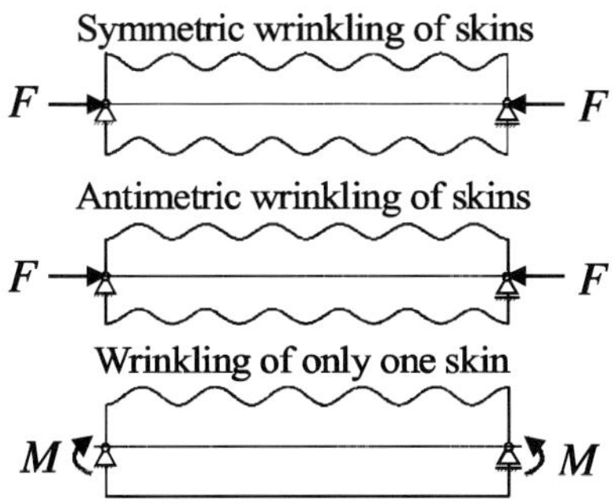

Fig. 1. Local stability cases

Due to consideration of quadratic core thickness and linear core shear strain as well as longitudinal core deformation an improved prediction of the wrinkling behavior is succeeded even for thick cores [8] and thin skins.

2 Assumptions

Sandwich (SW)-structures are considered to be plane and surface-like with the following properties:

- three-layered and symmetric with respect to the midplane
- the core material is softer than the skin material ${}^{c}E << {}^{s}E$, both are considered homogeneously
- the SW-skins are flexible according to stretching and bending but rigid to shear (corresponding to Kirchhoff).

3 Kinematic and static quantities of the SW-skins

The following kinematic modelling of the SW-continuum is based on parameters, which are defined with regard to the midsurface. In detail these parameters are the position vector $\boldsymbol{r}$ and the director $\boldsymbol{d}$ as well as two intensity coefficients α_{z1} and α_{z2} mainly describing the linear and quadratic part of the strain in z- direction including the associated core cross-section warping.

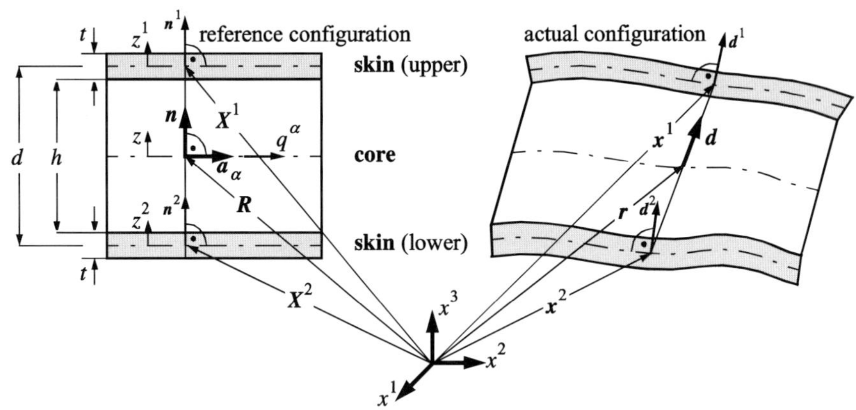

Fig. 2. On kinematic description of the SW-skins and the midsurface

As geometric SW-parameters are introduced:

t, h: thickness of SW-skin resp. SW-core, $d = h + t$: distance between skin midsurfaces.

Further, quantities marked with a top right index refers to the skins in general, ψ^{i}, $i = 1{,}2$: whereas the upper skin is marked by $i = 1$ and the lower skin by $i = 2$.

In case of a missing top right index, ψ is defined with regard to geometric midsurface.

Using Lagrangean coordinates q^{α} of the midsurface for the description of a material point in the reference configuration by $X^{i}(q^{\alpha})$ and in the actual configuration by $x^{i}(q^{\alpha})$, the vectors of the geometric midsurface (Fig. 2) are defined as follows:

$$\boldsymbol{R}(q^{\alpha}) = \frac{1}{2}(\boldsymbol{X}^{1}+\boldsymbol{X}^{2}) \text{ resp. } \boldsymbol{r}(q^{\alpha}) = \frac{1}{2}(\boldsymbol{x}^{1}+\boldsymbol{x}^{2}) \tag{1}$$

$$\boldsymbol{n}(q^{\alpha}) = \frac{1}{d}(\boldsymbol{X}^{1}-\boldsymbol{X}^{2}),\ |\boldsymbol{n}| = 1 \text{ resp. } \boldsymbol{d}(q^{\alpha}) = \frac{1}{d}(\boldsymbol{x}^{1}-\boldsymbol{x}^{2}),\ |\boldsymbol{d}| \neq 1. \tag{2}$$

According to Fig. 2 and Eqs. (1, 2) the exact description of the skin midsurfaces is given as follows:

$$\boldsymbol{X}^{i}(q^{\alpha}) = \boldsymbol{R} \pm \frac{d}{2}\boldsymbol{n} \text{ resp. } \boldsymbol{x}^{i}(q^{\alpha}) = \boldsymbol{r} \pm \frac{d}{2}\boldsymbol{d}. \tag{3}$$

Introducing and splitting the Nabla operator according to

$$\nabla = \nabla_{2} + \boldsymbol{n}\frac{\partial}{\partial z} \text{ in: normal part } \nabla_{n} = \boldsymbol{n}\frac{\partial}{\partial z} \text{ and planar part } \nabla_{2} = \boldsymbol{a}^{\alpha}\frac{\partial}{\partial q^{\alpha}}, \tag{4}$$

where $\boldsymbol{a}^{\alpha}$ is the reciprocal base for $\boldsymbol{a}_{\alpha} = \partial \boldsymbol{R}/\partial q^{\alpha} = \boldsymbol{R}_{,\alpha}$ with respect to the generally curvilinear coordinates q^{α}, results in:

$$\nabla_{2} \circ \boldsymbol{R} = \boldsymbol{a}^{\alpha} \circ \boldsymbol{R}_{,\alpha} = \boldsymbol{E}_{2} \mathrel{\widehat{=}} \delta_{\alpha\cdot}^{\cdot\beta}\boldsymbol{a}^{\alpha} \circ \boldsymbol{a}_{\beta}. \tag{5}$$

$\boldsymbol{E}_2$ represents the planar unit tensor and $\circ$ terms the dyadic or tensorial product, respectively. If cartesian coordinates $q^{\alpha} = X^{\alpha}$ are used, then it holds especially $\nabla_{2} = \boldsymbol{e}_{\alpha}\partial/\partial X^{\alpha}$ and $\boldsymbol{a}^{\alpha} = \boldsymbol{a}_{\alpha} = \boldsymbol{e}_{\alpha}$, where $\boldsymbol{e}_{\alpha}$ is the orthonormal base. Further developments are based on the introduction of the gradients of the midsurface vectors, which are planar with respect to the reference and three-dimensional with regard to the actual configuration:

$$\begin{aligned} \boldsymbol{F} &= \operatorname{Grad}^{\mathrm{T}} \boldsymbol{r} = (\nabla_{2} \circ \boldsymbol{r})^{\mathrm{T}} = (\boldsymbol{r}_{,\alpha} \circ \boldsymbol{a}^{\alpha})^{\mathrm{T}} \text{ and} \\ \boldsymbol{G} &= \operatorname{Grad}^{\mathrm{T}} \boldsymbol{d} = (\nabla_{2} \circ \boldsymbol{d})^{\mathrm{T}} = (\boldsymbol{d}_{,\alpha} \circ \boldsymbol{a}^{\alpha})^{\mathrm{T}}. \end{aligned} \tag{6}$$

Corresponding to [3,5] the particular Greenean skin membrane- and skin bending strains results in:

$$\boldsymbol{D}^{i} = \frac{1}{2}(\boldsymbol{F}^{i\mathrm{T}} \cdot \boldsymbol{F}^{i} - \boldsymbol{E}_{2}) = \frac{1}{2}\left\{\left(\boldsymbol{F}^{\mathrm{T}} \cdot \boldsymbol{F} - \boldsymbol{E}_{2} + \frac{d^{2}}{4}\boldsymbol{G}^{\mathrm{T}} \cdot \boldsymbol{G}\right) \pm \frac{d}{2}\left(\boldsymbol{F}^{\mathrm{T}} \cdot \boldsymbol{G} + \boldsymbol{G}^{\mathrm{T}} \cdot \boldsymbol{F}\right)\right\} \tag{7}$$

$$\begin{aligned} \overset{<\kappa>}{\boldsymbol{\kappa}}{}^{i} &= -\frac{1}{2}\Big\{\boldsymbol{d} \cdot \left(\boldsymbol{r} \circ \nabla_{2} \circ \nabla_{2}\right) + \left(\nabla_{2} \circ \nabla_{2} \circ \boldsymbol{r}\right) \cdot \boldsymbol{d} \pm \\ &\pm \frac{d}{2}\Big[\boldsymbol{d} \cdot \left(\boldsymbol{d} \circ \nabla_{2} \circ \nabla_{2}\right) + \left(\nabla_{2} \circ \nabla_{2} \circ \boldsymbol{d}\right) \cdot \boldsymbol{d}\Big]\Big\} \end{aligned} \tag{8}$$

Defining as resultants of the 2nd Piola-Kirchhoff stresses

$$\boldsymbol{n}^{i} = \int_{z^{i}=0}^{t} \boldsymbol{S}^{i} dz^{i} \text{ and } \boldsymbol{m}^{i} = \int_{z^{i}=0}^{t} z^{i} \boldsymbol{S}^{i} dz^{i},\ (i = 1, 2), \tag{9}$$

the **midsurface-based SW-quantities** are determined from rearrangement:

- SW-membrane force- and SW-membrane strain tensor (see (7) and (9))

$$N = n^1 + n^2 \qquad \text{(Fig. 3)} \tag{10}$$

$$D = \frac{1}{2}(D^1 + D^2) = \frac{1}{2}\left[F^{\mathrm{T}} \cdot F - E_2 + \frac{d^2}{4} G^{\mathrm{T}} \cdot G\right] \tag{11}$$

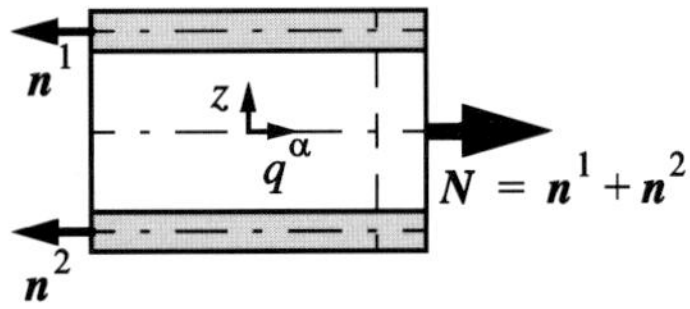

Fig. 3. Membrane force

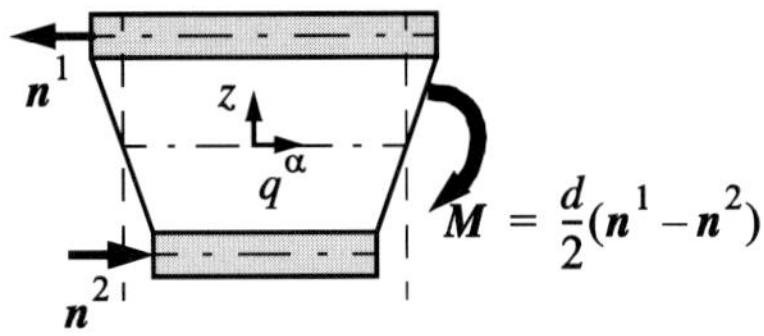

Fig. 4. (Membrane-)Moment

- SW-moment- and SW-bending strain tensor (see (7) and (9))

$$M = \frac{d}{2}\left(n^1 - n^2\right) \qquad \text{(Fig. 4)} \tag{12}$$

$$\kappa = \frac{1}{d}\left(D^1 - D^2\right) = \frac{1}{2}\left(F \cdot G^{\mathrm{T}} + G \cdot F^{\mathrm{T}}\right) \tag{13}$$

- SW-skin bending moment- and SW-skin bending strain-sum-tensor (see (8) and (9))

$${}^{s}m_S = m^1 + m^2 \tag{14}$$

$$\overset{<\kappa>}{\kappa}_S = \frac{1}{2}\left(\overset{<\kappa>1}{\kappa} + \overset{<\kappa>2}{\kappa}\right) = -\frac{1}{2}\left\{ d \cdot \left(r \circ \nabla_2 \circ \nabla_2\right) + \left(\nabla_2 \circ \nabla_2 \circ r\right) \cdot d \right\} \tag{15}$$

- SW-skin bending moment- and SW-skin bending strain-difference-tensor (see (8) and (9))

$${}^{s}m_D = m^1 - m^2 \tag{16}$$

$$\overset{<\kappa>}{\kappa}_D = \frac{1}{2}\left(\overset{<\kappa>1}{\kappa} - \overset{<\kappa>2}{\kappa}\right) = -\frac{1}{2}\left\{\frac{d}{2}\left[d \cdot \left(d \circ \nabla_2 \circ \nabla_2\right) + \left(\nabla_2 \circ \nabla_2 \circ d\right) \cdot d\right]\right\} \tag{17}$$

4 Kinematic and static variables of the SW-core

Different preliminary considerations have shown [1,3,7,8], that the short-wave wrinkling problem, mainly appearing for very thin skins and a thick core, requires an improved description of the SW-core concerning transversal, shear and longitudinal stiffness. Therefore an extended kinematic approach for the core deformation in comparison to [2,3,5] is developed which provides a quadratic thickness stretch (represented by α_{z1} and α_{z2}) and a linear shear strain in z - direction each as well as a consideration of associated cross-section warping. The derivation of this compatible displacement field is firstly realised for the SW-bar considering small displacements (1st order theory, see Fig. 5 for illustration). A generalisation in terms of the components of the non-linear Green's strain tensor works analogously to [2,3,5] by describing the strain quantities in a coordinate-invariant way using 8 SW-DOF $\boldsymbol{r}$, $\boldsymbol{d}$, α_{z1}, α_{z2} and an application of (11,13,15,21) as follows:

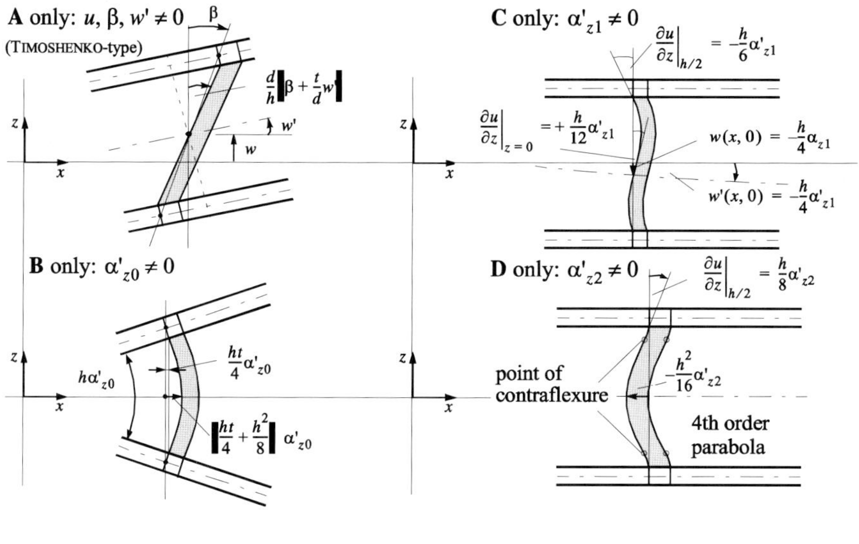

Fig. 5. Contribution of warping

$$ {}^{c}\boldsymbol{D}(q^{\alpha},z)={}^{c}\boldsymbol{D}_0 + z\frac{d}{h}(\boldsymbol{\kappa} - \frac{t}{d}\overset{<K>}{\boldsymbol{\kappa}_S}) + \sum_{i=0}^{2} g(z^{i+2})_{,i}\nabla_2 \circ \nabla_2 \alpha_{zi} \tag{18} $$

$$ {}^{c}\varepsilon_{zz}(q^{\alpha},z)=\alpha_{z0}+\frac{z}{h/2}\alpha_{z1}+(\frac{z^2}{h^2/6}-\frac{1}{2})\alpha_{z2} \tag{19}$$

$$ {}^{c}\boldsymbol{e}_S(q^{\alpha},z)=\frac{d}{2h}(\boldsymbol{F}^{\mathrm{T}}\cdot\boldsymbol{d}-\frac{h^2}{6d}\nabla_2\alpha_{z1})+\frac{z}{8}\nabla_2\alpha_{z2}$$

$$ \text{where } {}^{c}\boldsymbol{D}_0=\frac{1}{2}(\boldsymbol{F}^{\mathrm{T}}\cdot\boldsymbol{F}-\boldsymbol{E}_2)\,,\ \alpha_{z0}=\frac{d}{2h}(\boldsymbol{d}\cdot\boldsymbol{d}-1) \tag{20}$$

and with the warping functions

$$ g_0(z^2)=\frac{1}{2}[\frac{th}{2}+(\frac{h^2}{4}-z^2)]\,;\ g_1(z^3)=\frac{z}{3h}(\frac{h^2}{4}-z^2) \tag{21}$$

$$ g_2(z^4)=\frac{1}{2h^2}(\frac{h^2}{4}-z^2)(z^2-\frac{h^2}{2})\,.$$

The definition of generalized core-membrane, core-shear and core-transversal stress resultants according to

$$ \overset{c\,<l>}{\boldsymbol{n}}=\int z^{l}\,{}^{c}S_{\alpha\beta}dz\boldsymbol{e}_{\alpha}\circ\boldsymbol{e}_{\beta}\,,\quad \overset{c\,<l>}{\boldsymbol{Q}}=\int z^{l}S_{z\alpha}dz\boldsymbol{e}_{\alpha}\,,\quad \overset{c\,<l>}{t}=\int z^{l}S_{zz}dz \tag{22}$$

results in $\overset{c\,<0>}{\boldsymbol{n}}={}^{c}\boldsymbol{n}$, $\overset{c\,<1>}{\boldsymbol{n}}={}^{c}\boldsymbol{m}$, $\overset{c\,<2>}{\boldsymbol{n}}$, $\overset{c\,<3>}{\boldsymbol{n}}$, $\overset{c\,<4>}{\boldsymbol{n}}$ as well as in $\overset{c\,<0>}{\boldsymbol{Q}}={}^{c}\boldsymbol{Q}$, $\overset{c\,<1>}{\boldsymbol{Q}}$ and $\overset{c\,<0>}{t}$, $\overset{c\,<1>}{t}$, $\overset{c\,<2>}{t}$.

5 The nonlinear equilibrium and boundary conditions

By the use of the abbreviations

$$ \overline{\boldsymbol{N}}=\boldsymbol{N}+{}^{c}\boldsymbol{n}\ ;\ \overline{\boldsymbol{M}}=\boldsymbol{M}+\frac{d}{h}{}^{c}\boldsymbol{m} \tag{23}$$

$$ \overline{\boldsymbol{m}}_S={}^{s}\boldsymbol{m}_S-\frac{t}{h}{}^{c}\boldsymbol{m}\quad;\quad \overline{\boldsymbol{m}}_D=\frac{d}{2}{}^{s}\boldsymbol{m}_D-\frac{d}{h}(\frac{th}{4}+\frac{h^2}{8}){}^{c}\boldsymbol{n}+\frac{d}{2h}\overset{c\,<2>}{\boldsymbol{n}} \tag{24}$$

$$ \tilde{\boldsymbol{n}}=\frac{h^2}{16}{}^{c}\boldsymbol{n}-\frac{3}{8}\overset{c\,<2>}{\boldsymbol{n}}+\frac{1}{2h^2}\overset{c\,<4>}{\boldsymbol{n}}\quad;\quad \tilde{\boldsymbol{m}}=-\frac{h}{12}{}^{c}\boldsymbol{m}+\frac{1}{3h}\overset{c\,<3>}{\boldsymbol{n}}\,, \tag{25}$$

from the principle of virtual work with regard to [3,5] the 1st Piola-Kirchhoff-tensors which are needed for a satisfaction of equilibrium

$$ \overline{\boldsymbol{N}}^{1P}=\overline{\boldsymbol{N}}\cdot\boldsymbol{F}^{\mathrm{T}}+\overline{\boldsymbol{M}}\cdot\boldsymbol{G}^{\mathrm{T}}+\nabla_2\cdot(\overline{\boldsymbol{m}}_S\circ\boldsymbol{d}) \tag{26}$$

$$ \overline{\boldsymbol{M}}^{1P}=\overline{\boldsymbol{M}}\cdot\boldsymbol{F}^{\mathrm{T}}+\frac{d^2}{4}\boldsymbol{N}\cdot\boldsymbol{G}^{\mathrm{T}}+\nabla_2\cdot(\overline{\boldsymbol{m}}_D\circ\boldsymbol{d}) \tag{27}$$

according to the variations $\delta\boldsymbol{r}$, $\delta\boldsymbol{d}$, $\delta\alpha_{z1}$, $\delta\alpha_{z2}$ the field equations for force, momentum and transversal equilibrium are obtained.

$$\delta \boldsymbol{r}: \qquad \nabla_2 \cdot \left(\overline{\boldsymbol{N}}^{1P} + \frac{d}{h} {}^c\boldsymbol{Q} \circ \boldsymbol{d} \right) + \boldsymbol{p}_0 = \boldsymbol{0} \tag{28}$$

$$\delta \boldsymbol{d}: \qquad \nabla_2 \cdot \left(\overline{\boldsymbol{M}}^{1P} \right) - \frac{d}{h} {}^c\boldsymbol{Q} \cdot \boldsymbol{F}^T - \frac{d}{h} {}^{c}\overset{<0>}{t}\, \boldsymbol{d} + \overline{\boldsymbol{m}}_S \cdot\cdot (\nabla_2 \circ \nabla_2 \circ \boldsymbol{r}) + \overline{\boldsymbol{m}}_D \cdot\cdot (\nabla_2 \circ \nabla_2 \circ \boldsymbol{d}) + \boldsymbol{p}_1 = \boldsymbol{0} \tag{29}$$

$$\delta \alpha_{z1}: \qquad \nabla_2 \cdot \left(-\frac{h}{6} {}^c\boldsymbol{Q} + \nabla_2 \cdot \widetilde{\boldsymbol{m}} \right) - \frac{2}{h} {}^{c}\overset{<1>}{t} = 0 \tag{30}$$

$$\delta \alpha_{z2}: \qquad \nabla_2 \cdot \left(\frac{1}{4} {}^{c}\overset{<1>}{\boldsymbol{Q}} + \nabla_2 \cdot \widetilde{\boldsymbol{n}} \right) + \frac{1}{2} {}^{c}\overset{<0>}{t} - \frac{6}{h^2} {}^{c}\overset{<2>}{t} = 0 \ . \tag{31}$$

The considerations described above include the associated boundary conditions:

$$\delta \boldsymbol{r}: \quad \boldsymbol{p}^b = \boldsymbol{e}_\perp \cdot \left(\overline{\boldsymbol{N}}^{1P} + \frac{d}{h} {}^c\boldsymbol{Q} \circ \boldsymbol{d} \right) + \frac{\partial}{\partial s} (\overline{m}_{S_{\perp b}} \boldsymbol{d}) \qquad \frac{\partial \delta \boldsymbol{r}}{\partial s_\perp}: \ \overline{\boldsymbol{m}}_S^b = \boldsymbol{e}_\perp \circ \boldsymbol{e}_\perp \cdot\cdot (-\overline{\boldsymbol{m}}_S \circ \boldsymbol{d}) \tag{32}$$

$$\delta \boldsymbol{d}: \quad \boldsymbol{m}^b = \boldsymbol{e}_\perp \cdot \overline{\boldsymbol{M}}^{1P} + \frac{\partial}{\partial s} (\overline{m}_{D_{\perp b}} \boldsymbol{d}) \qquad \frac{\partial \delta \mathrm{d}}{\partial s_\perp}: \ \overline{\mathrm{m}}_D^b = \mathrm{e}_\perp \circ \mathrm{e}_\perp \cdot\cdot (-\overline{\mathrm{m}}_D \circ \mathrm{d}) \tag{33}$$

$$\partial \alpha_{z1}: \quad \begin{aligned} {}^c q_{z1}^h &= \boldsymbol{e}_\perp \cdot \left(-\frac{h}{6} {}^c\boldsymbol{Q} + \nabla_2 \cdot \widetilde{\boldsymbol{m}} \right) - \frac{\partial}{\partial s} (\widetilde{m}_{\perp b}) \\ \frac{\partial \delta \alpha_{z1}}{\partial s_\perp}&: \quad -\widetilde{m}_{\perp\perp}^b = \boldsymbol{e}_\perp \circ \boldsymbol{e}_\perp \cdot\cdot (-\widetilde{\boldsymbol{m}}) \end{aligned} \tag{34}$$

$$\partial \alpha_{z2}: \quad {}^c m_{z2}^h = \boldsymbol{e}_\perp \cdot \left(\frac{1}{4} {}^{c}\overset{<1>}{\boldsymbol{Q}} + \nabla_2 \cdot \widetilde{\boldsymbol{n}} \right) - \frac{\partial}{\partial s} (\widetilde{n}_{\perp b}) \qquad \frac{\partial \delta \alpha_{z2}}{\partial s_\perp}: \ -\widetilde{n}_{\perp\perp}^b = \boldsymbol{e}_\perp \circ \boldsymbol{e}_\perp \cdot\cdot (-\widetilde{\boldsymbol{n}}) \tag{35}$$

The 1st P.K.-tensors (26,27) and the expression ${}^c\boldsymbol{Q} \circ \boldsymbol{d}$ are dyades (as well as $\boldsymbol{F}^{\mathrm{T}}$ and $\boldsymbol{G}^{\mathrm{T}}$), which are planar with respect to the reference and three-dimensional with regard to the actual configuration according to an arbitrary, e.g. cartesian base. Thus the terms $\overline{\boldsymbol{m}}_S \circ \boldsymbol{d}$ resp. $\overline{\boldsymbol{m}}_D \circ \boldsymbol{d}$ are corresponding to [2x3x3] tensors of 3rd order. A comparison with [2,3,5] clearly shows the enlargement due to the consideration of the core warping (additional terms ${}^c\psi$) and the additional Eqs. (31) due to the 8th DOF.

6 Wrinkling instabilities of the SW plate-like bar

Single-side skin wrinkling due to $\overset{\circ}{M} = (d/2)P$

As an example (see also [9]), the critical wrinkling stress for the bending case is calculated using a second order specialization. The parameters are ${}^{s}E/{}^{c}E = 480$, ${}^{s}\nu = 0{,}3$, ${}^{c}\nu \equiv 0$, as well as $d/t = 10$ for thick skins (I) and $d/t = 40$ for thin skins (II). In Fig. 7,8 the results of the presented theory (A) are compared to results from (B) [3] and to results received from a detailed FE-plain-strain analysis.

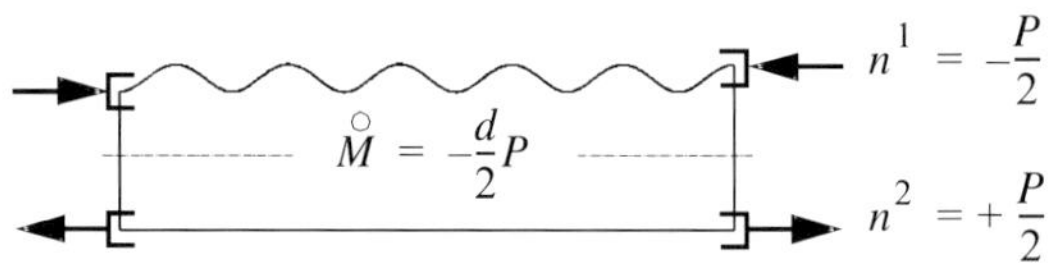

Fig. 6. One-side skin wrinkling

Fig. 7 shows a good match for case I (thick skins) with a closer approximation of A. The minimum failure load appears in case of half wave lengths of about 1.5-fold *d*. For case II (thin skins, Fig. 8) A results in a stiffer and B a weaker solution with regard to the reference solution while A shows the closer correlation.

Comparing case II (thin skins) with the usual estimation formula [6,9] for the wrinkling stress

$$ {}^{s}\sigma_{xx_k} \approx C \sqrt[3]{{}^{s}E \cdot {}^{c}E \cdot {}^{c}G} \qquad \text{where} \qquad C \approx 0{,}85 \tag{36} $$

the parameter mentioned above becomes approximately

$$ {}^{s}\sigma_{x_k} / {}^{s}E \approx 0{,}0110 \tag{37} $$

which is in good accordance with the reference values of the FEM-computation and the results from the presented theory A.

7 Summary

The geometrically nonlinear theory for plane SW-structures using 8 kinematic degrees of freedom ($\boldsymbol{r}$, $\boldsymbol{d}$, α_{z1}, α_{z2}) described is valid for arbitrary displacements and rotations with the restriction to moderate strains. From the detailed consideration of the core deformation with respect to cross-section warping as well as transversal and shear stiffness inclusive independent skin bendings, a comprehensive and closed description of the global and local (load application, skin wrinkling, ...) SW-structural behaviour becomes possible. The nonlinear equations are including simple specialisations with respect to the 1st or 2nd order theory. From the more improved core description of 8 DOF - theory a more extended scope of application up to wrinkling half wave lengths of ca. $a/d > 0{,}3$ is achieved. That means that a good description of very thin-skinned SW-structures becomes possible. Furthermore, important questions e.g. concerning the interface stresses (peeling- resp. shear stresses) between core and skins can be answered more precisely.

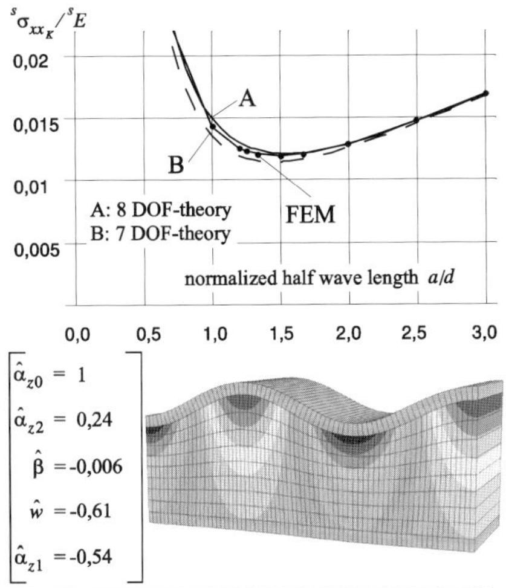

I: thick skins	B	A	FEM
min ${}^s\sigma_{xx_K}/{}^sE$	0,0115	0,0121	0,0119
$\left\lvert\frac{X-FEM}{FEM}\right\rvert$ [%]	3,4	1,7	-
at: a/d	1,45	1,54	1,50

Fig. 5. Case I-thick skins

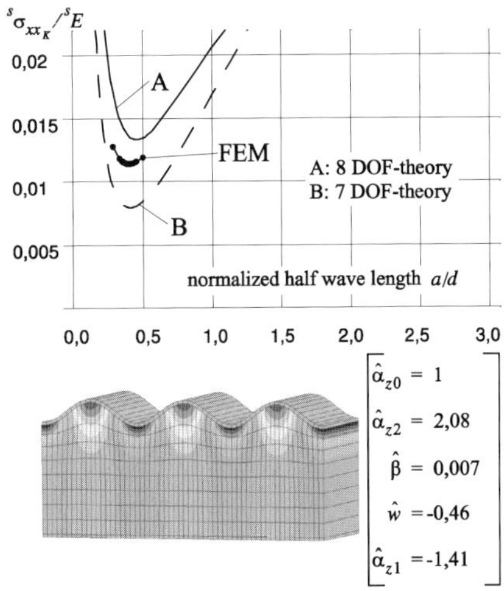

II: thin skins	B	A	FEM
min ${}^s\sigma_{xx_K}/{}^sE$	0,0079	0,0133	0,0114
$\left\lvert\frac{X-FEM}{FEM}\right\rvert$ [%]	30,7	16,6	-
at: a/d	0,41	0,46	0,40

Fig. 6. Case II- thin skin

References

1. Golze M (2000) Untersuchungen zum Knitterverhalten (lokale Instabilitäten) ebener Sandwichtragwerke unter Heranziehung einer verfeinerten Theorie mit quadratischer Dickenverzerrung. Diplomarbeit, Technische Universität Berlin, BTU Cottbus
2. Kühhorn A (1993) Nichtlineare Theorie und finites Element für Sandwichtragwerke zur Beschreibung des globalen und lokalen (Knittern) Versagens. ZAMM 73: 438-443
3. Kühhorn A (1991) Geometrisch nichtlineare Theorie für Sandwichschalen unter Einbeziehung des Knitterphänomens, Reihe 18, Nr 100, VDI-Verlag, Düsseldorf
4. Kühhorn A, Golze M: An improved sandwich theory for a better prediction of the wrinkling phenomenon, WCCM V, http://wccm.tuwien.ac.at

5. Kühhorn A, Schoop H (1992) A nonlinear theory for sandwich shells including the wrinkling phenomenon. Arch Appl Mech 62: 413-427
6. Stamm K, Witte H (1974) Sandwichkonstruktionen - Berechnung, Fertigung, Ausführung. Springer, Wien
7. Vonach WK, Rammerstorfer FG (2000) Wrinkling of thick orthotropic sandwich plates under general loading conditions. Arch Appl Mech 70: 338-348
8. Vonach WK, Rammerstorfer FG (2000) The effects of in-plane core stiffness on the wrinkling behavior of thick sandwiches. Acta Mech 141: 1-10
9. Wiedemann J (1986) Leichtbau: Band 1 – Elemente. Springer, Berlin

On the Finite Element Analysis of Layered Structures With Thin Piezoelectric Actuators and Sensors

R. Lammering and S. Mesecke-Rischmann

Institute of Mechanics, University of the Federal Armed Forces Hamburg, Germany

1 Introduction

Thin piezoelectric layers bonded to the surfaces of a host structure and working as actuators or sensors are easy to manufacture and effective e.g. for vibration attenuation in combination with a control concept. Hence, well established design tools like the finite element method have to be extended in order to enable a proper structural design. Especially, the coupled electromechanical constitutive equations and the above mentioned typical constructions have to be taken into account.

In recent years various finite plate and shell elements with the above mentioned features have been formulated, cf. (Benjeddou 2000), where displacements, rotations and electric potentials have been introduced as independent variables. The corresponding variational principle can be interpreted as a two-field variational functional and indicates that linear or quadratic shape functions are required for their approximation. But in experiments the measurement of continuous voltage output of distributed electroded piezoelectric patches on a passive host structure is impossible. Moreover, the only non-vanishing electric field component is obtained in thickness direction. So constant shape functions for the electric potential with respect to the directions within the shell surface have been implemented, see also (Hwang and Park 1993). These independent variables are subsequently eliminated on the element level, so that the system's degrees of freedom only include mechanical degrees of freedom. These elements can easily be linked with purely mechanical ones and are proven very time efficient in numerical examples. Also linear shape functions for the electric potential are implemented. Here the system of equations contains mechanical and electrical degrees of freedom, cf. (Mesecke and Lammering 2002). In the past, hybrid-mixed methods have been used successfully in purely mechanical analyses, i.e. in order to avoid locking phenomena, cf. (Simo and Rifai 1990), or to enable a simple implementation of nonlinear constitutive equations, cf. (Ghandi and Hagood 1997). Analogously to these well known hybrid finite element formulations in elastodynamics, similar techniques are applied to electromechanically coupled problems. A new variational principle for electromechanical systems is derived augmenting the total potential energy in the Hu-Washizu functional further. This three-field variational functional is achieved by introducing the electric field-potential relation in the formulation via a Lagrange multiplier. Based on this variational principle, electromechanically coupled hybrid finite elements are obtained in which the dielectric displacement serves as an independent variable, cf. (Sze 1999), and can be condensed on the element level, so that it does not enter the system of equations, cf. (Mesecke and Lammering 2002). The performance of these new finite element formulations is tested in a numerical example in which the piezoelectric material works as sensor.

2 Governing equations for coupled electromechanical systems

Starting point for the formulation of weak forms and corresponding finite elements for piezoelectric material is the balance of momentum

$$\mathrm{Div}\,\boldsymbol{\sigma} + \rho_0 \left(\boldsymbol{b} - \frac{\partial^2 \boldsymbol{u}}{\partial t^2} \right) = 0, \tag{2.1}$$

where $\boldsymbol{\sigma}$ denotes the symmetric stress tensor, $\rho_0 \boldsymbol{b}$ are the specific body forces and $\rho_0\, \partial^2\mathbf{u}/\partial t^2$ are the specific inertial forces. The charge conservation condition for an insulating material is formulated as

$$\mathrm{Div}\,\mathbf{D} = 0 \tag{2.2}$$

with the dielectric displacement $\mathbf{D}$. The electromechanical coupling for piezoelectric material is described by the linear constitutive relations as

$$\begin{aligned} \boldsymbol{\sigma} &= \mathbf{C}:\mathbf{E}_{\mathrm{m}} - \mathbf{e}\cdot\mathbf{E}_{\mathrm{el}} \\ \mathbf{D} &= \mathbf{e}^{\mathrm{T}}:\mathbf{E}_{\mathrm{m}} + \boldsymbol{\varepsilon}\cdot\mathbf{E}_{\mathrm{el}} \end{aligned} \tag{2.3}$$

The elasticity tensor $\mathbf{C}$ is used as in Hooke's law and its material constants are evaluated at constant electric field $\mathbf{E}_{\mathrm{el}}$. The purely electric material behaviour is formulated through the permittivity $\boldsymbol{\varepsilon}$ measured at constant mechanical strain $\mathbf{E}_{\mathrm{m}}$. The electromechanical coupling is obtained through the piezoelectric modulus $\mathbf{e}$.

Using the mechanical strains and the dielectric displacement as independent variables a conjugate formulation of the linear constitutive relations is derived

$$\begin{aligned} \boldsymbol{\sigma} &= \mathbf{C}_{\mathrm{d}}:\mathbf{E}_{\mathrm{m}} - \mathbf{h}\cdot\mathbf{D} \\ \mathbf{E}_{\mathrm{el}} &= -\,\mathbf{h}^{\mathrm{T}}:\mathbf{E}_{\mathrm{m}} + \boldsymbol{\beta}\cdot\mathbf{D} \end{aligned} \tag{2.4}$$

where $\mathbf{C}_{\mathrm{d}}$ denotes the elasticity tensor at constant dielectric displacement $\mathbf{D}$, $\boldsymbol{\beta}$ the impermittivity tensor at constant mechanical strain $\mathbf{E}_{\mathrm{m}}$ and $\mathbf{h}$ the piezoelectric modulus.

There exist natural and essential mechanical and electrical boundary conditions for the displacements $\mathbf{u} = \bar{\mathbf{u}}$ on ∂B_u and the traction $\bar{\mathbf{t}} = \boldsymbol{\sigma}\cdot\mathbf{n}$ on ∂B_σ with the unit outward normal vector $\mathbf{n}$ and for the electric potential $\Phi = \bar{\Phi}$ on ∂B_Φ and the charge density $\bar{d} = \mathbf{D}\cdot\mathbf{n}$ on the surface ∂B_D.

Linear strain-displacement and electric field-electric potential relations are assumed

$$\mathbf{E}_m = \frac{1}{2}\left(Grad\,\mathbf{u} + Grad^T\mathbf{u}\right),\ \delta\mathbf{E}_m = \frac{1}{2}\left(Grad\,\delta\mathbf{u} + Grad^T\delta\mathbf{u}\right) \tag{2.5}$$

$$\mathbf{E}_{el} = -\mathrm{Grad}\Phi, \quad \delta\mathbf{E}_{el} = -\mathrm{Grad}\delta\Phi. \tag{2.6}$$

3 Two-field weak form and elements

The weak form of equilibrium is derived via multiplying Eq. (2.1) by the virtual displacement field $\delta\mathbf{u}$ and Eq. (2.2) by the virtual electric potential $\delta\Phi$. The subsequent summation, integration over the body B and applying Gauss's Theorem yields

$$\begin{aligned}\delta\mathcal{G}(\mathbf{u},\delta\mathbf{u},\Phi,\delta\Phi) &= \int_B \delta\mathbf{E}_m : \boldsymbol{\sigma}\,dV + \int_B \delta\mathbf{E}_{el}\cdot\mathbf{D}\,dV \\ &+ \int_B \rho_0\frac{\partial^2\mathbf{u}}{\partial t^2}\cdot\delta\mathbf{u}\,dV - \int_{\partial B_\sigma}\delta\mathbf{u}\cdot\bar{\mathbf{t}}\,dA - \int_B \delta\mathbf{u}\cdot\rho_0\mathbf{b}\,dV - \int_{\partial B_D}\delta\Phi\cdot\bar{d}\,dA = 0.\end{aligned} \tag{3.1}$$

Introducing the constitutive Eq. (2.3), the well known two-field variational formulation is obtained:

$$\begin{aligned}&\int_B \delta\mathbf{E}_m:(\mathbf{C}:\mathbf{E}_m + \mathbf{e}\cdot Grad\Phi)\,dV + \int_B \rho_0\frac{\partial^2\mathbf{u}}{\partial t^2}\cdot\delta\mathbf{u}\,dV - \int_{\partial B_\sigma}\delta\mathbf{u}\cdot\bar{\mathbf{t}}\,dA - \int_B \delta\mathbf{u}\cdot\rho_0\mathbf{b}\,dV = 0\\ &\int_B Grad\,\delta\Phi\cdot\left(\mathbf{e}^T:\mathbf{E}_m - \boldsymbol{\varepsilon}\cdot Grad\Phi\right)dV - \int_{\partial B_D}\delta\Phi\bar{d}\,dA = 0.\end{aligned} \tag{3.2}$$

The following finite element formulation is based on the isoparametric piezoelectric shell element of the Mindlin/Reissner type, see (Mesecke and Lammering 2002). The 4-noded sandwich structure consists of two piezoelectric layers with a core of conventional material. The displacements $\underline{u}$ are approximated by bilinear shape functions and the electric potential $\underline{\Phi}$ is approximated by constant and in addition also by bilinear shape functions. $\underline{N}$ denotes the matrix of shape functions and $\underline{B}$ the matrix of their derivatives. The indices u and Φ refer to the mechanical and electrical part, respectively

$$\begin{aligned}&\underline{u} = \underline{N}_u\hat{\underline{u}},\quad \delta\underline{u} = \underline{N}_u\delta\hat{\underline{u}},\qquad \underline{E}_m = \underline{B}_u\hat{\underline{u}},\qquad \delta\underline{E}_m = \underline{B}_u\delta\hat{\underline{u}},\\ &\underline{\Phi} = \underline{N}_\Phi\hat{\underline{\Phi}},\ \delta\underline{\Phi} = \underline{N}_\Phi\delta\hat{\underline{\Phi}},\ Grad\underline{\Phi} = \underline{B}_\Phi\delta\hat{\underline{\Phi}},\ Grad\delta\underline{\Phi} = \underline{B}_\Phi\delta\hat{\underline{\Phi}}.\end{aligned} \tag{3.3}$$

The weak form of equilibrium with $\underline{u}$ and $\underline{\Phi}$ as independent variables is now written on the element level

$$\delta G\left(\underline{\hat{u}},\delta\underline{\hat{u}},\underline{\hat{\Phi}},\delta\underline{\hat{\Phi}}\right)=\begin{bmatrix}\delta\underline{\hat{u}}_e^T\\ \delta\underline{\hat{\Phi}}_e^T\end{bmatrix}\begin{bmatrix}\int\limits_{Ve}\underline{B}_u^T\underline{C}\underline{B}_u dV & \int\limits_{Ve}\underline{B}_u^T\underline{e}\underline{B}_\Phi dV\\ \int\limits_{Ve}\underline{B}_\Phi^T\underline{e}^T\underline{B}_u dV & -\int\limits_{Ve}\underline{B}_\Phi^T\underline{\varepsilon}\underline{B}_\Phi dV\end{bmatrix}\begin{bmatrix}\underline{\hat{u}}_e\\ \underline{\hat{\Phi}}_e\end{bmatrix}$$

$$+\begin{bmatrix}\delta\underline{\hat{u}}_e^T\\ \delta\underline{\hat{\Phi}}_e^T\end{bmatrix}\begin{bmatrix}\int\limits_{Ve}\rho_0\underline{N}_u^T\underline{N}_u dV & \underline{0}\\ \underline{0} & \underline{0}\end{bmatrix}\begin{bmatrix}\partial^2\underline{\hat{u}}_e/\partial t^2\\ \underline{\hat{\Phi}}_e\end{bmatrix}-\begin{bmatrix}\delta\underline{\hat{u}}_e^T\\ \delta\underline{\hat{\Phi}}_e^T\end{bmatrix}\begin{bmatrix}\int\limits_{Ve}\underline{N}_u^T\rho_0\underline{b}dV+\int\limits_{A_\sigma}\underline{N}_u^T\underline{\bar{t}}\,dA\\ \int\limits_{A_D}\underline{N}_\Phi^T\underline{\bar{d}}\,dA\end{bmatrix}=0. \tag{3.4}$$

The system of equations that has to be solved yields

$$\begin{bmatrix}\underline{K}_{uu} & \underline{K}_{u\Phi}\\ \underline{K}_{\Phi u} & -\underline{K}_{\Phi\Phi}\end{bmatrix}\begin{bmatrix}\underline{\hat{u}}_e\\ \underline{\hat{\Phi}}_e\end{bmatrix}+\begin{bmatrix}\underline{M}_{uu} & \underline{0}\\ \underline{0} & \underline{0}\end{bmatrix}\begin{bmatrix}\partial^2\underline{\hat{u}}_e/\partial t^2\\ \underline{\hat{\Phi}}_e\end{bmatrix}=\begin{bmatrix}\underline{F}_u\\ \underline{F}_\Phi\end{bmatrix}. \tag{3.5}$$

Elimination of the electric potential on the element level gives

$$(\underline{K}_{uu}+\underline{K}_{u\Phi}\underline{K}_{\Phi\Phi}^{-1})\underline{\hat{u}}_e+\underline{M}_{uu}\partial^2\underline{\hat{u}}_e/\partial t^2=\underline{F}_u+\underline{K}_{\Phi\Phi}^{-1}\underline{F}_\Phi\,. \tag{3.6}$$

As it can be seen from Eq. (3.2), a C^0-continuity is required for the displacements as well as for the electric potential. However, when thin piezoelectric plates are bonded to plate and shell surfaces, the only non-vanishing electric field component is obtained by the electric potential gradient in thickness direction. The electric potential within the shell surface can be neglected. Moreover, in practice the measurement of a continuous voltage output is impossible for a distributed electrode. Therefore, constant shape functions have been introduced as well in piezoelectric shell formulations and successfully applied to various examples, see e.g. (Mesecke and Lammering 2002, Hwang and Park 1993).

4 Multi-field weak form and element

Multi-field weak forms are used for the construction of purely mechanical finite elements which do not lock and possess other performance enhancements. The most general formulation is the Hu-Washizu variational principle. Analogously, weak forms for the electro-mechanical problems are derived, where the displacement, electric potential, and, additionally, the dielectric displacement serve as degrees of freedom. The electric field-electric potential relation is cast into the variational principle as a constraint. The resulting variational principle reads with δG from Eq. (3.1) where $\delta\mathbf{D}$ is the virtual dielectric displacement and $\delta\mathbf{E}_{el}$ is the virtual electric field

$$\delta G^{*}\left(\mathbf{u},\delta\mathbf{u},\Phi,\delta\Phi,\mathbf{D},\delta\mathbf{D}\right)=\delta G\left(\mathbf{u},\delta\mathbf{u},\Phi,\delta\Phi\right)$$
$$+\int_{B}\delta\mathbf{D}\cdot\left(\mathbf{E}_{el}+Grad\Phi\right)dV+\int_{B}\mathbf{D}\cdot\left(\delta\mathbf{E}_{el}+Grad\delta\Phi\right)dV=0. \tag{4.1}$$

So the electric field $\delta\mathbf{E}_{el}$ is computed from Eq. (2.4) and not from Eq. (2.6) and therefore now serves as a dependent variable. After some straightforward calculations, the principle results in a three-field variational formulation:

$$\begin{aligned}
&\int_{B}\delta\mathbf{E}_{m}:(\mathbf{C}_{d}:\mathbf{E}_{m}-\mathbf{h}\cdot\mathbf{D})dV+\int_{B}\rho_{0}\left(\frac{\partial^{2}\mathbf{u}}{\partial t^{2}}-\mathbf{b}\right)\delta\mathbf{u}\,dV-\int_{\partial B_{\sigma}}\delta\mathbf{u}\cdot\bar{\mathbf{t}}\,dA=0\\
&\int_{B}Grad\delta\Phi\cdot\mathbf{D}dV-\int_{\partial B_{D}}\delta\Phi\,\bar{d}\,dA=0\\
&\int_{B}\delta\mathbf{D}\cdot\left(-\mathbf{h}^{T}:\mathbf{E}_{m}+\boldsymbol{\beta}\cdot\mathbf{D}+Grad\Phi\right)dV=0.
\end{aligned} \tag{4.4}$$

In addition to the Eq. (3.3), linear shape functions are also chosen for the approximation of the dielectric displacement field and its variation:

$$\underline{D}=\underline{N}_{D}\hat{\underline{D}},\qquad \delta\underline{D}=\underline{N}_{D}\delta\hat{\underline{D}}\,. \tag{4.5}$$

Here, the index D refers to the dielectric displacement. Neglecting inertia terms for brevities sake one obtains therewith

$$\bigcup_{e=1}^{n_e}\left\{\begin{bmatrix}\delta\hat{\underline{u}}_e^T\\ \delta\hat{\underline{\Phi}}_e^T\\ \delta\hat{\underline{D}}_e^T\end{bmatrix}\begin{bmatrix}\int_{Ve}\underline{B}_u^T\underline{C}_d\underline{B}_u dV & \underline{0} & -\int_{Ve}\underline{B}_u^T\underline{h}\underline{N}_D dV\\ \underline{0} & \underline{0} & \int_{Ve}\underline{B}_\Phi^T\underline{N}_D dV\\ -\int_{Ve}\underline{N}_D^T\underline{h}^T\underline{B}_u dV & \int_{Ve}\underline{N}_D^T\underline{B}_\Phi dV & \int_{Ve}\underline{N}_D^T\underline{\beta}\underline{N}_D dV\end{bmatrix}\begin{bmatrix}\hat{\underline{u}}_e\\ \hat{\underline{\Phi}}_e\\ \hat{\underline{D}}_e\end{bmatrix}\right.$$
$$\left.-\begin{bmatrix}\delta\hat{\underline{u}}_e^T\\ \delta\hat{\underline{\Phi}}_e^T\\ \delta\hat{\underline{D}}_e^T\end{bmatrix}\begin{bmatrix}\int_{Ve}\underline{N}_u^T\rho_0\underline{b}dV+\int_{A_\sigma}\underline{N}_u^T\underline{\bar{t}}dA\\ \int_{A_D}\underline{N}_\Phi^T\bar{d}\,dA\\ \underline{0}\end{bmatrix}\right\}=0. \tag{4.6}$$

Introduction of mnemonic abbreviations for the sub-matrices gives the following set of equations

$$\begin{bmatrix}\underline{K}_{uu} & \underline{0} & -\underline{K}_{uD}\\ \underline{0} & \underline{0} & \underline{K}_{\Phi D}\\ -\underline{K}_{Du} & \underline{K}_{D\Phi} & \underline{K}_{DD}\end{bmatrix}\begin{bmatrix}\hat{\underline{u}}\\ \hat{\underline{\Phi}}\\ \hat{\underline{D}}\end{bmatrix}=\begin{bmatrix}\underline{F}_u\\ \underline{F}_\Phi\\ \underline{0}\end{bmatrix}. \tag{4.7}$$

In analogy to the hybrid methods the system of equations can be reduced by elimination of the dielectric displacement $\hat{\underline{D}}$:

$$\begin{bmatrix} \underline{K}_{uu} - \underline{K}_{uD}\underline{K}_{DD}^{-1}\underline{K}_{Du} & \underline{K}_{uD}\underline{K}_{DD}^{-1}\underline{K}_{D\Phi} \\ \underline{K}_{\Phi D}\underline{K}_{DD}^{-1}\underline{K}_{Du} & -\underline{K}_{\Phi D}\underline{K}_{DD}^{-1}\underline{K}_{D\Phi} \end{bmatrix} \begin{bmatrix} \hat{\underline{u}} \\ \hat{\underline{\Phi}} \end{bmatrix} = \begin{bmatrix} \underline{F}_u \\ \underline{F}_\Phi \end{bmatrix}. \tag{4.8}$$

5 Example

For the numerical verification of the three different finite element formulations from Sects. 3 and 4 a 4-node Mindlin/Reissner shell element with linear shape functions for the displacements and selective reduced integration of the shear terms was chosen, see (Mesecke and Lammering 2002).

A rectangular plate with bonded piezoelectric patches with opposite polarisation on both sides, see Figure 5.1, and simply supported edges is under consideration. The geometry and material data for the system is given in Table 5.1.

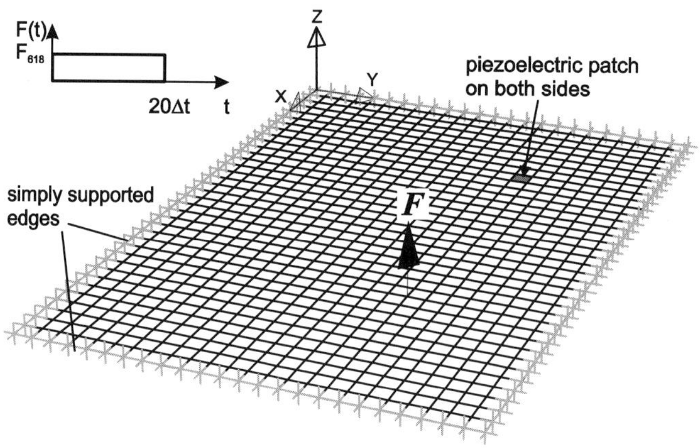

Fig. 5.1. FE-mesh of the plate with piezoelectric patches on both sides, load

Table 5.1. Geometry and material data

Geometry data		Material data	
plate	piezoelectric film	plate	piezoelectric film
$L_x = 0.90$ m	$L_{x,piezo} = 0.04$ m	$E = 1.57 \cdot 10^{11}$ N/m²	$E = 1.57 \cdot 10^{11}$ N/m ²
$B_y = 0.60$ m	$B_{y,piezo} = 0.04$ m	$\nu = 0.0$	$\nu_{film} = 0.0$
$h = 0.004$ m	$h_{piezo} = 0.0002$ m	$\rho = 2700$ kg/m³	$\rho_{film} = 0$ kg/m³
$L_1 = 0.20$ m		$\eta = 0.01$	$e = -47.1$ N/(Vm)
$B_2 = 0.40$ m			$\varepsilon = 3.54 \cdot 10^{-8}$ F/m

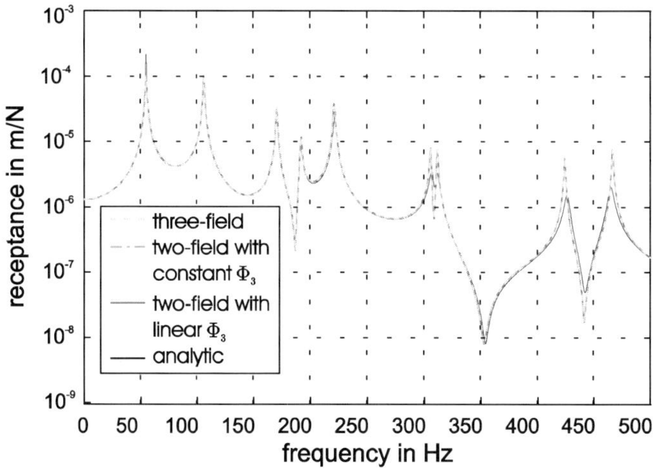

Fig. 5.2. Receptance

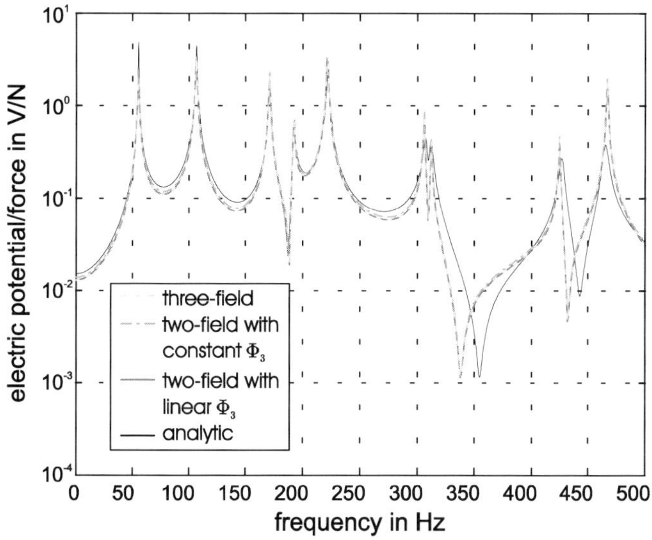

Fig. 5.3. Fourier transformed of the electric potential

For the transient calculation a low Rayleigh damping is assumed and a time increment of Δt=1.59E-5 s chosen. The system is loaded by a force F = 1.0 N at (x_f,y_f) = (0.625 m,0.4 m) for 20 time steps and the decay of the oscillations is observed after the load is removed. For a representative node (x_p,y_p) = (0.2 m,0.4 m) the results are transformed in the frequency domain and the three different formulations are compared for the sensor operation mode. The receptance and the electric potential/force are shown in Figures 5.2 and 5.3. The analytical solution shown is given by (Vogl et al. 2001).

All the formulations agree very well, although the damping coefficients have only been fitted for the first eigenfrequency, which is at about f = 55 Hz. Also for the electric potential all three formulations match each other. Moreover the shape of these curves are similar to the corresponding ones obtained from the displacements. This indicates that the sensor signal reproduces the plates mode of vibration correctly.

6 Concluding remarks

Coupled electromechanical problems and their analysis by the use of the finite element method are in the focus of this paper. Beside the two-field formulation in which the displacements and the electric potential serve as unknowns, multi-field formulations and related elements have been presented and successfully tested.

When the material behaviour cannot be described through linear approximations properly, for the small and also for the high signal ranges, then multi-field formulations and related elements could be an alternative for these systems. So further investigations will be concerned with non-linear material behaviour and with the inf-sup condition as a basic mathematical criterion that determines whether a mixed finite element discretisation is stable and convergent.

References

Benjeddou A (2000) Advances in piezoelectric finite element modeling of adaptive structural elements: a survey. Comp Struct 76:347-363

Ghandi K, Hagood NW (1997) A finite element model for phase transitions in nonlinear electro-mechanically coupled material. In: Varadan VV (ed) Proceedings of SPIE 3039. SPIE, Washington, pp 97-112

Hwang WS, Park HC (1993) Finite element modeling of piezoelectric sensors and actuators. AIAA J 31:930-937

Mesecke S, Lammering R (2002) Hybrid finite element formulations for layered structures with piezoelectric material. In: Mang HA., Rammerstorfer FG, Eberhardsteiner J (eds) Proceedings of the Fifth World Congress on Computational Mechanics (WCCM V). Vienna University of Technology, Austria, ISBN 3-9501554-0-6,

Simo JC, Rifai MS (1990) A class of mixed assumed strain methods and the method of incompatible modes. Int J Num Meth Eng 29:1595–1638

Sze KY (1999) Hybrid finite element models for piezoelectric materials. J Sound Vibr 226: 519-547

Vogl B, Herold S, Wahl F (2001) Vibroakustische Sensorik zur Minimierung der Schallabstrahlung schwingender Strukturen. Techn Mech 21: 167-178

Free Waves in Incoherent Slabs

G. Lancioni

Università di Roma “Tor Vergata”, Dipartimento di Ingegneria Civile, Roma, Italy

1 Introduction

Standard plate theories [6-9] deal with *coherent* plates, that is to say, plates made of a linearly elastic material which (either is isotropic or) has a preferred response axis **c** parallel to the normal **z** to the plate's cross section: typically, the material is transversely isotropic with respect to **c**, or monoclinic with respect to a plane normal to **c**, and is so arranged in the plate that $\mathbf{c}\times\mathbf{z}=0$. *Incoherent* plates are those for which $sin\,\theta=|\mathbf{c}\times\mathbf{z}|\in\,]0,1[$. Incoherence - the smallest amount of it – destroys the prime source of simplicity of classical plate theories, i.e., the orthogonality in energy of the membrane and flexure regimes.

Incoherent plates have been considered in [2-4]. The basic idea of [2] is, roughly speaking, that incoherence would simply force a user of any hierarchical plate theory to choose a polynomial in the thickness coordinate of an order higher than for the coherent case to achieve the same approximation. The approach of [3,4] is based on the conjecture that polynomials in the oblique coordinate along the material-symmetry axis **c** would perhaps work better. Here such conjecture is tested in the case of a *freely-vibrating* infinite slab. Explicit solutions are available for this dynamic problem in three-dimensional linear elasticity [1, 12]. These solutions can be used as benchmarks to evaluate and compare the predictions of different plate theories. We consider two of such two-dimensional theories, both of the first order.

For p a typical point of the slab, we write either

$$p=x+\zeta\,\mathbf{z}\quad\text{or}\quad p=x+\xi\,\mathbf{c}\tag{1}$$

where points x and y are, respectively, the orthogonal and oblique projections of p on the cross section, and where ζ and ξ denote the thickness coordinates taken along **z** and **c**. For the displacement field, we use the representation

$$\mathbf{u}(x,\zeta,t)=\mathbf{u}_0(x,t)+\zeta\,\mathbf{u}_1(x,t)\tag{2}$$

in the first theory, and the representation

$$\tilde{\mathbf{u}}(x,\xi,t)=\tilde{\mathbf{u}}_0(x,t)+\xi\,\tilde{\mathbf{u}}_1(x,t)\tag{3}$$

in the second. The former kinematics is a slight generalization of Reissner Mindlin's; the latter is tailored to account for incoherence, and we wish to estimate to what extent it does. To obtain the governing equations, we insert arbitrary velocity fields consistent with either the one or the other of (3) and (2) into the three-dimensional virtual work equation, and then integrate through thickness. We also derive and use, in both cases, an explicit formula for the elasticity tensor as a function of the incoherence angle θ. In the following we refer to the theory based on (3) ((2)) as to Theory A (B).

Just as expected, no matter which of the two approximate theories we use, we find that incoherence couples the vibration modes in all possible ways: separation between flexure and thickness waves is lost (for coherent electroelastic plates, such vibrations are analyzed in, respectively, [5] and [11]). We check the accuracy of the two theories by comparing the dispersion curves they yield with those obtained from the parent three-dimensional theory. We find that the theory based on the *Ansatz* (2) guarantees a better matching when incoherence is strong ($\sin\theta$ close to 1) and transverse isotropy weak (that is to say, the material response along **c** differs little from the response in the isotropy plane orthogonal to **c**); the situation reverses when incoherence is weak ($\sin\theta$ close to 0) and the material's stiffness is sensibly smaller in the isotropy plane than in the transverse direction.

2 Incoherent slab

2.1 Geometry

Let us consider a *slab-like region* Ξ in three-dimensional Euclidean space, with

$$\Xi=\{\ p=x+\zeta\mathbf{z}\ |\ x\in\pi_z,\ \zeta\in I_z\ \} \tag{4}$$

(cf. the first alternative in (1)). Here π_z is the slab's middle-plane, **z** the unit vector normal to π_z, and $I_z = (-\varepsilon, +\varepsilon)$ an interval. Since the correspondence $p\leftrightarrow(x,\zeta)$ is one-to-one, we identify Ξ with the Cartesian product $\pi_z\times I_z$. In particular, the upper and lower faces of Ξ are $\pi_z^{\pm} = \pi_z\times\{\pm\varepsilon\}$. Finally, we choose an origin point $o\in\pi_z$ and introduce the orthonormal Cartesian frame $(o;\ \mathbf{c}_1,\ \mathbf{c}_2,\ \mathbf{z})$, so that the position vector of the typical point $x\in\pi_z$ with respect to the origin is $x-o=x_\alpha\mathbf{c}_\alpha,\ \alpha=1,2$.

With a view toward considering the slab as a transversely isotropic material body with fibers aligned in a direction oblique to the middle plane, and without any loss of generality, in the cone of axis **z** and half-opening the *incoherence angle* θ we choose the unit vector **c** obtained by rotating **z** around $\mathbf{c}_2$ counterclockwise by θ, that is, $\mathbf{c} = \mathbf{Q}(\theta)\mathbf{z}$, where

$$\mathbf{Q}(\theta)=\mathbf{I}+\sin\theta\mathbf{W}+(1-\cos\theta)\mathbf{W}^2,\ \ \mathbf{W}=2skw(\mathbf{c}_1\otimes\mathbf{z})\ .$$

We then introduce the oblique frame (o; $\mathbf{c}_1$, $\mathbf{c}_2$, $\mathbf{z}$), and represent a point $p\in\Xi$ as in the second of (1). Note that

$$y = x + \zeta \tan\theta\, \mathbf{c}_1 , \qquad \zeta = \xi / \cos\theta ; \tag{5}$$

hence, the map $(y, \xi)\leftrightarrow(x, \zeta)$ is one-to-one. We can identify Ξ with the Cartesian product $\pi_z\times I_c$, where $I_c := (-\varepsilon/\cos\theta, \varepsilon/\cos\theta)$. In particular, the upper and lower faces of Ξ are the sets $\pi_z^{\pm} = \pi_z\times\{\pm\varepsilon/\cos\theta\}$, and the *fiber* through a point $y\in\pi_z$ is the set $\{ y \} \times I_c$.

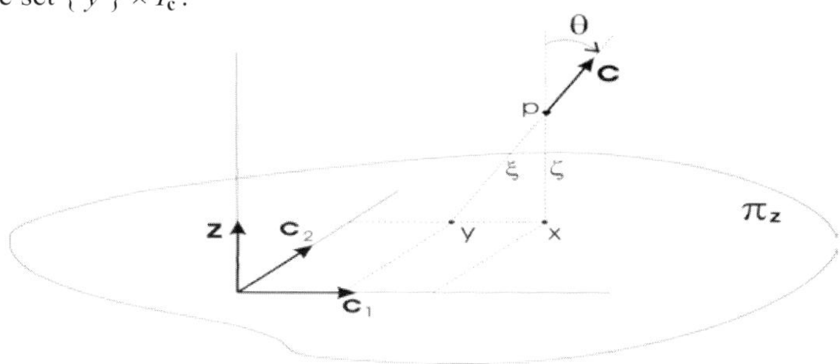

Fig. 1. Oblique geometry

2.2 Elasticity tensor

The space limitations set for this writing force us to do no more than delineating the procedure used in [3,4] to construct the *elasticity tensor* $\mathbf{C}(\theta)$ for a linearly elastic material being transversely isotropic with respect to the axis $\mathbf{c}$.

We start from the elasticity tensor $\mathbf{C}_0 = \mathbf{C}(0)$ for a transversely isotropic material in the form given in section 16 of [10], with $\mathbf{z}$ the preferred axis. We then rotate $\mathbf{C}_0$ by means of $\mathbf{Q}(\theta)$, the fourth-order tensor $\mathbf{Q}(\theta)$ which operates as follows on any second-order tensor $\mathbf{A} : \mathbf{Q}(\theta)[\mathbf{A}] = \mathbf{Q}(\theta)\mathbf{A}\mathbf{Q}^T(\theta)$; so as to obtain

$$\mathbf{C}(\theta) = \mathbf{Q}(\theta)\mathbf{C}_0\mathbf{Q}^T(\theta) . \tag{6}$$

For an extended representation of $\mathbf{C}(\theta)$ we refer the reader to [3] or [4]. Here we only point out that $\mathbf{C}(\theta)$ turns out to reflect the response of a special monoclinic material characterized by the symmetry transformation $\mathbf{R}_{\mathbf{c}_2}^{\pi}$, the rotation of π around $\mathbf{c}_2$, with each of its 13 material moduli a function of the 5 moduli of $\mathbf{C}_0$ and of the incoherence angle θ.

3 Evolution equations

Free vibrations are in order in a three-dimensional slab when the contact loads on the upper and lower faces are null and the only distance load are inertial. The virtual work equation reads:

$$\int_{\Xi} \mathbf{S}\cdot\nabla\mathbf{v} = -\int_{\Xi}\rho \;\; \ddot{\mathbf{u}}\cdot\mathbf{v}. \tag{7}$$

with $\mathbf{S}$ the stress, ρ the mass density, $\ddot{\mathbf{u}}$ the acceleration and $\mathbf{v}$ an arbitrary test field. We deduce the balance equations of the two-dimensional theories A and B by thickness integration of (7), when written for test fields consistent with, respectively, the kinematics (3) and (2).

Here we formulate the two theories A and B based on the kinematic assumptions (3) and (2) respectively. It is shown that theory B descends from theory A.

Theory A. To evaluate the gradient of the displacement field (3) we express it in terms of the variables (x, ζ, t):

$$\tilde{\mathbf{u}}(y(x,\zeta),\xi(\zeta),t) = \tilde{\tilde{\mathbf{u}}}(x,\zeta,t;\theta) = \tilde{\mathbf{u}}_0(y(x,\zeta),t) + \xi(\zeta)\tilde{\mathbf{u}}_1(y(x,\zeta),t) \tag{8}$$

where the maps $(x, \zeta)\rightarrow y(x, \zeta)$ and $\zeta\rightarrow\xi(\zeta)$ depend on the incoherence angle θ as specified by (5). A use of the chain rule yields

$$\nabla\tilde{\mathbf{u}} = ({}^s\nabla\tilde{\mathbf{u}}_0 + \xi\, {}^s\nabla\tilde{\mathbf{u}}_1)\nabla y + \tilde{\mathbf{u}}_1 \otimes \nabla y, \tag{9}$$

where ${}^s\nabla\tilde{\mathbf{u}}_i = (\partial\tilde{\mathbf{u}}_i / \partial y_\alpha)\otimes\mathbf{c}_\alpha$, $\nabla y = \mathbf{c}_\alpha \otimes \mathbf{c}_\alpha - \tan\theta\,\mathbf{c}_1 \otimes \mathbf{z}$ and $\nabla\xi = (\cos\theta)^{-1}\mathbf{z}$ ($i = 0,1$ and $\alpha = 1,2$). For test fields whose gradient has the form (3.3), starting from the virtual work Eq. (7) and using the divergence theorem and a standard localization argument, it is easy to obtain the following balance equations of theory A

$$\begin{gathered}\mathrm{div}\mathbf{M}_0 - \tan\theta\nabla(\mathbf{M}_0\mathbf{z})\mathbf{c}_1 = 2\varepsilon\rho\ddot{\tilde{\mathbf{u}}}_0, \\ \mathrm{div}\mathbf{M}_1 - \frac{1}{\cos\theta}\mathbf{M}_0\mathbf{z} - \tan\theta\nabla(\mathbf{M}_1\mathbf{z})\mathbf{c}_1 = \frac{2}{3\cos^2\theta}\varepsilon^3\rho\ddot{\tilde{\mathbf{u}}}_1,\end{gathered} \tag{10}$$

to hold in $\pi_z \times [0, t_0]$, where the two-dimensional stress descriptors

$$\mathbf{M}_i(y,t) = \int_{I_c}\xi^i\mathbf{S}\cos\theta\, d\xi, \quad i = 0,\, 1 \tag{11}$$

give a global account of the stress distribution along the oblique fiber $\{y\}\times I_c$ at time t. Note that here thickness integration is performed along the oblique axis $\mathbf{c}$.

To arrive at the desired evolution equations, the stress descriptors $\mathbf{M}_i$ ($i = 0,1$) must be expressed in terms of the kinematic unknowns $\mathbf{u}_i$ by the use of the constitutive and compatibility relations. Combining (11) with $\mathbf{S} = \mathbf{C}(\theta)\left[sym\nabla\tilde{\mathbf{u}}\right]$ and (6), we find

$$\mathbf{M}_i(y,t)=\int_{I_c}\{\sum_{j=0}^{1}(\xi^{i+j}\mathbf{C}(\theta)[\nabla\tilde{\mathbf{u}}_j-\tan\theta(\nabla\tilde{\mathbf{u}}_j)\mathbf{c}_1\otimes\mathbf{z}])+\\
+\frac{\xi^i}{\cos\theta}\mathbf{C}(\theta)[\tilde{\mathbf{u}}_1\otimes\mathbf{z}]\}\cos\theta d\xi. \tag{12}$$

The resulting evolution equations are written in full in [11] and [12]; in those works one can also find an "oblique" representation for the displacement more general than the first-order representation (3).

Theory B. Just as kinematics (3) reduces to (2) when the incoherence angle is null, the balance equations of theory B follow by setting $\theta = 0$ into equations (10). One finds:

$$\mathrm{div}\mathbf{M}_0 = 2\varepsilon\rho\ddot{\mathbf{u}}_0, \qquad \mathrm{div}\mathbf{M}_1-\mathbf{M}_0\mathbf{z}=\frac{2}{3}\varepsilon^3\rho\,\ddot{\mathbf{u}}_1, \tag{13}$$

where the stress descriptors $\mathbf{M}_i = \int\zeta^i\mathbf{S}d\zeta$ account for the stress distribution along fibers normal to the middle-plane π_z. To obtain the corresponding evolution equations, we write the stress in the integrands of the descriptors $\mathbf{M}_i$ in terms of the elasticity tensor and the strain field: $\mathbf{S}=\mathbf{C}(\theta)\left[sym\nabla\tilde{\mathbf{u}}\right]$, so that

$$\mathbf{M}_i = \int_{I_c}\{\sum_{j=0}^{1}(\xi^{i+j}\mathbf{C}(\theta)[\nabla\mathbf{u}_j])+\xi^i\mathbf{C}(\theta)[\mathbf{u}_1\otimes\mathbf{z}]\}.$$

Note that theory A features a double dependence on the incoherence angle, namely, in the kinematics and in the elastic response. Only the latter dependence is present in theory B, whose kinematics (2), while does not reflect the "obliquity" of the material response, has the advantage of being affine along the shortest segment crossing the slab and thus, in a sense, is the less constrained one among all kinematics of the first order in a thickness variable.

4 Free-wave propagation

The vibrations we wish to study are *plane progressive waves propagating in the direction* $\mathbf{c}_1$ with wavelength k, frequency f, and, hence, velocity $c=f/k$. Accordingly, for the corresponding solutions of systems (10) and (13) we assume the following representations:

$$\tilde{\mathbf{u}}_i = g(x,t)\tilde{\mathbf{U}}_i, \qquad \mathbf{u}_i = g(x,t)\mathbf{U}_i, \qquad i=0,1, \tag{14}$$

where $\tilde{\mathbf{U}}_i$ and $\mathbf{U}_i$ are constant amplitude vectors and the wave shape is $g(x,t)=\exp(i(k\mathbf{c}_1\cdot(x-o)-ft))$. Substitution of $(14)_1$ into the evolution Eqs. (10) yields two linear algebraic systems, namely,

$$\tilde{\mathbf{L}}(k_\varepsilon,f_\varepsilon)\tilde{\mathbf{a}}=\mathbf{0}, \quad \tilde{\mathbf{I}}(k_\varepsilon,f_\varepsilon)\tilde{\mathbf{b}}=\mathbf{0}. \tag{15}$$

Here the unknowns are the "vectors"

$$[\tilde{\mathbf{a}}]=[\tilde{\mathbf{U}}_0\cdot\mathbf{z},\tilde{\mathbf{U}}_0\cdot\mathbf{c}_1,\varepsilon\tilde{\mathbf{U}}_1\cdot\mathbf{z},\varepsilon\tilde{\mathbf{U}}_1\cdot\mathbf{c}_1]^T \quad \text{and} \quad [\tilde{\mathbf{b}}]=[\tilde{\mathbf{U}}_0\cdot\mathbf{c}_2,\varepsilon\,\tilde{\mathbf{U}}_1\cdot\mathbf{c}_2]^T,$$

the matrices $\tilde{\mathbf{L}}(k_\varepsilon,f_\varepsilon)$ and $\tilde{\mathbf{I}}(k_\varepsilon,f_\varepsilon)$, whose detailed form is to be found in [11], depend on the dimensionless parameters $k_\varepsilon=\varepsilon k$, $f_\varepsilon=\varepsilon(\rho/\mu)^{1/2}f$. For nontrivial solutions to exist, we require that

$$\det\tilde{\mathbf{L}}(k_\varepsilon,f_\varepsilon)=0\,, \qquad \det\tilde{\mathbf{I}}(k_\varepsilon,f_\varepsilon)=0\,. \tag{16}$$

We point out that, as a consequence of our choice of the propagation direction $\mathbf{c}_1$, the wave motion we study splits into an irrotational oscillation in the coordinate plane $x_2 = 0$ and a solenoidal oscillation directed along the x_2-axis; the former is determined by system $(15)_1$, the latter by $(15)_2$. Separation of flexure and membrane waves is lost for an incoherent slab. Moreover, incoherence couples dilatational and isochoric waves at cut-off.

If we substitute $(14)_2$ into (13), we get two linear algebraic systems completely analogous to the systems (15):

$$\det\bar{\mathbf{L}}(k_\varepsilon,f_\varepsilon)=0\,, \qquad \det\bar{\mathbf{I}}(k_\varepsilon,f_\varepsilon)=0\,. \tag{17}$$

In the following we restrict our attention to nontrivial irrotational oscillations in the plane $x_2 = 0$, a case in which, depending on what theory we deal with, either conditions $(16)_1$ or conditions $(17)_1$ prevail, with their left side a fourth-order polynomial both in k_ε^2 and f_ε^2.

5 Two examples

For the purpose both to compare the predictions of the two-dimensional theories we have sketched and to check those predictions against the exact consequences of the three-dimensional theory, we solve Eqs. $(16)_1$ and $(17)_1$ numerically for two significant and different choices of the coherence and isotropy parameters:

1. Strong incoherence & weak anisotropy. Consider a ceramic slab (we use the constitutive moduli of lead titanate-zirconate P1-88 from Table 1 [3]), with an incoherence angle $\theta = 30^\circ$. For such a material the anisotropy is *weak* in the sense that the stiffness modulus of fibers along **c** (120 GPa) is close to that of fibers in the isotropy plane (130 GPa). The pairs $(f_\varepsilon, k_\varepsilon)$ which satisfy $(16)_1$ (theory A) and $(17)_1$ (theory B) are plotted in Fig. 2 (a) (dashed and dotted lines, respectively[1].

The accuracy of the two theories is checked by comparing the four branches of each theory with the first four *dispersion branches* (solid line) given by the parent three-dimensional theory (we refer to [3,4] for a detailed study and to [1] and [11]

[1] We confined our investigation to real wave numbers because we are interested only in *progressive* waves. When k_ε is imaginary, the waves are *standing* and decay exponentially in space.

for a study of wave propagation in isotropic and anisotropic slabs, respectively). We notice that the dotted curves match the corresponding three-dimensional branches better than the dashed ones. This feature is preserved when the incoherence angle increases. In Fig. 2 (b), for $\theta = 60^\circ$, the distance between dashed and solid branches increases, whereas the distance between dotted and solid curves stays essentially the same.

We conclude that, when incoherence is strong, theory B works better than theory A, at least as far as yielding accurate predictions for this type of propagating waves.

2. Weak incoherence & strong anisotropy. We now take $\theta = 60^\circ$, and adjust the constitutive moduli from the previous example so as to get a material much stiffer in the direction **c** than that in any direction in the plane orthogonal to **c**. Precisely, we divide by 50 the moduli λ and μ, which determine the stiffness of material fibers in the plane orthogonal to **c**, whereas we multiply by 50 the moduli τ_1 and η (τ_1 measures the stiffness of a fiber along **c**; η is the shear modulus relative to any plane through **c**); finally, we multiply by 1.3 the modulus τ_2 measuring the Poisson effect relative to any pair of orthogonal directions (**c**, **d**). Note that these adjustments preserve the positive-definiteness of the stored-energy density.

The dispersion branches for such a slab are plotted in Fig. 2 (c)[2], again with dashed lines for theory A and dotted lines for theory B, and compared with the dispersion branches given by the three-dimensional theory (solid lines). Clearly, the dashed curves match the corresponding solid curves better than the dotted ones.

Hence, when incoherence is weak and anisotropy strong (that is to say, the material's stiffness is substantially smaller in the isotropy plane than in the transverse direction), theory A yields more accurate predictions than theory B.

Acknowledgements

I thank Professor Paolo Podio-Guidugli for his valuable comments and suggestions and for his useful help with the drafting of the paper.

References

1. Achenbach JD (1980) Wave propagation in elastic solids. 3 edn, North Holland , New York
2. Batra RC, Vidoli S (2000) Derivation of plate and rod equations for a piezoelectric body from a mixed three-dimensional variational principle. J Elasticity 59: 23-50
3. Lancioni G (2002) Dinamica di piastre incoerenti. Tesi di dottorato

[2] To have a clear drawing of the dispersion branches we plot two figures characterized by different frequency ranges: pairs ($f_\varepsilon \in (0,10)$) (Fig. 2 (c)1) and ($f_\varepsilon \in (40,180)$) (Fig. 2 (c)2).

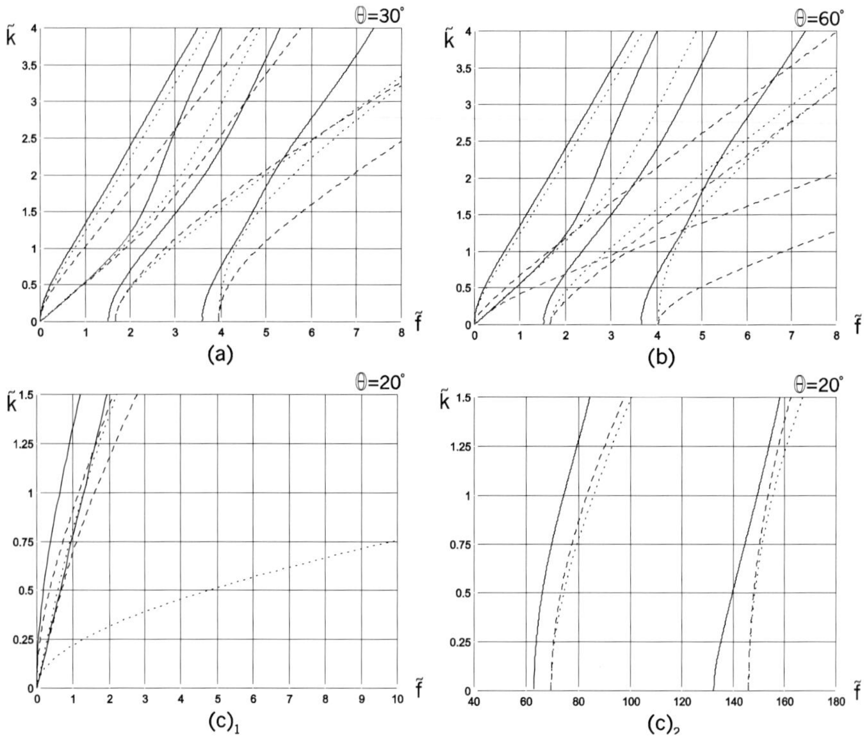

Fig. 2. Dispersion branches: theory A ~ dashed line; theory B ~ dotted line; 3D theory ~ solid line. (a) $\theta = 30°$; (b) $\theta = 60°$; (c) $\theta = 20°$, modified material moduli.

4. Lancioni G, Podio-Guidugli P (2003) forthcoming
5. Lancioni G, Tomassetti G (2002) Flexure waves in electroelastic plates, Wave motion 35: 257-269
6. Lekhnitskii SG (1987) Anisotropic plates. Gordon and Breach Science Publishers, New York
7. Love AEH (1927) A treatise on the mathematical Theory of Elasticity. 4th edn. Cambridge University Press, (Dover Publications Edition, 1944)
8. Mindlin RD (1951) Thickness-Shear and Flexural Vibrations of Cristal Plates. J Applied Physics 22-3
9. Nikodem Z, Lee PC (1974) Approximate theory of vibration of crystal plates at high-frequency. J Solids Structures 10: 177-196
10. Podio-Guidugli P (2000) A Primer in Elasticity. Kluwer
11. Podio-Guidugli P, Tomassetti G (2001) Thickness waves in electroelastic plates. Wave motion 34: 175-191
12. Rogerson AG, Kossovich LY (1999) Approximation of the dispersion relation for an elastic plate composed of strongly anisotropic elastic material. J Sound Vib 225: 283-305

Inelastic Behaviour of Stepped Square Plates

J. Lellep and A. Mürk

Institute of Applied Mathematics, Tartu University, Tartu, Estonia

1 Introduction

The most of the work in dynamic plasticity is dedicated to the dynamic behaviour of beams and axisymmetric plates (see N. Jones [2], T. X. Yu and F. Chen [4], W. Stronge [3]). The only exact theoretical solution on dynamic response of a non- axisymmetric plastic plate is obtained by A. Cox and L. M. Morland [1]. In the present paper the method of mode form motions is used in the case of small deflections of stepped square plates subjected to impulsive loadings.

2 Formulation of the problem and governing equations

Consider a square plate with a side length $2L$ (Fig. 1). The plates with and without cutout will be studied. It is assumed that the plate is subjected to the initial impact loading. We assume that the initial kinetic energy K_0 is given whereas the initial transverse velocity field may be unknown. The dynamic response of the plate will be studied within the concept of a rigid- plastic body.

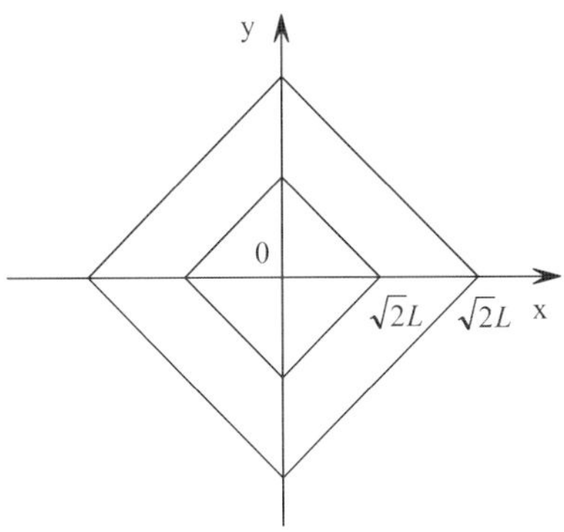

Fig. 1. Square plate

Let the plate thickness be piece wise constant, e.g.

$$h = h_j , \tag{1}$$

for $(x, y) \in D_j$; $j = 0, \ldots, n$. We restrict our attention to the concentric case when the inner and outer boundaries of regions D_j are squares. Let the boundaries of regions of constant thickness intersect x- and y- axis at points $\alpha_0 \sqrt{2}L$, $\alpha_1 \sqrt{2}L$, …, $\alpha_{n+1} \sqrt{2}L$. Here

$$\alpha_0 = \frac{a}{L}, \; \alpha_{n+1} = 1 \tag{2}$$

provided $2a$ is the length on the internal edge of the plate. Note that in the case of a full plate without cutout $\alpha_0 = a = 0$.

It is assumed that the initial kinetic energy is high enough to cause plastic strains. Moreover, we assume that elastic stains are small in comparison with plastic strains so that elastic counterparts of strains can be neglected. The material of plates is an isotropic homogeneous material which can be treated as an ideal plastic material obeying Johansen's yield condition (Fig. 2). Here M_1, M_2 stand for the principal moments

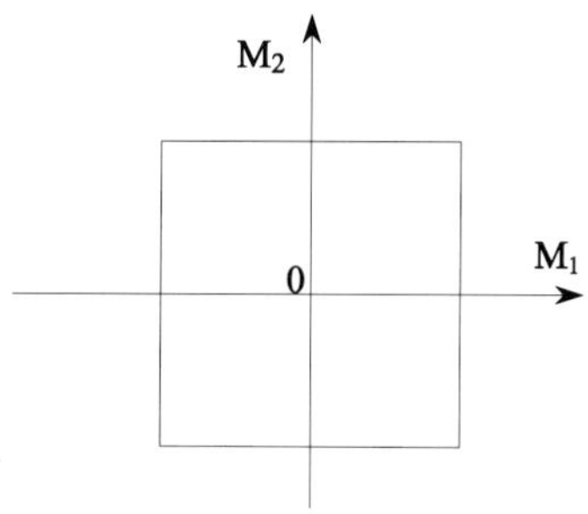

Fig. 2. Johansen's yield condition

The equation of motion can be presented as

$$\frac{\partial^2 M_x}{\partial x^2} + 2\frac{\partial^2 M_{xy}}{\partial x \partial y} + \frac{\partial^2 M_y}{\partial y^2} = \mu\, h \frac{\partial^2 W}{\partial t^2}. \tag{3}$$

The curvatures corresponding to (5) have the form

$$\kappa_x = -\frac{\partial^2 W}{\partial x^2}, \kappa_y = -\frac{\partial^2 W}{\partial y^2}, \kappa_{xy} = -\frac{\partial^2 W}{\partial x \partial y}. \tag{4}$$

It is reasonable to assume that the deformation process is symmetrical with respect to coordinate axis. Thus we can restrict our attention to the first quadrant only. Following A. Cox and L. Morland [1] we introduce a new variable

$$z = \frac{1}{\sqrt{2}L}(x+y). \tag{5}$$

Evidently, in the first quadrant $0 \le z \le 1$. The method of mode form motions will be used in the present paper. This method was suggested by J. Martin and P. Symonds. It was established that the approximate theoretical predictions are in good correlation with exact solutions and experimental results.

Perhaps the simpliest kinematically admissible transverse velocity distribution is

$$\frac{\partial W}{\partial t} = V(t)\frac{1-z}{1-\alpha}, \tag{6}$$

for a plate with cutout. If there is no cutout one can take $\alpha = 0$. Here $V(t)$ stands for an unknown function of time.

3 Determination of the acceleration

Substituting (1) and (6) in (3) leads to the equation

$$\frac{\partial^2 M_x}{\partial x^2} + 2\frac{\partial^2 M_{xy}}{\partial x \partial y} + \frac{\partial^2 M_y}{\partial y^2} = \frac{\mu h_j \dot{V}}{1-\alpha}(1-z), \tag{7}$$

which holds good in the region D_j ($j = 0,\dots,n$). Let us interprete D_j as the region in the first quadrant where $z \in (\alpha_j, \alpha_{j+1})$.

In (7) $\dot{V}$ stands for the acceleration of the free edge of the plate ($z = \alpha$) or of the central point of the plate ($z = 0$) in the case of the plate without cutout. It is assumed herein that the outer edge of the plate is simply supported or clamped whereas the inner edge is free (in the case of the plate with cutout).

Boundary conditions at the edge can be obtained from the relation for the moment with respect to an inclined edge [1]

$$M_n = M_x \sin^2 \beta + M_y \cos^2 \beta + M_{xy} \sin 2\beta, \tag{8}$$

where β stands for the angle of inclination with respect to the x-axis. Similar relation can be derived for the shear force as well

$$Q_n = Q_x \cos \beta + Q_y \sin \beta. \tag{9}$$

We are looking for the solution of (7) in the form

$$\begin{aligned} M_x &= M_j + x^2 F_j(z), \\ M_y &= M_j + y^2 F_j(z), \\ M_{xy} &= xyF_j(z), \end{aligned} \tag{10}$$

for $(x,y)\in D_j$ ($j=0,...,n$). Here M_j stands for the limit moment for the part D_j of the plate. Thus $M_j=\sigma_0 h_j^2/4$, σ_0 being the yield stress of the material. Functions F_j ($j=0,...,n$) in (10) are unknown functions of z which have to be defined so that the Eq. (7) is satisfied in each region D_j ($j=0,...,n$). Substituting (10) in (7) gives after appropriate transformations

$$z^2\frac{\partial^2 F_j}{\partial z^2}+6z\frac{\partial^2 F_j}{\partial z}+6F_j=\frac{\mu h_j\dot{V}(t)}{1-\alpha}(1-z), \tag{11}$$

for $z\in D_j$ ($j=0,...,n$). The Eq. (13) has a solution

$$F_j=\frac{\mu h_j\dot{V}}{12(1-\alpha)}(2-z)+\frac{B_j}{z^2}+\frac{C_j}{z^3}, \tag{12}$$

B_j and C_j being arbitrary constants.

Consider the plate with clamped outer edge and free inner edge in a greater detail. At the outer edge one has $M_n\big|_1=-M_n$ whereas at the free edge $M_n\big|_\alpha=0$, $Q_n\big|_\alpha=0$. Thus according to (8) - (10) boundary conditions can be presented as

$$F_0(\alpha)=-\frac{M_0}{\alpha^2L^2}, \tag{13}$$

$$F_0'(\alpha)=\frac{3M_0}{\alpha^3L^2}.$$

At the outer edge one has

$$F_n(1)=-\frac{2M_n}{L^2}. \tag{14}$$

Due to the continuity of M_n and Q_n at $z=\alpha_j$ ($j=0,...,n$) following requirements must be met by the set functions F_j

$$\begin{aligned}F_j(\alpha_{j+1}-0)&=F_{j+1}(\alpha_{j+1}+0)+\frac{M_{j+1}-M_j}{\alpha_j^{\,2}L^2},\\ F_j'(\alpha_{j+1}-0)&=F_{j+1}'(\alpha_{j+1}+0)-3\frac{M_{j+1}-M_j}{\alpha_j^{\,3}L^2}.\end{aligned} \tag{15}$$

For determination of $2n+3$ unknown constants B_j, C_j ($j=1,...,n$) and $\dot{V}$ one has $2n+3$ requirements which are given by (13) – (15). The solution of the non- linear system (13) – (15) in general case is quite complicated.
However, in the particular case when $n=1$ the system takes the form

$$\frac{\mu h_0 \dot{V}}{12(1-\alpha)}(2-\alpha)+\frac{B_0}{\alpha^2}+\frac{C_0}{\alpha^3}=-\frac{M_0}{\alpha^2 L^2},$$

$$\frac{\mu h_0 \dot{V}}{12(1-\alpha)}+\frac{2B_0}{\alpha^3}+\frac{3C_0}{\alpha^4}=-\frac{3M_0}{\alpha^3 L^2},$$

$$\frac{\mu h_1 \dot{V}}{12(1-\alpha)}+B_1+C_1=-\frac{2M_1}{L^2}, \tag{16}$$

$$\frac{\mu h_0 \dot{V}}{12(1-\alpha)}(2-\theta)+\frac{B_0}{\theta^2}+\frac{C_0}{\theta^3}=\frac{\mu h_1 \dot{V}}{12(1-\alpha)}(2-\theta)+\frac{B_1}{\theta^2}+\frac{C_1}{\theta^3},$$

$$\frac{\mu h_0 \dot{V}}{12(1-\alpha)}+\frac{2B_0}{\theta^3}+\frac{C_0}{\theta^4}=\frac{\mu h_1 \dot{V}}{12(1-\alpha)}+\frac{2B_1}{\theta^3}+\frac{3C_1}{\theta^4}.$$

In (16) the notation $\theta=\alpha_1$ is used . From (16) one readily obtains that

$$\dot{V}=\frac{12(1-\alpha)\left[M_0(\alpha-\theta)+M_1(\theta-2)\right]}{\mu L^2\left[(h_0-h_1)(3\theta^4-8\theta^3+6\theta^2)-h_0(3\alpha^4-8\alpha^3+6\alpha^2)-h_1\right]}. \tag{17}$$

4 Residual deflections

As it might be expected the acceleration of the free edge is a constant (see (17)). Thus one can easily integrate the acceleration which gives

$$\dot{W}(\alpha)=V(t)=\dot{V}t+V_0. \tag{18}$$

Integrating once more one obtains

$$\dot{W}(\alpha,t)=\frac{\dot{V}}{2}t^2+V_0 t, \tag{19}$$

when deriving (18) and (19) initial conditions $\dot{W}(\alpha,0)=V_0$, $W(\alpha,0)=0$ have been taken into account.

It follows from (18) that the motion ceases at the moment

$$t_1=-\frac{V_0}{\dot{V}}, \tag{20}$$

and the maximal residual deflection is

$$W_1=W(\alpha,t_1)=-\frac{V_0^2}{2\dot{V}}. \tag{21}$$

Note that we are employing the method of mode form motions. Thus the velocity distribution (6) holds good at each moment of time during the motion of the plate. Therefore, the initial kinetic energy of a plate with a single step can be calculated as

$$K_0 = \frac{2\mu V}{(1-\alpha)^2} \left\{ \int_0^{\sqrt{2}\alpha L} dx \int_{-x+\sqrt{2}\alpha L}^{-x+\sqrt{2}\theta L} h_0 (1-z)^2 dy + \int_{\sqrt{2}\alpha L}^{\sqrt{2}\theta L} dx \int_0^{-x+\sqrt{2}\theta L} h_0 (1-z)^2 dy + \right.$$
$$\left. + \int_0^{\sqrt{2}\theta L} dx \int_{-x+\sqrt{2}\theta L}^{-x+\sqrt{2}L} h_1 (1-z)^2 dy + \int_{\sqrt{2}\theta L}^{\sqrt{2}L} dx \int_0^{-x+\sqrt{2}L} h_1 (1-z)^2 dy \right\}. \tag{22}$$

If the number of steps in the thickness exceeds unity the relation (22) must be accommodated appropriately. It easily follows from (22) that the initial velocity can be presented as

$$V_0^2 = \frac{12K_0 (1-\alpha)^2}{4\mu L^2 \left\{ (h_1 - h_0)(1-\theta)^3 (3\theta+1) + h_0 (1-\alpha)^3 (3\alpha+1) \right\}}. \tag{23}$$

Substituting (17) and (23) in (21) leads to the maximal residual displacement

$$W_1 = \frac{K_0 (1-\alpha)}{8\left[M_0 (\theta-\alpha) + M_1 (2-\theta) \right]}. \tag{24}$$

It follows from (24) that in the case of a plate of constant thickness h_* the maximal deflection is

$$W_1 = \frac{K_0 (1-\alpha)}{8M_* (2-\alpha)}.$$

Here M_* stands for the limit moment of the plate of constant thickness h_* ,e.g. $M_* = \sigma_0 h_*^2 / 4$.

5 Numerical results

Maximal residual deflections for plates with a single step are depicted in Fig. 3 - 5. The results presented in Fig. 3 correspond to plates simply supported at the edge whereas Fig. 4 is associated with the clamped plates.

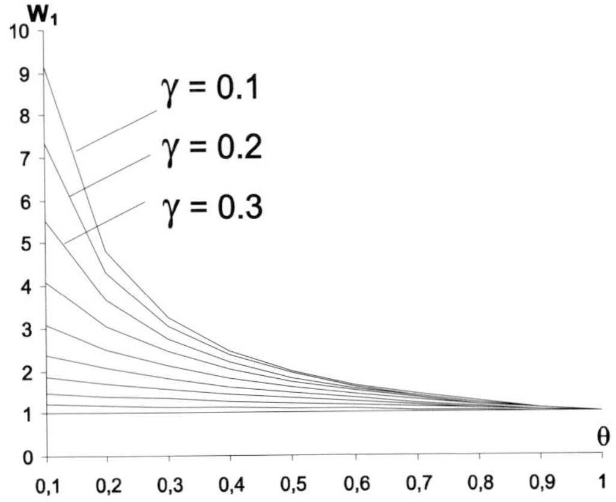

Fig. 3. Maximal deflections of a simply supported plate

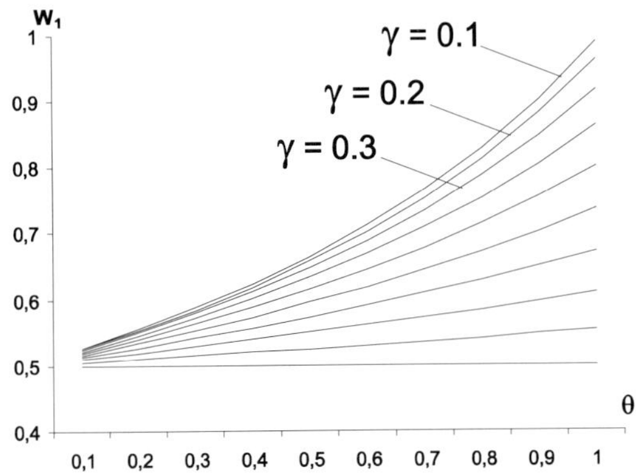

Fig. 4. Deflections of a clamped plate

The results of Fig. 5 are obtained for clamped plates with cutout. In the latter case the transverse deflection attains its maximal value at the internal edge of the plate. It can be seen from Fig. 3- 5 that the maximal residual deflection tends to that of the reference plate of constant thickness when the step tends to the outer edge (Fig. 3) or to the center of the plate (Fig. 4) or to the inner edge of the plate (Fig. 5).

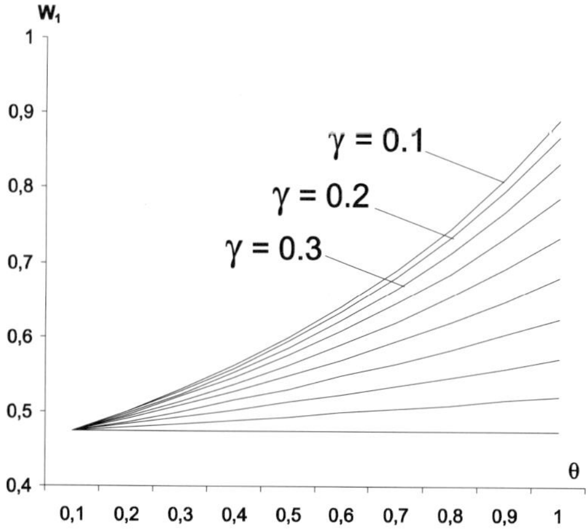

Fig. 5. Deflections of a plate with cutout

6 Concluding remarks

A method for prediction of the behaviour of rigid- plastic square plates under initial impulsive loading has been developed. The plates under consideration have piece wise constant thickness and obey Johansen's yield condition.

Numerical analysis showed that the method of mode form motions leads to results which are close to the exact solution in case of plates of constant thickness. The statical admissibility of solutions for stepped plates was established numerically.

References

1. Cox AD, Morland LM (1959) Dynamic plastic deformations of simply supported square plates. J Mech Phys Solids 7: 229-241.
2. Jones N (1989) Structural Impact. Cambridge University Press
3. Stronge W, Yu TX (1993) Dynamic models of structural plasticity. Springer Verlag, London
4. Yu TX, Chen FL (2000) Failure of plastic structures under intense dynamic loading: modes criteria and thresholds. Int J Mech Sci 42: 1537-1554

An Assessment of Classical and Refined Plate Theories Derived From the Principle of Virtual Displacements

J. Meenen

BASF Aktiengesellschaft, WLF/FF Materials Engineering Ludwigshafen, Germany

1 Introduction

Classical and refined plate theories can be derived either by postulating the existence of a Cosserat surface (direct approach) or by reducing a set of three-dimensional equations of continuum to a two-dimensional system (reduction techniques). In the first case, it is a priori assumed that the plate can be represented by a two-dimensional surface, whose deformation is described by translations and additional kinematical degrees of freedom. The additional degrees of freedom must in a later stage be physically interpreted, f. e. as mid-plane rotations. Constitutive relations are formulated between these kinematical degrees of freedom and corresponding kinetic degrees of freedom, which are in most cases physically interpreted as forces and moments.

Table 1. Derivation of Classical and Refined Plate Theories

Reduction Techniques		Direct Approach
3D Continuum Theory		2D Cosserat Surface
Mathem. Techniques: Asympt. Integration Series Expansion	Hypotheses: Virtual Displacements Virtual Forces	Degrees of Freedom: Forces/Moments, Displacements/Rotations
Convergence Behaviour complicated equations	Boundary Conditions Incompatibilities	Physical Interpretation Constitutive Equations

The reduction techniques can be subdivided into theories using mathematical techniques such as the method of asymptotic expansions, and into theories which are based on physically motivated assumptions or series expansions. The introduction of assumptions or series expansions into the principle of virtual displacements is perhaps the most common way to derive plate theories and has in the past been used for a significant number of different plate theories. A review of the relevant literature can be found for example in [1]. For laminated plates, the assumptions and series expansions can either be formulated for the whole plate thickness (single-layer theories) or for each layer separately (discrete-layer theories). These refined plate theories usually involve more than five degrees of freedom. Simplified theories with only five degrees of freedom can be derived by imposing constraints on the distribution of stresses or strains through the plate thickness. Examples of these simplified theories are the discrete-layer theory of di Sciuva [2] where the transverse shear stress is constant over the plate thickness, and the single-layer plate theory of Bhimarradi and Stevens [3] where the condition of shear-free plate surfaces is fulfilled.

2 Derivation of a general theory

It is rather straightforward to show that the kinematical assumptions of most refined plate theories can be written by

$$\begin{aligned} \bar{u}_\alpha &= \sum_{q=0}^{K_1} \bar{u}_\alpha^q(X_\alpha)\phi^q(X_3) + \sum_{q=0}^{K_2} \bar{u}_{3,\alpha}^q(X_\alpha)\psi^q(X_3), \\ \bar{u}_3 &= \sum_{q=0}^{K_2} \bar{u}_3^q(X_\alpha)\chi^q(X_3), \end{aligned} \tag{1}$$

where ϕ^q, ψ^q and χ^q describe the distribution of the displacements u_α and u_3 as a function of the plate thickness coordinate X_3 in the reference configuration. If we assume that the displacements are small and that the rotations of the plate normal are moderate, the first Piola-Kirchhoff-tensor Σ_{ij}^I can be approximated by

$$\begin{aligned} \Sigma_{i\alpha}^I &= \Sigma_{i\alpha} + \Sigma_{ik}\frac{\partial u_\alpha}{\partial X_k} \approx \Sigma_{i\alpha}, \\ \Sigma_{i3}^I &= \Sigma_{i3} + \Sigma_{ik}\frac{\partial u_3}{\partial X_k} \approx \Sigma_{i3} + \Sigma_{i\alpha}\frac{\partial u_3}{\partial X_\alpha} \end{aligned}$$

and the Green-Lagrange strain tensor G_{ij} can be approximated by

$$\begin{aligned} G_{\alpha\beta} &= \frac{1}{2}(u_{\alpha,\beta} + u_{\beta,\alpha} + u_{k,\alpha}u_{k,\beta}) \approx \frac{1}{2}(u_{\alpha,\beta} + u_{\beta,\alpha} + u_{3,\alpha}u_{3,\beta}), \\ G_{\alpha 3} &= \frac{1}{2}(u_{\alpha,3} + u_{3,\alpha} + u_{k,\alpha}u_{k,3}) \approx \frac{1}{2}(u_{\alpha,3} + u_{3,\alpha}), \\ G_{33} &= u_{3,3} + \frac{1}{2}u_{k,3}u_{k,3} \approx u_{3,3} \end{aligned}$$

these approximations will be termed the von Kármán stress and strain tensors. In all equations, Einstein's summation convention is applied. Greek indices take the values 1 and 2 and represent the in-plane directions of the plate, whereas lower arabic indices run from 1 to 3.

The equations of equilibrium corresponding to this displacement field can be derived from the principle of virtual displacements in the reference configuration. Defining geometrically linear stress resultants by

$$
\begin{aligned}
R^{\phi^q}_{\alpha\beta} &:= \int_{-h/2}^{h/2} \Sigma_{\alpha\beta}\phi^q \mathrm{d}X_3, \qquad & Q^{\phi^q}_{\alpha} &:= \int_{-h/2}^{h/2} \Sigma_{\alpha 3}\phi^{q\prime} \mathrm{d}X_3,\\
R^{\psi^q}_{\alpha\beta} &:= \int_{-h/2}^{h/2} \Sigma_{\alpha\beta}\psi^q \mathrm{d}X_3, \qquad & Q^{\psi^q}_{\alpha} &:= \int_{-h/2}^{h/2} \Sigma_{\alpha 3}(\psi^{q\prime}+\chi^q) \mathrm{d}X_3,\\
N^{\chi^q} &:= \int_{-h/2}^{h/2} \Sigma_{33}\chi^{q\prime} \mathrm{d}X_3
\end{aligned}
$$

and geometrically non-linear stress resultants by

$$
R^{\chi^q\chi^r}_{\alpha\beta} := \int_{-h/2}^{h/2} \Sigma_{\alpha\beta}\chi^q\chi^r \mathrm{d}X_3, \qquad Q^{\chi^q\chi^r}_{\alpha} := \int_{-h/2}^{h/2} \Sigma_{\alpha 3}\chi^{q\prime}\chi^r \mathrm{d}X_3
$$

leads to the equations of equilibrium in terms of stress resultants

$$
\begin{aligned}
-R^{\phi^q}_{\alpha\beta,\beta} + Q^{\phi^q}_{\alpha} &= F^{\phi^q}_{\alpha} + P^{\phi^q}_{\alpha},\\
R^{\psi^q}_{\alpha\beta,\alpha\beta} - Q^{\psi^q}_{\alpha,\alpha} + N^{\chi^q} &= (R^{\chi^q\chi^r}_{\beta\alpha}\,\bar{u}^r_{3,\alpha})_{,\beta} - Q^{\chi^q\chi^r}_{\alpha}\,\bar{u}^r_{3,\alpha} + F^{\chi^q}_3 - F^{\psi^q}_{\alpha,\alpha}\\
&\quad + P^{\chi^q}_3 - P^{\psi^q}_{\alpha,\alpha}
\end{aligned}
$$

In these equations, $P^{\phi^q}_{\alpha}$, $P^{\psi^q}_{\alpha}$ and $P^{\chi^q}_3$ are resultants of the loads p_i acting onto the top and the bottom surface of the plate

$$
\begin{aligned}
P^{\phi^q}_{\alpha} &:= \left[p_\alpha\phi^q(X_3)\right]^{h/2}_{X_3=-h/2}, \qquad & P^{\psi^q}_{\alpha} &:= \left[p_\alpha\psi^q(X_3)\right]^{h/2}_{X_3=-h/2},\\
P^{\chi^q}_3 &:= \left[p_3\chi^q(X_3)\right]^{h/2}_{X_3=-h/2}
\end{aligned}
$$

$F^{\phi^q}_{\alpha}$, $F^{\psi^q}_{\alpha}$ and $F^{\chi^q}_3$ are resultants of the volume loads ρf_i

$$
F^{\phi^q}_{\alpha} := \int_{-h/2}^{h/2} \rho f_\alpha \phi^q(X_3)\mathrm{d}X_3, \qquad F^{\psi^q}_{\alpha} := \int_{-h/2}^{h/2} \rho f_\alpha \psi^q(X_3)\mathrm{d}X_3,
$$

$$F_3^{\chi^q} := \int_{-h/2}^{h/2} \rho f_3 \chi^q(X_3) \mathrm{d}X_3$$

The natural boundary conditions are described in [8]. The geometrically non-linear resultants can be replaced by their geometrically linear counterparts if the conditions

$$\chi^q \chi^r = \alpha^{qrs} \phi^s + \beta^{qrs} \psi^s, \qquad \chi^{q\prime} \chi^r = \gamma^{qrs} \phi^{s\prime} + \eta^{qrs} (\psi^{s\prime} + \chi^s) \tag{2}$$

are fulfilled. In this case, it can be assured that the system of equations is determined after introducing appropriate constitutive equations.

3 Constitutive hypothesis

For the derivation of plate constitutive equations, a linear relationship between the second Piola-Kirchhoff stress tensor and the Green-Lagrange strain tensor is used

$$\Sigma_{ij} = C_{ijkl} G_{kl},$$

which can for moderate rotations be approximated by the von Kármán strain tensor. To reduce this three-dimensional formula to a two-dimensional plate constitutive equation, is is assumed that the distribution of strains through the plate thickness can be approximately calculated from the displacement field (1). Assuming furthermore that the conditions (2) are fulfilled leads to

$$\begin{aligned} G_{\alpha\beta} &= \epsilon^q_{\alpha\beta} \phi^q + \kappa^q_{\alpha\beta} \psi^q, \\ G_{\alpha 3} &= \frac{1}{2} \left[\bar{u}^q_\alpha \phi^{q\prime} + \bar{u}^q_{3,\alpha} (\psi^{q\prime} + \chi^q) \right], \\ G_{33} &= \bar{u}^q_3 \chi^{q\prime}, \end{aligned}$$

where the abbreviations

$$\begin{aligned} \epsilon^q_{\alpha\beta} &= \frac{1}{2} (\bar{u}^q_{\alpha,\beta} + \bar{u}^q_{\beta,\alpha}) + \alpha^{qrs} \bar{u}^r_{3,\alpha} \bar{u}^s_{3,\beta}, \\ \kappa^q_{\alpha\beta} &= \bar{u}^q_{3,\alpha\beta} + \beta^{qrs} \bar{u}^r_{3,\alpha} \bar{u}^s_{3,\beta} \end{aligned}$$

have been used. Introducing this strain field into the definitions of the stress resultants leads to the constitutive plate equations

$$\begin{aligned} R^{\phi^q}_{\alpha\beta} &= \mathcal{A}^{qr}_{\alpha\beta\gamma\delta} \epsilon^r_{\gamma\delta} + \mathcal{B}^{qr}_{\alpha\beta\gamma\delta} \kappa^r_{\gamma\delta} + \mathcal{C}^{qr}_{\alpha\beta} \bar{u}^r_3, \\ R^{\psi^q}_{\alpha\beta} &= \mathcal{B}^{rq}_{\alpha\beta\gamma\delta} \epsilon^r_{\gamma\delta} + \mathcal{D}^{rq}_{\alpha\beta\gamma\delta} \kappa^r_{\gamma\delta} + \mathcal{E}^{rq}_{\alpha\beta} \bar{u}^r_3, \end{aligned}$$

$$
\begin{aligned}
Q_\alpha^{\phi^q} &= \mathcal{F}_{\alpha\beta}^{qr}\bar{u}_\beta^r + \mathcal{G}_{\alpha\beta}^{qr}\bar{u}_{3,\beta}^r,\\
Q_\alpha^{\psi^q} &= \mathcal{G}_{\alpha\beta}^{rq}\bar{u}_\beta^r + \mathcal{H}_{\alpha\beta}^{rq}\bar{u}_{3,\beta}^r,\\
N^{\chi^q} &= \mathcal{C}_{\gamma\delta}^{rq}\epsilon_{\gamma\delta}^r + \mathcal{E}_{\gamma\delta}^{qr}\kappa_{\gamma\delta}^r + \mathcal{L}^{qr}\bar{u}_3^r
\end{aligned}
$$

with the plate stiffness

$$
\begin{aligned}
\mathcal{A}_{\alpha\beta\gamma\delta}^{qr} &= \int_{-h/2}^{h/2} \phi^q C_{\alpha\beta\gamma\delta}\phi^r \mathrm{d}X_3, & \mathcal{B}_{\alpha\beta\gamma\delta}^{qr} &= \int_{-h/2}^{h/2} \phi^q C_{\alpha\beta\gamma\delta}\psi^r \mathrm{d}X_3,\\
\mathcal{C}_{\alpha\beta}^{qr} &= \int_{-h/2}^{h/2} \phi^q C_{\alpha\beta 33}\chi^{r\prime} \mathrm{d}X_3, & \mathcal{D}_{\alpha\beta\gamma\delta}^{qr} &= \int_{-h/2}^{h/2} \psi^r C_{\alpha\beta\gamma\delta}\psi^q \mathrm{d}X_3,\\
\mathcal{E}_{\alpha\beta}^{qr} &= \int_{-h/2}^{h/2} \chi^{q\prime} C_{\alpha\beta 33}\psi^r \mathrm{d}X_3, & \mathcal{F}_{\alpha\beta}^{qr} &= \int_{-h/2}^{h/2} \phi^{q\prime} C_{\alpha 3\beta 3}\phi^{r\prime} \mathrm{d}X_3,\\
\mathcal{G}_{\alpha\beta}^{qr} &= \int_{-h/2}^{h/2} \phi^{q\prime} C_{\alpha 3\beta 3}(\psi^{r\prime} + \chi^r) \mathrm{d}X_3, & \mathcal{L}^{qr} &= \int_{-h/2}^{h/2} \chi^{q\prime} C_{3333}\chi^{r\prime} \mathrm{d}X_3,\\
\mathcal{H}_{\alpha\beta}^{qr} &= \int_{-h/2}^{h/2} (\psi^{q\prime} + \chi^q) C_{\alpha 3\beta 3}(\psi^{r\prime} + \chi^r)) \mathrm{d}X_3 & &
\end{aligned}
$$

The equilibrium of stresses leads to the field equations

$$
\begin{aligned}
-\mathcal{A}_{\alpha\beta\gamma\delta}^{qr}\epsilon_{\gamma\delta,\beta}^r - \mathcal{B}_{\alpha\beta\gamma\delta}^{qr}\kappa_{\gamma\delta,\beta}^r - \left(\mathcal{C}_{\alpha\beta}^{qr} - \mathcal{G}_{\alpha\beta}^{qr}\right)\bar{u}_{3,\beta}^r + \mathcal{F}_{\alpha\beta}^{qr}\bar{u}_\beta^r &= F_\alpha^{\phi^q} + P_\alpha^{\phi^q},\\
\mathcal{B}_{\alpha\beta\gamma\delta}^{rq}\epsilon_{\gamma\delta,\alpha\beta}^r + \mathcal{D}_{\alpha\beta\gamma\delta}^{rq}\kappa_{\gamma\delta,\alpha\beta}^r + \left(2\mathcal{E}_{\alpha\beta}^{rq} - \mathcal{H}_{\alpha\beta}^{rq}\right)\kappa_{\alpha\beta}^r + \left(\mathcal{C}_{\alpha\beta}^{rq} - \mathcal{G}_{\alpha\beta}^{rq}\right)\epsilon_{\alpha\beta}^r,\\
+\mathcal{L}^{qr}\bar{u}_3^r = (R_{\beta\alpha}^{\chi^q\chi^r}\bar{u}_{3,\alpha}^r)_{,\beta} - Q_\alpha^{\chi^q\chi^r}\bar{u}_{3,\alpha}^r + F_3^{\chi^q} - F_{\alpha,\alpha}^{\psi^q} + P_3^{\chi^q} - P_{\alpha,\alpha}^{\psi^q}
\end{aligned}
$$

4 Numerical assessment with reference solutions

Plate theories derived by reduction techniques are obviously approximations of the underlying three-dimensional continuum mechanics, since in the course of their derivation, series expansions or hypotheses are introduced or higher order terms are neglected. Plate theories involving a Cosserat surface seem at the first sight free of any assumption or approximation, since their derivation follows a purely two-dimensional approach. It is however impossible to identify the coefficients of the two-dimensional constitutive equations of Cosserat plate theories, as long as the additional degrees of freedom have not been given a certain physical meaning. Since at that stage assumptions have to be introduced, Cosserat plate theories become approximations of three-dimensional continuum mechanics as soon as they are used to solve any plate problem.

frag replacements

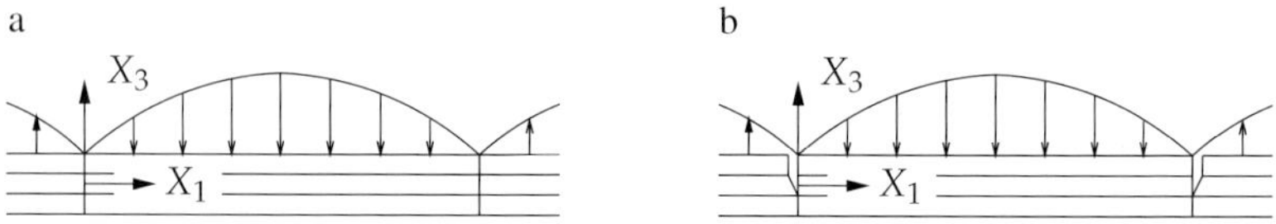

Fig. 1. Simply supported boundary conditions. a Antiperiodic boundary conditions. b fractured boundary

The accuracy of classical and refined plate theories is often assessed by comparing their results with analytical reference solutions, such as Pagano's solution for the simply supported plate strip [4]. Koiter and Simmonds [5] have shown that the global error of classical plate theories is of the order of $(h/L)^2$ for regular boundary conditions, which do not involve a boundary layer. For irregular boundary conditions, the global error increases to the order of (h/l) if the edge disturbances are of the same order of magnitude as the regular stresses and if the length of the boundary layer is approximately equal to the plate thickness. Any assessment should therefore be performed with classes of statically equivalent reference solutions. Such a class of solutions can be constructed from the two different boundary conditions of a three-ply plate strip shown in Fig. 1, which are for a first order plate theory both simply supported. An analytical solution of this problem can be found by using a modified Fourier transformation technique [6, 7].

The transverse shear stress in the middle of the top layer as a function of the X_1-coordinate for the three-ply plate of Fig. 1 is shown in Fig. 2 for different values of E_l/E_T. In this example, all three plies are transversely isotropic, with the axis of symmetry parallel to the X_1-axis. All stiffnesses of the middle layer are of a factor of 10 smaller than the stiffnesses of the outer layers, E_T is equal to G_{LT}, and ν_{LT} and ν_{TT} take the values of 0.25. For comparison, the results for an isotropic plate are also shown.

It turns out that the stresses inside the boundary layer (boundary condition b in Fig. 1) differ significantly from the stresses of the antiperiodic case (boundary condition a in Fig. 1 and dashed line in Fig. 2) and that the size of the boundary layer increases with increasing values of E_L/E_T. To describe this behaviour, refined plate theories with more than five unknowns are required. Simplified plate theories have got the same number of unknowns as classical plate theories and are therefore not able to describe this boundary layer. They are however well suited for the interior of the plate where the boundary layer stresses have vanished, since they need less computational effort.

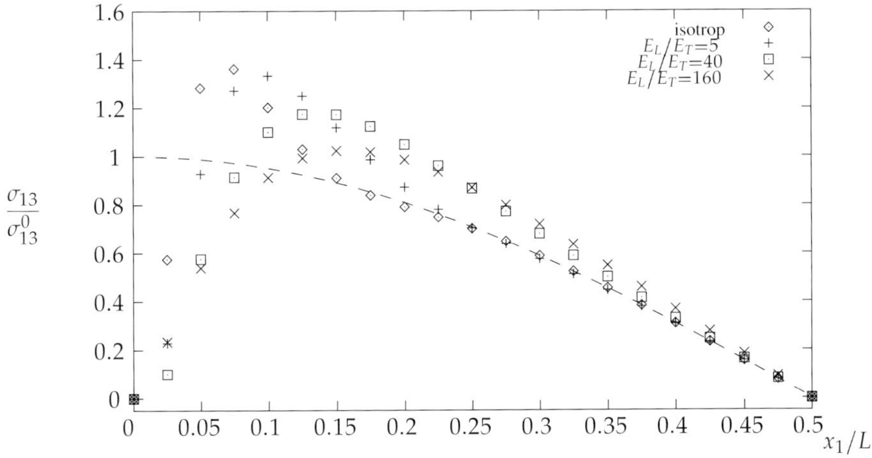

Fig. 2. Decay of boundary disturbances for different values of E_L/E_T

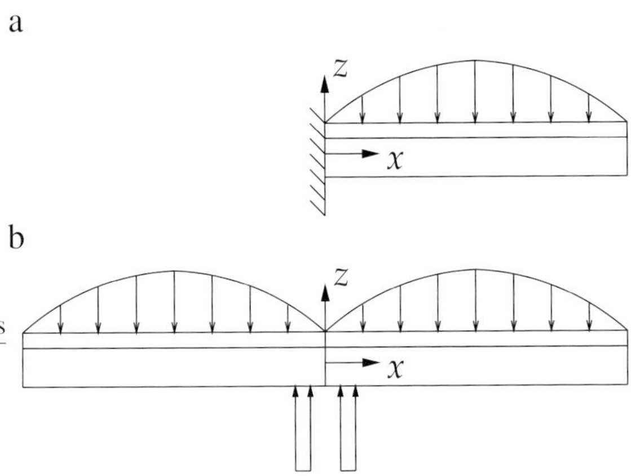

PSfrag replacements

Fig. 3. Clamped plate. a Plate problem. b Reference solution

For the two-ply plate clamped at $X_1 = 0$ and free at $X_1 = 4L$ shown in Fig. 3a [9], the distribution of the shear stresses over the plate thickness for different values of X_1/L are shown in Fig. 4. As a reference solution, the plate strip shown in Fig. 3b with periodic boundary conditions was used, which leads to similar results except in a region close to $X_1 = 0$.

The displacement field of the refined plate theory of Soldatos and Watson [9] contains six unknowns and has been adapted to yield the same results as Pagano's solution for the antiperiodic plate strip. The transverse displacements on top and bottom of the plate coincide very well with the results of the reference solution, but for an accurate description of the shear stresses, additional unknowns are required (see Fig. 4).

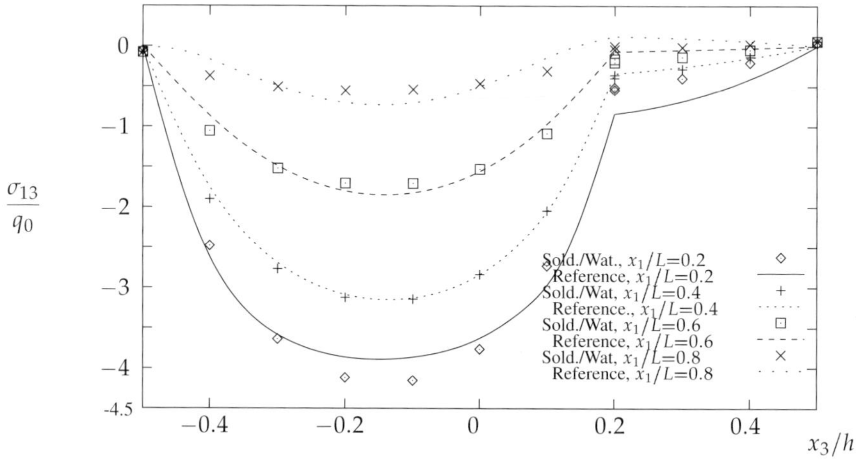

Fig. 4. Distribution of transverse shear stresses σ_{13}

References

1. Noor AK, Burton WS (1989) Assessment of shear deformation theories for multilayered composite plates. Applied Mechanics Reviews 42(1):1-13
2. di Sciuva M (1984) Bending, vibration and buckling of simply supported thick multilayered anisotropic plates. Meccanica dei solidi 118:279-295
3. Bhimarradi A, Stevens LK (1984) A higher order theory for the free vibration of orthotropic, homogeneous and laminated rectangular plates. Journal of Applied Mechanics 51:195-198
4. Pagano NJ (1969) Exact solutions for composite laminates in cylindrical bending. Journal of Composite Materials 3:398–411
5. Koiter WT, Simmonds JG (1972) Foundations of Shell Theory. WTHD 40, Technische Hoogeschool Delft
6. Meenen J, Altenbach H (1998) Statically equivalent solutions for the assessment of refined plate theories. Mech Comp Mat 34(4):461-476
7. Meenen J (2000) A Reference solution for the numerical assessment of refined plate theories including boundary layers and edge disturbances. Technische Mechanik 20(3):275-282
8. Meenen J, Altenbach H (2000) A consistent deduction of von Kármán-type plate theories from threedimensional non-linear continuum mechanics. Acta Mechanica 147:1-17
9. Soldatos KP, Watson P (1997) Accurate stress analysis of laminated plates combining a two-dimensional theory with the exact three-dimensional solution for simply supported edges. Mathematics and Mechanics of Solids 2(4):459-489

Stochastic Finite Element Analysis of Plates. Theory and Application.

M. Mestrovic

Faculty of Civil Engineering, University of Zagreb, Croatia

1 Introduction

The assumption that structures have deterministic material properties, is implicitly involved in the most calculation of standard finite element analysis of the structures. The material properties of real structures have uncertainties, which have to be considered in the structural analysis. The uncertainties of the structure are than considered through the increase of the safety factors using deterministic analysis.

The stochastic analysis is applied on the problem of the bending of the thin plate with random material properties. The stochastic analysis of the plates is concentrated on the influence of the material uncertainties on the response deflection. The elastic modulus is assumed to constitute homogenous, two-dimensional stochastic field. It means that the response deflection is than also stochastic field. It leads to stochastic differential operator, bilinear form in the variational formulation of thin plate bending problem. The bilinear form is continuous and coercive for the each random value of elastic modulus. According Lax-Milgram theorem, [1], the unique solution of the variational problem for the response deflection exists for the each realization of input stochastic field, [4].

The stochastic finite element formulation is used to find the influence of the randomness of elastic modulus on the randomness of the response deflection. The randomness of input quantity is involved in the finite element analysis by local average method, [5]. The stochastic input field is approximated with only one random variable per each finite element. The stochastic stiffness matrix is expressed as sum of deterministic and stochastic part. The response variability is calculated using the first-order Taylor expansion approximation of the variability response function, [2].

2 Variability of input quantities

We consider a plate with a spatially varying elastic modulus $E(x,y)$. Elastic modulus is assumed to constitute a homogenous two-dimensional random field in following form

$$E^{(e)}(\xi,\eta) = E_0\left[1 + r^{(e)}(\xi,\eta)\right] \quad , \tag{1}$$

where E_0 is expectation of elastic modulus and $r^{(e)}(\xi,\eta)$ is homogenous two-dimensional random field with expectation zero. This random field is represented with its variance σ_{rr} and autocorrelation function

$$R_{rr}(\zeta) = \mathrm{E}\left[r(\mathbf{x}+\zeta)r(\mathbf{x})\right] \quad . \tag{2}$$

3 Finite element formulation

Standard finite element formulation of thin plate bending problem is

$$\mathbf{Kw} = \mathbf{q} \quad . \tag{3}$$

Involving the randomness, the formulation for stochastic analysis is

$$(\mathbf{K}_0 + \Delta\mathbf{K})\mathbf{w} = \mathbf{q} \quad , \tag{4}$$

where $\mathbf{K}_0$ is deterministic stiffness matrix and $\Delta\mathbf{K}$ is stochastic part of stiffness matrix. The displacement vector is than approximated, with assumption that variance is sufficiently small , with

$$w = \sum_{i=1}^{\infty}\left(-\mathbf{K}_0^{-1}\Delta\mathbf{K}\right)^i \mathbf{K}_0^{-1}\mathbf{q} \approx \left(\mathbf{I} - \mathbf{K}_0^{-1}\Delta\mathbf{K}\right)\mathbf{w}_0 \quad . \tag{5}$$

The rectangular bicubic finite element will be used to derive the stochastic stiffness matrix. This element has 4 degrees of freedom per each node $\left(w, w_x, w_y, w_{xy}\right)$ for a total of 16 degrees of freedom per element. The element stiffness matrix for the bending of the thin plate is given by

$$\mathbf{K}^{(e)} = \int_{(e)}\left(\mathbf{B}^{(e)}\right)^{\mathrm{T}} \mathbf{D}^{(e)}\mathbf{B}^{(e)} d\xi d\eta \quad , \tag{6}$$

where $D^{(e)}$ is the constitutive matrix

$$\mathbf{D}^{(e)} = \frac{E^{(e)}t^3}{12(1-\nu^2)}\begin{bmatrix} 1 & \nu & 0 \\ \nu & 1 & \\ 0 & 0 & \dfrac{1-\nu}{2} \end{bmatrix} . \tag{7}$$

The constitutive matrix is also stochastic field which can be written as

$$\mathbf{D}^{(e)} = \mathbf{D}_0^{(e)}\left[1 + r^{(e)}(x)\right] . \tag{8}$$

4 Local average method

Local average method is used to represent the stochastic field approximation by only one random variable per each finite element. The stochastic part of element stiffness matrix is expressed as

$$\Delta\mathbf{K}^{(e)} = \alpha^{(e)}\mathbf{K}_0^{(e)} , \tag{9}$$

where the local average $\alpha^{(e)}$ is defined with

$$\begin{aligned} \alpha^{(e)} &= \frac{1}{L_x^{(e)}L_y^{(e)}}\int_0^{L_x^{(e)}}\int_0^{L_y^{(e)}} r^{(e)}\left(\xi_e,\eta_e\right)d\xi_e d\eta_e \\ &= \frac{1}{L_x^{(e)}L_y^{(e)}} X_{00}^{(e)} . \end{aligned} \tag{10}$$

The response vector is approximated by its first-order expansion around the mean value of random integral in local average

$$\begin{aligned} \mathbf{w} &= \mathbf{w}_0 + \sum_{(e)=1}^{N^{(e)}} X_{00}^{(e)}\left[\frac{\partial\mathbf{w}}{\partial X_{00}^{(e)}}\right]_{\mathrm{E}} \\ &= \mathbf{w}_0 - \sum_{(e)=1}^{N^{(e)}} \mathbf{K}_0^{-1}\left[\frac{\partial\mathbf{w}}{\partial X_{00}^{(e)}}\right]_{\mathrm{E}} \mathbf{w}_0 X_{00}^{(e)} \\ &= \mathbf{w}_0 - \sum_{(e)=1}^{N^{(e)}} \frac{1}{L_x^{(e)}L_y^{(e)}}\mathbf{K}_0^{-1}\mathbf{K}_0^{(e)}\mathbf{w}_0 X_{00}^{(e)} . \end{aligned} \tag{11}$$

Using the first-order approximation, we can evaluate the first-order approximation of covariance matrix, [2,3]

$$\mathrm{Cov}\left[\mathbf{w},\mathbf{w}\right] = \sum_{(e),(f)=1}^{N^{(e)}} \frac{1}{L_x^{(e)}L_y^{(e)}L_x^{(f)}L_y^{(f)}}\mathbf{K}_0^{-1}\mathbf{K}_0^{(e)}\mathbf{w}_0\mathbf{w}_0^{\mathrm{T}}\left(\mathbf{K}_0^{(f)}\right)^{\mathrm{T}} \cdot \mathrm{E}\left[X_{00}^{(e)}X_{00}^{(f)}\right] \tag{12}$$

Taking into account Wiener-Khintchine transformation, [5], between autocorrelation function and spectral-density function $S_{rr}(\kappa_x, \kappa_y)$, variance of the response vector can be written as

$$\operatorname{var}[\mathbf{w}] = \int_{-\infty}^{\infty} \int_{-\infty}^{\infty} S_{rr}(\kappa_x, \kappa_y) \mathrm{VRF}_{\mathrm{LA}}(\kappa_x, \kappa_y) d\kappa_x d\kappa_y \quad , \tag{13}$$

where the vector $\mathrm{VRF}_{\mathrm{LA}}(\kappa_x, \kappa_y)$ is the first-order approximation of the variability response function expressed as

$$\begin{aligned} &\mathrm{VRF}_{\mathrm{LA}}(\kappa_x, \kappa_y) = \\ &\sum_{(e)} \sum_{(f)} \frac{1}{L_x^{(e)} L_y^{(e)} L_x^{(f)} L_y^{(f)}} \operatorname{diag}\left(\mathbf{K}_0^{-1} \Delta \mathbf{K}_0^{(e)} \mathbf{w}_0\right) \mathbf{K}_0^{-1} \Delta \mathbf{K}_0^{(f)} \mathbf{w}_0 \cdot \\ &\cdot \Big[\left(CI^{(e)} CI^{(f)} + SI^{(e)} SI^{(f)} \right) \cos\left(\Delta x_{f_e} \kappa_x + \Delta y_{f_e} \kappa_y \right) - \\ &- \left(SI^{(e)} CI^{(f)} - CI^{(e)} SI^{(f)} \right) \sin\left(\Delta x_{f_e} \kappa_x + \Delta y_{f_e} \kappa_y \right) \Big] \quad , \end{aligned} \tag{14}$$

where Δx_{f_e} and Δy_{f_e} are coordinate differences between centroids of finite elements and

$$CI^{(e)} = \int_0^{L_x^{(e)}} \cos(\kappa_x \xi_e + \kappa_y \eta_e) d\xi_e d\eta_e \tag{15}$$

$$SI^{(e)} = \int_0^{L_x^{(e)}} \sin(\kappa_x \xi_e + \kappa_y \eta_e) d\xi_e d\eta_e \quad . \tag{16}$$

For specific degree of freedom w_i, the variance is bounded

$$\operatorname{var}[w_i] \le \sigma_{rr}^2 \, \mathrm{VRF}_{\mathrm{LA}}^i \left(\kappa_x^*, \kappa_y^* \right) \quad , \tag{17}$$

where (κ_x^*, κ_y^*) is the point at which the variability response function takes its maximum value.

5 Mathematical formulation

Let (Ω_ω, F, p) be a probability space, where Ω_ω is the set of outcomes, the values of flexural rigidity of plates F $\sigma-$algebra of events and $p: F \to [0,1]$ is a probability measure. Consider stochastic boundary value problem: for given $q \in L^2(\Omega)$, find $w: \Omega \to \mathbf{R}$, such that

$$\begin{cases} \bar{D}\Delta w) = q \ \text{in}\, \Omega \\ \quad w = w_0 \ \text{on}\, \partial\Omega \end{cases} , \tag{18}$$

where Ω is bounded domain in $\mathbf{R}^2$ and $\bar{D}$ real stochastic function. Let v is bounded stochastic function in $L^2(\Omega\times\Omega_\omega)$

$$v:\Omega\times\Omega_\omega\to\mathbf{R},\quad \|v\|<\infty\quad, \tag{19}$$

with norm

$$\|v\|^2=\mathrm{E}\left[\int_\Omega v^2(x,y,\cdot)dxdy\right] \tag{20}$$

For stochastic function $\bar{D}$, we will assume that $\bar{D}$ is bounded and strictly positive, $\exists\alpha_1,\alpha_2>0$ such that

$$0<\alpha_1\le\bar{D}(x,y)\equiv D(x,y,\omega)\le\alpha_2<\infty\quad. \tag{21}$$

In order to rewrite boundary problem in variational form, let us consider the space

$$V=\left\{w:\Omega\times\Omega_\omega\to\mathbf{R}:\|w\|v<\infty,\quad w\Big|_{\partial\Omega\times\Omega_\omega}=w_0\right\}\quad. \tag{22}$$

The function $D(x,y,\omega)$ is bounded, what leads to well defined norm $\|\cdot\|_V$ with

$$\|w\|_V^2=E\left[\int_\Omega D(x,y,\omega)\left[(1-\nu)\left|\nabla(\nabla w)\right|^2+\nu\left|\Delta w\right|^2\right]dxdy\right]\quad, \tag{23}$$

where all needed differentiation are with respect only to plane coordinates x, y. The inner product that induces this norm is given by the bilinear form $a:V\times V\to\mathbf{R}$,

$$\begin{aligned}&a_\omega\left(w,w'\right)=\mathrm{E}\cdot\\&\cdot\left[\int_\Omega D(x,y,\omega)\left[(1-\nu)\nabla\left(\nabla w\left(x,y,\omega\right)\right)\cdot\nabla\left(\nabla w'\left(x,y,\omega\right)\right)+\right.\right.\\&\left.\left.+\nu\Delta w\left(x,y,\omega\right)\Delta w'\left(x,y,\omega\right)\right]dxdy\right]\;.\end{aligned} \tag{24}$$

Obviously, V is Hilbert space of stochastic function. Bilinear form $a_\omega(w,w')$ is continuous and coercive

$$\begin{aligned}&\left|a_\omega\left(w,w'\right)\right|\le\|w\|_V\,\|w'\|_V\\&\forall w,w'\in V\end{aligned} \tag{25}$$

$$\begin{aligned}&\|w\|_V^2\le a_\omega\left(w,w\right)\\&\forall w\in V\end{aligned}\quad. \tag{26}$$

For every $q\in L^2(\Omega)$, there exists a constant C such that

$$\left| \mathrm{E}\left[\int_{\Omega} qwd\Omega \right] \right| \leq C \|q\|_{L^2(\Omega)} \|w\|_V \quad . \tag{27}$$

Then, in view of the Lax-Milgram theorem, for every $q \in L^2(\Omega)$, there exists a unique weak solution $w \in V$ of stochastic boundary problem satisfying

$$\begin{gathered} a_{\omega}(w, w') = \mathrm{E}\left[\int_{\Omega} qwd\Omega \right] \quad . \\ \forall w' \in V \end{gathered} \tag{28}$$

6 Numerical examples

The first example is taken same as in [2], where the numerical calculation was provided with different finite element (12 degrees of freedom). The simply supported square plate, divided to 64 finite elements, is considered under antisymmetric loading (Fig. 1.).

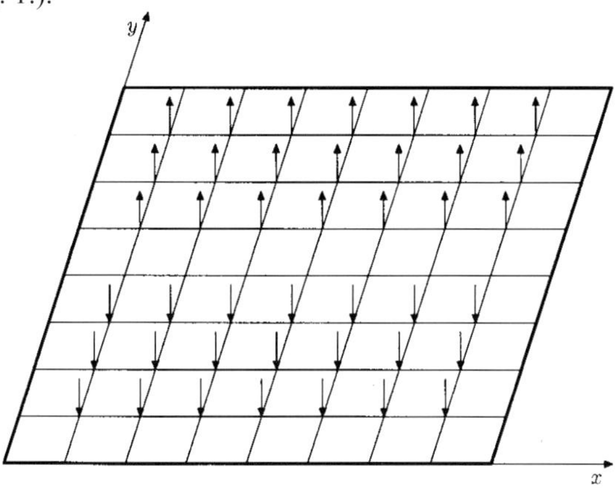

Fig. 1 Square plate under antisymmetric loading

The coefficient of variation is calculated for point $A = (L/4, L/4)$ with fixed value κ_x as a function of κ_y, (Fig. 2.).

The spectral-distribution-free upper bounds is computed larger than coefficient of variation of input elastic modulus as follows

$$\begin{gathered} \mathrm{COV}\left[w_{\mathrm{A}} \right] \leq 1.2\sigma_{ff} \\ \kappa^{*}_{x,\mathrm{A}} = 0.0 \\ \kappa^{*}_{x,\mathrm{A}} = \pm 0.39 \end{gathered} \tag{29}$$

Second example is taken to show the influence of boundary conditions on the variability offresponse deflection. The plate is with same loading and geometry as in first example. The numerical calculation is provided with all clamped edges and graphically compared with solution for simply supported plate in first example, (Fig. 3).

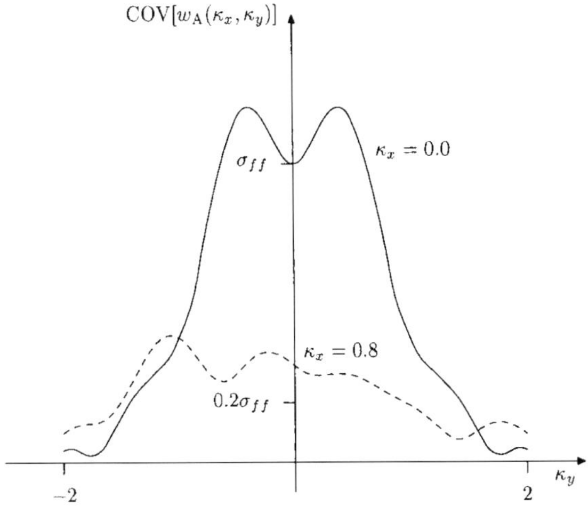

Fig. 2. Coefficient of variation as function of κ_y for $\kappa_x = 0.0$ and $\kappa_x = 0.8$

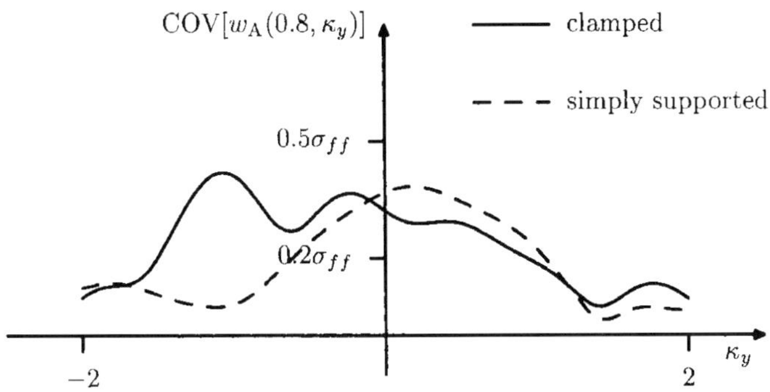

Fig. 3. Coefficient of variation as function of κ_y for $\kappa_x = 0.8$

Third example is unit simply supported square plate under uniform loading calculated with different number of finite elements, (Fig. 4.). Coefficient of variation is calculated for the deflection at the center of the plate and graphically plotted for value $\kappa_x = 0.0$ as function of κ_y. The sequences of numerical results, calculated by different number of finite elements, show the quadratic convergence of the coefficient of variation evaluated by stochastic finite element analysis.

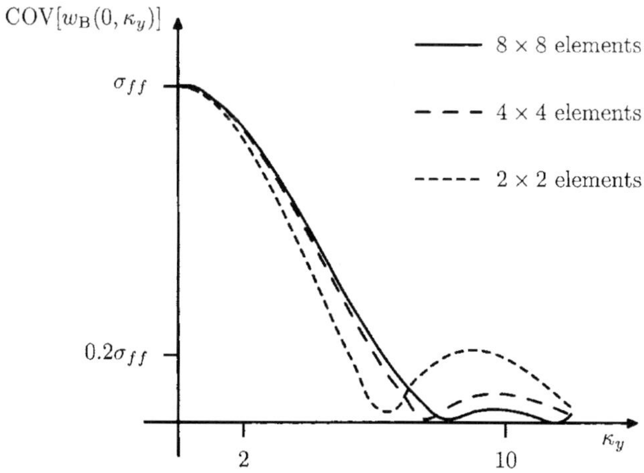

Fig. 4. Coefficient of variation as function of mesh size and κ_y for $\kappa_x = 0.0$

7 Conclusions

The concept of variability response function based on the local average method was extended to the thin plate bending problem. The spectral-distribution-free upper bounds of response variability has been computed. The quadratic convergence of the sequences of the numerical results calculated by increasing the number of finite elements has been shown. The coefficient of variation of the response deflection can become larger than the coefficient of variation of input random material property. The mathematical formulation of stochastic boundary value problem of the bending of thin plate is given. The existence and uniqueness of weak solution has been proved according the Lax-Milgram theorem.

References

1. Brenner SC, Scott LR (1996) The mathematical theory of finite element methods, Springer
2. Graham L, Deodatis G (1998) Structural safety, Vol. 20, pp. 167–188
3. Mestrovic M (2001) Structural safety and reliability, proceedings of eighth international conference, ICOSSAR'01, Corotis et al (Eds.), Swets & Zeitlinger
4. Stoepler K (1998) Stochastische Finite Elemente zur Lösung von Differentialgleichungen mit zufälligen Koeffizienten, Herbert Utz Verlag
5. VanMarcke E (1983) Random fields: analysis and synthesis, MIT Press, Web Edition by rare book services, Princeton University, (1998)

On Nonlinear Theory of Rigid-Flexible Shells Without the Kirchhoff Hypotheses

E. I. Mikhailovskii and A. V. Yermolenko

Syktyvkar State University, Syktyvkar, Russia

As a development and a specification of some variants of nonlinear shell theory suggested in [1,3-5,7,9,11], we deduce the equations of shell mechanics without using the Kirchhoff hypotheses. The longitudinal deformation of normal fibres is taken into account without increasing the order of differential equilibrium equations. The transversal shears are approximated according to a linear law. Using the generalized forces and moments, the equilibrium equations are reduced to a canonical form, i.e. as they look in linear theory. It is shown that a nonlinear boundary problem of shell mechanics if the boundary conditions are expressed exclusively in terms of the displacements, is generally incorrect in sense of Lagrange principle,

1. Suppose that the radius-vector

$$\overset{\circ}{\mathbf{R}}(\alpha,\xi)=\overset{\circ}{\mathbf{r}}(\alpha)+\xi\,\overset{\circ}{\mathbf{n}}(\alpha),\ \alpha=(\alpha^1,\alpha^2)\in\overset{\circ}{\Omega},\ \xi\in[-\tfrac{1}{2}\overset{\circ}{h},\tfrac{1}{2}\overset{\circ}{h}]\,, \tag{1.1}$$

after deformation of the shell turns into

$$\mathbf{R}(\alpha,\xi)=\mathbf{r}(\alpha)+\lambda_\zeta(\alpha)[\xi+\tfrac{1}{2}\xi^2\kappa_\zeta(\alpha)]\mathbf{n}(\alpha)+\xi\psi_\beta(\alpha)\mathbf{r}^\beta(\alpha)\ , \tag{1.2}$$

where $\mathbf{r}=\overset{\circ}{\mathbf{r}}+\mathbf{u}$, $\mathbf{u}=u^\alpha\,\overset{\circ}{\mathbf{r}}_\alpha+w\overset{\circ}{\mathbf{n}}$; the quantities characterizing the initial (undeformed) configuration are denoted by "$\circ$".

The parameters λ_ξ,κ_ξ describe longitudinal deformation of transverse fibres, $\lambda_\xi=h/\overset{\circ}{h}$. The functions $\psi^i(\alpha),\ i=1,2$ are transversal shears.

In addition to (1.1) – (1.2), the following geometric assumptions will be used:

1. the curvilinear coordinates α^1,α^2 are orthogonal in initial configuration of the middle surface, i.e. $\overset{\circ}{a}_{12}=\overset{\circ}{\mathbf{r}}_1\cdot\overset{\circ}{\mathbf{r}}_2=0$;
2. the shell is thin, i.e.

$$\xi b_{<ij>}\equiv\xi b_{ij}/\sqrt{\overset{\circ}{a}_{ii}\overset{\circ}{a}_{jj}}<<1\,, \tag{1.3}$$

($b_{<ij>}$ are physical components of the curvature tensor of the middle surface);

3. the components of the Green-Lagrange strain tensor depend linearly on shell thickness;
4. the transversal shears are approximated according to linear law;

5. the convertibility of the function $\lambda_\xi(\alpha)$ can be ignored.

The assumptions (3) – (5) and (1.3) are natural for rigid-flexible shells, i.e. shells made of hard compressible material (steel, for example). Rigid-flexible shells work under large displacements and large angles of rotation, but with small deformations. Another type of shells is soft-flexible shells, i.e. shells made of soft incompressible material (elastic, for example). Large displacements of soft-flexible shells are related to both large angles of rotation and large elastic deformations.

Using (2) – (5), the components of the Green-Lagrange strain tensor $\mathring{\mathbf{\Gamma}} = \gamma_{ij}^{\xi}\, \mathring{\mathbf{R}}^i \otimes \mathring{\mathbf{R}}^j$ can be obtained by formulas

$$\gamma_{ij}^{\xi} = \gamma_{ij} + \xi(\kappa_{ij} + \mu_{ij}), \quad \gamma_{i3}^{\xi} = \frac{1}{2}\psi_i, \quad \gamma_{33}^{\xi} = \frac{1}{2}(\lambda_\xi - 1) + \xi\lambda_\xi^2\kappa_\xi, \tag{1.4}$$

where

$$\gamma_{ij} = \frac{1}{2}(a_{ij} - \mathring{a}_{ij}), \quad \kappa_{ij} = -\lambda_\xi b_{ij} + \mathring{b}_{ij}, \quad \mu_{ij} = \frac{1}{2}(\nabla_i\psi_j + \nabla_j\psi_i), \tag{1.5}$$

(a pair of roman subscript and the same superscript means summing from 1 to 3, a pair of greek subscript and superscript – summing from 1 to 2).

To describe the stressed-deformed state, we use the elastic potential of the second order standard material (STM-2). For STM-2, the law of elasticity has the form [2]

$$\{\mathbf{F}^{-1} \cdot J\mathbf{\Sigma} \cdot \mathbf{F}^{-1*}\} = 2\mu\mathring{\mathbf{\Gamma}} + \lambda I_{\mathring{\Gamma}}\mathbf{1},$$

($\{\mathbf{F}^{-1} \cdot J\mathbf{\Sigma} \cdot \mathbf{F}^{-1*}\} = J\sigma^{kl}\, \mathring{\mathbf{R}}_k \otimes \mathring{\mathbf{R}}_l$ is the Piola-Kirchhoff stress tensor; $\mathbf{F}$ is the double gradient tensor of motion [10]; $\mathbf{\Sigma}$ is the Cauchy tensor of true stresses; J is Jacobian of the transformation from Lagrangian to Eulerian coordinates; $I_{\mathring{\Gamma}}$ is the first main invariant of the tensor $\mathring{\mathbf{\Gamma}}$; $\mathbf{1}$ is the unit tensor; λ, μ are the Lame elastic constants; "$\cdot$" is the scalar product) or (in terms of corresponding tensors)

$$J\sigma^{ij} = (\lambda\, \mathring{g}^{ij}\mathring{g}^{kl} + 2\mu\, \mathring{g}^{ik}\mathring{g}^{jl})\gamma_{kl}^{\xi}. \tag{1.6}$$

Using relation (1.3), Eq. (1.6) can be written in the form

$$J\sigma^{ij} = (\lambda\, \mathring{a}^{ij}\mathring{a}^{\alpha\beta} + 2\mu\, \mathring{a}^{i\alpha}\mathring{a}^{j\beta})\gamma_{\alpha\beta}^{\xi} + \lambda\, \mathring{a}^{ij}\gamma_{33}^{\xi},$$

$$J\sigma^{33} = \lambda\, \mathring{a}^{\alpha\beta}\gamma_{\alpha\beta}^{\xi} + (\lambda + 2\mu)\gamma_{33}^{\xi}, \quad J\sigma^{i3} = 2\mu\, \mathring{a}^{i\beta}\gamma_{\beta 3}^{\xi}. \tag{1.7}$$

The parameters λ_ξ, κ_ξ are defined from the boundary conditions on the shell surfaces

$$J\sigma^{33}(\tfrac{1}{2}\mathring{h}) = q_n^{+}, \quad J\sigma^{33}(-\tfrac{1}{2}\mathring{h}) = q_n^{-}.$$

After same easy transformations, we obtain

$$\lambda_\xi^2 = 1 - \frac{2\lambda}{\lambda+2\mu}\mathring{a}^{\alpha\beta}\gamma_{\alpha\beta} + \frac{2m_n}{(\lambda+2\mu)\mathring{h}},$$

$$\lambda_\xi^2\kappa_\xi = -\frac{\lambda}{\lambda+2\mu}\mathring{a}^{\alpha\beta}(\kappa_{\alpha\beta}+\mu_{\alpha\beta}) + \frac{q_n}{(\lambda+2\mu)\mathring{h}}, \tag{1.8}$$

where

$$q_n = q_n^+ - q_n^-, \quad m_n = \frac{1}{2}(q_n^+ + q_n^-)\mathring{h}. \tag{1.9}$$

Using $(1.4)_3$, $(1.7)_2$ and (1.8), we get

$$\gamma_{33}^\xi = -\frac{\lambda}{\lambda+2\mu}\mathring{a}^{\alpha\beta}\gamma_{\alpha\beta}^\xi + \frac{m_n+\xi q_n}{(\lambda+2\mu)\mathring{h}}, \qquad J\sigma^{33} = \frac{1}{\mathring{h}}(m_n+\xi q_n). \tag{1.10}$$

Forces and moments are defined by the following relations: (ν is the Poisson ratio).

$$S^{ij} = \int_{-\mathring{h}/2}^{\mathring{h}/2}(J\sigma^{ij} - \frac{\nu}{1-\nu}\mathring{a}^{ij}J\sigma^{33})d\xi, \quad T_{\cdot n}^i = \lambda_\xi\int_{-\mathring{h}/2}^{\mathring{h}/2}J\sigma^{i3}d\xi,$$

$$M^{ij} = \lambda_\xi\int_{-\mathring{h}/2}^{\mathring{h}/2}(J\sigma^{ij} - \frac{\nu}{1-\nu}\mathring{a}^{ij}J\sigma^{33})\xi d\xi. \tag{1.11}$$

Taking into account the Eqs (1.7), (1.10) and (1.11), the determinative equations of elasticity of rigid-flexible shell can be written in the form

$$S^{ij} = CA^{ij,\alpha\beta}\gamma_{\alpha\beta}, \quad T_{\cdot n}^i = \lambda_\xi\mu\mathring{h}\mathring{a}^{i\beta}\psi_\beta, \tag{1.12}$$

$$M^{ij} = M_K^{ij} + M_T^{ij}, \quad M_K^{ij} = \lambda_\xi DA^{ij,\alpha\beta}\kappa_{\alpha\beta}, \quad M_T^{ij} = \lambda_\xi DA^{ij,\alpha\beta}\mu_{\alpha\beta},$$

where

$$A^{ij,\alpha\beta} = \mathring{a}^{i\alpha}\mathring{a}^{j\beta} + \nu\mathring{c}^{i\alpha}\mathring{c}^{j\beta}, \quad C = \frac{E\mathring{h}}{1-\nu^2}, \quad D = \frac{E\mathring{h}^3}{12(1-\nu^2)},$$

E is the Young's modulus, $\mathring{c}^{ij}$ are components of the discriminant tensor of the middle surface.

Assuming for simplicity that the load on side surface of shell is absent, the Lagrange variational equation can be written in the form

$$\delta U = A(\delta\mathbf{R}), \tag{1.13}$$

where

$$\delta U = \int_{\mathring{\Omega}}(\int_{-\mathring{h}/2}^{\mathring{h}/2}\delta\Phi d\xi)d\mathring{\Omega},$$

$$A(\delta\mathbf{R}) = \int_\Omega(\mathbf{q}^+\cdot\delta\mathbf{R}^+ - \mathbf{q}^-\cdot\delta\mathbf{R}^-)d\Omega,$$

$$\delta\Phi = \{\mathbf{F}^{-1}\cdot J\mathbf{\Sigma}\cdot\mathbf{F}^{-1*}\} : \delta\mathring{\mathbf{\Gamma}} = J\sigma^{\alpha\beta}\delta\gamma_{\alpha\beta} + 2J\sigma^{\alpha3}\delta\gamma_{\alpha3} + J\sigma^{33}\delta\gamma_{33}, \tag{1.14}$$

(":" is the double scalar product).

Note that Eq. (1.13) is not a variational principle, because the expression $A(\delta\mathbf{R})$ is generally not a variation of any functional. This is connected mainly with dependence of load on displacement. If $A(\delta\mathbf{R}) = dV$, then the load is *conservative* and the functional V is *a potential of surface load.* In particular, the weight load is conservative. It can be shown that evenly divided following pressure is conservative under some additional conditions.

2. Now we give, omitting some details, the deduction of equilibrium equations. Using (1.4), (1.10) and (1.11), one can write

$$\delta U = I_1 + I_2, \tag{2.1}$$

where

$$I_1 = \int_{\mathring{\Omega}} (S^{\alpha\beta}\delta\gamma_{\alpha\beta} + \lambda_\xi^{-1} M^{\alpha\beta}\delta\kappa_{\alpha\beta}) d\mathring{\Omega},$$

$$I_2 = \int_{\mathring{\Omega}} (M^{\alpha\beta}\delta\kappa_{\alpha\beta} + T^{\alpha}_{\cdot n}\delta\psi_\alpha)\lambda_\xi^{-1} d\mathring{\Omega}. \tag{2.2}$$

Using the integration by parts and the formulas

$$\delta\lambda_\xi = -\frac{\nu}{1-\nu}\mathring{a}^{\alpha\beta}\mathbf{r}_\alpha \cdot \delta\mathbf{r}_\beta,$$

$$\delta(\lambda_\xi \kappa_\xi) = -\frac{\nu}{1-\nu}\mathring{a}^{\alpha\beta}(\delta\kappa_{\alpha\beta} + \delta\mu_{\alpha\beta}),$$

$$\delta\kappa_{\alpha\beta} = -\lambda_\xi \delta b_{\alpha\beta} + \frac{\nu}{1-\nu}\mathring{a}^{\nu\mu} b_{\alpha\beta}\mathbf{r}_\nu \cdot \delta\mathbf{r}_\mu, \qquad \delta b_{\alpha\beta} = \nabla_\alpha(\mathbf{n}\cdot\delta\mathbf{r}_\beta) + b^\nu_\alpha \mathbf{r}_\nu \cdot \delta\mathbf{r}_\beta,$$

obtained previously, we get

$$I_1 = \sum_{k=1}^{3} J_k - \int_{[\Omega]} \partial_\beta[\sqrt{\mathring{a}}\, T^{\beta\alpha}\mathbf{r}_\alpha + \sqrt{a}\nabla_\alpha (A M^{\alpha\beta})\mathbf{n}]\cdot\delta\mathbf{r} d\alpha^1 d\alpha^2, \tag{2.3}$$

where $[\Omega]$ is topological mapping of the domains $\mathring{\Omega}$ and Ω on the plane α^1, α^2

$$T^{ij} = S^{ij} - b^j_\alpha M^{i\alpha} + \frac{\nu}{1-\nu}\mathring{a}^{ij} b_{\alpha\beta} M^{\alpha\beta},$$

$$\mathring{a} = \mathring{a}_{11}\mathring{a}_{22},\ a = a_{11}a_{22} - a_{12}a_{21}, \qquad A = \sqrt{\mathring{a}}\,/\sqrt{a},$$

$$J_1 = \oint_{\partial\mathring{\Omega}} \mathring{\nu}_\beta T^{\alpha\beta}\mathbf{r}_\alpha \cdot \delta\mathbf{r} d\mathring{s}_t,\ \ J_2 = -\oint_{\partial\mathring{\Omega}} \mathring{\nu}_\alpha M^{\alpha\beta}\mathbf{n}\cdot\delta\mathbf{r}_\beta d\mathring{s}_t,$$

$$J_3 = \oint_{\partial\mathring{\Omega}} \mathring{\nu}_\beta A^{-1}\nabla_\alpha (A M^{\alpha\beta})\mathbf{n}\cdot\delta\mathbf{r} d\mathring{s}_t,\ \ \mathring{\nu}_i = \mathring{\mathbf{v}}\cdot\mathring{\mathbf{r}}_i. \tag{2.4}$$

Similarly, according to assumption (ε), we can write

$$I_2 = \int_{\mathring{\Omega}} [-A^{-1}\nabla_\beta (A M^{\alpha\beta}) + T^\alpha_{\cdot n}]\delta\psi_\alpha \lambda_\xi^{-1} d\mathring{\Omega} + K_1, \tag{2.5}$$

where

$$K_1 = \oint_{\partial\mathring{\Omega}} \mathring{\nu}_\beta M^{\alpha\beta}\delta\psi_\alpha \lambda_\xi^{-1} d\mathring{s}_t\,. \tag{2.6}$$

Taking into account our assumptions, expression for the work of external forces on respective displacements can be written in the form

$$A(\delta\mathbf{R}) = \int_{\Omega}[\tilde{\mathbf{q}}\cdot\delta\mathbf{r} + h_\lambda^2\nabla_\alpha(\mathring{a}^{\alpha\beta}q_n)]\delta\psi_\beta d\Omega + \sum_{k=4}^{6} J_k + K_2 . \tag{2.7}$$

Here

$$\begin{gathered}
\tilde{\mathbf{q}} = \mathbf{q} + \frac{\lambda_\xi h_\lambda^2}{\sqrt{a}}\partial_\alpha[\sqrt{a}\nabla_\beta(\mathring{a}^{\alpha\beta}q_n)\mathbf{n}] + \\
\frac{\nu}{1-\nu}\frac{1}{\sqrt{a}}\partial_\alpha(\sqrt{a}m_n\mathring{a}^{\alpha\beta}\mathbf{r}_\beta), \\
J_4 = -\lambda_\xi h_\lambda^2 \oint_{\partial\mathring{\Omega}} A^{-1}\mathring{\nu}_\alpha\nabla_\beta(\mathring{a}^{\alpha\beta}q_n)\mathbf{n}\cdot\delta\mathbf{r}\, d\mathring{s}_t, \\
J_5 = \lambda_\xi h_\lambda^2 \oint_{\partial\mathring{\Omega}} A^{-1}\mathring{\nu}_\alpha\mathring{a}^{\alpha\beta}q_n\mathbf{n}\cdot\delta\mathbf{r}_\beta\, d\mathring{s}_t, \\
J_6 = -\frac{\nu}{1-\nu}\oint_{\partial\mathring{\Omega}} A^{-1}\mathring{\nu}_\alpha\mathring{a}^{\alpha\beta}m_n\mathbf{r}_\beta\cdot\delta\mathbf{r}\, d\mathring{s}_t, \\
h_\lambda^2 = \frac{\nu\mathring{h}^2}{8(1-\nu)} .
\end{gathered} \tag{2.8}$$

Equating to zero the multipliers at $\delta\mathbf{r}$, $\delta\psi_i$, $i=1,2$ in the transformed equation (1.13) (see (2.1) – (2.3), (2.5) and (2.7)), we obtain

$$\begin{gathered}
\partial_\beta[\sqrt{a}(\tilde{T}^{\alpha\beta}\mathbf{r}_\alpha + \mathbf{n}\nabla_\alpha\tilde{M}^{\alpha\beta})] + \sqrt{a}\mathbf{q} = \mathbf{0}, \\
\tilde{T}^i_{\cdot n} = \nabla_\beta\tilde{M}^{\beta i},\ i=1,2,
\end{gathered} \tag{2.9}$$

where

$$\tilde{T}^{ij} = AT^{ij} + \frac{\nu}{1-\nu}\mathring{a}^{ij}m_n,\quad \tilde{T}^i_{\cdot n} = AT^i_{\cdot n},\quad \tilde{M}^{ij} = AM^{ij} + \lambda_\xi h_\lambda^2\mathring{a}^{ij}q_n . \tag{2.10}$$

The Eqs. (2.10) can be written in the form

$$\nabla_\alpha\tilde{T}^{\alpha i} - b_\alpha^i\tilde{T}^\alpha_{\cdot n} + q^i = 0,\quad \nabla_\alpha\tilde{T}^\alpha_{\cdot n} + b_{\alpha\beta}\tilde{T}^{\alpha\beta} + q_n = 0,\quad \nabla_\alpha\tilde{M}^{\alpha j} - \tilde{T}^j_{\cdot n} = 0, \tag{2.11}$$

or (after substituting the normal shearing forces from third equation in first and second equations)

$$\nabla_\alpha\tilde{T}^{\alpha i} - b_\alpha^i\nabla_\beta\tilde{M}^{\beta\alpha} + q^i = 0,\qquad \nabla_\alpha\nabla_\beta\tilde{M}^{\alpha\beta} + b_{\alpha\beta}T^{\alpha\beta} + q_n = 0 . \tag{2.12}$$

The Eqs. (2.11) are identical (in appearance) to the Eqs. (6.84) [8] of equilibrium of the linear shell theory up to the sign "tilde". Therefore, the form (2.11) of equilibrium equations, taking into account transversal deformations, can be naturally called *canonical form*. If we linearize the Eqs. (2.11), then we obtain the canonical form of equilibrium equations of linear shell theory not using the Kirchhoff hypotheses. For example, if in the Eqs. (2.11) we accept

$$\tilde{M}^{ij} = h_\lambda^2 q_n\mathring{a}^{ij},\quad \tilde{T}^{ij} = \overset{*}{T}{}^{ij} + \frac{\nu}{1-\nu}m_n\mathring{a}^{ij},$$

then we obtain linear equations of membrane theory, made more exact by using transversal deformations:

$$\overset{\circ}{\nabla}_\alpha \overset{*}{T}{}^{\alpha i} + q^i + \underline{\frac{\nu}{1-\nu}\overset{\circ}{\nabla}_\alpha (m_n \overset{\circ}{a}{}^{\alpha i}) - h_\lambda^2 \overset{\circ}{b}{}^i_\alpha \overset{\circ}{\nabla}_\beta (q_n \overset{\circ}{a}{}^{\alpha\beta})} = 0 ,$$

$$\overset{\circ}{b}_{\alpha\beta} \overset{*}{T}{}^{\alpha\beta} + q_n + \underline{\frac{\nu}{1-\nu}\overset{\circ}{a}{}^{\alpha\beta}\overset{\circ}{b}_{\alpha\beta} m_n + h_\lambda^2 \overset{\circ}{\nabla}_\alpha \overset{\circ}{\nabla}_\beta (q_n \overset{\circ}{a}{}^{\alpha\beta})} = 0 . \tag{2.13}$$

The Eq. (2.13) differ from the classic ones (see e.g. (2.3) [8]) in the underlined terms, connected with load.

Remark. The Eqs. (1.11) do not allow one to determine, which of the stress tensor was used to calculate forces and moments: either the Kirchhoff tensor $J\Sigma = J\sigma^{ij}\mathbf{R}_i \otimes \mathbf{R}_j$ or the nominal stresses tensor $\{\mathbf{F}^{-1}\cdot J\Sigma\} = J\sigma^{ij}\mathbf{R}_i \otimes \mathbf{R}_j$ or the Piola-Kirchhoff tensor $\{\mathbf{F}^{-1}\cdot J\Sigma \cdot \mathbf{F}^{-1*}\} = J\sigma^{ij}\mathbf{R}_i \otimes \mathbf{R}_j$. Similarly, this cannot be established from expression of variation of elastic potential, because

$$\delta\Phi = \{\mathbf{F}^{-1}\cdot J\Sigma \cdot \mathbf{F}^{-1*}\} : \delta\overset{\circ}{\Gamma} = \{\mathbf{F}^{-1}\cdot J\Sigma\} : \delta\mathbf{F} = J\sigma^{ij}\delta\gamma_{ij} .$$

At the same time, the symbols ∇_k in the Eqs. (2.11), (2.12) are covariant derivatives, only when quantities $J\sigma^{ij}$ are the components of the tensor $J\Sigma$; otherwise ∇_k are just differential operators, which work as above-mentioned derivatives.

The Eqs. (2.11) without tildes are identical to equations of the quasi-Kirchhoff shell theory of K.F. Chernykh [3]. These equations can be written in the form

$$\hat{\nabla}_\alpha T^{\alpha i} - b^i_\alpha T^\alpha_{\cdot n} + A^{-1} q^i = 0 , \qquad \overset{\circ}{\nabla}_\alpha T^\alpha_{\cdot n} + b_{\alpha\beta} T^{\alpha\beta} + A^{-1} q_n = 0 ,$$

$$\hat{\nabla}_\alpha M^{\alpha i} - T^i_{\cdot n} = 0 ,$$

where $\hat{\nabla}_k, \overset{\circ}{\nabla}_l$ are covariant derivatives of components of double tensors

$$\mathbf{T} = T^{\alpha\beta}\, \overset{\circ}{\mathbf{r}}_\alpha \otimes \mathbf{r}_\beta, \;\; \mathbf{M} = M^{\alpha\beta}\, \overset{\circ}{\mathbf{r}}_\alpha \otimes (\mathbf{n}\times \mathbf{r}_\beta), \;\; \mathbf{T}_{\cdot n} = T^\alpha_{\cdot n}\, \overset{\circ}{\mathbf{r}}_\alpha \otimes \mathbf{n} , \tag{2.14}$$

i.e.

$$\hat{\nabla}_k \begin{bmatrix} T^{ij} \\ M^{ij} \end{bmatrix} = \partial_k \begin{bmatrix} T^{ij} \\ M^{ij} \end{bmatrix} + \overset{\circ}{\Gamma}{}^i_{k\alpha} \begin{bmatrix} T^{\alpha j} \\ M^{\alpha j} \end{bmatrix} + \Gamma^j_{k\alpha} \begin{bmatrix} T^{i\alpha} \\ M^{i\alpha} \end{bmatrix},$$

$$\overset{\circ}{\nabla}_k T^i_{\cdot n} = \partial_k T^i_{\cdot n} + \overset{\circ}{\Gamma}{}^i_{k\alpha} T^\alpha_{\cdot n} .$$

Namely the tensors (2.14) were used in [3] for obtaining the (2.10) type of equilibrium equations from the equilibrium conditions for shell element.

3. When conditions (2.9) hold, the Lagrange variational equation reduces to the following boundary equation:

$$\sum\nolimits_{k=1}^{3} (J_k - J_{k+3}) + K_1 - K_2 = 0 , \tag{3.1}$$

Skipping, because of limited space, the details of very technical deduction, we give the Eq. (3.1) in its final transformed form (assuming the contour $\partial\overset{\circ}{\Omega}$ does not contain nodal points):

$$\oint_{\partial\overset{\circ}{\Omega}}[Q_{\nu\nu}\delta u_{\nu}+Q_{\nu t}\delta u_{t}+Q_{\nu n}\delta w+(1+\tilde{\varepsilon})\tilde{M}_{\nu\nu}\delta\vartheta_{\nu}+(\lambda_{\xi}\boldsymbol{A})^{-1}\tilde{M}_{\nu\nu}\delta\psi_{\nu}+ \tag{3.2}$$
$$+(\lambda_{\xi}\boldsymbol{A})^{-1}\tilde{M}_{\nu t}\delta\psi_{t}-\underline{\tilde{M}_{\nu\nu}\tilde{\vartheta}_{\nu}\delta(\frac{du_{\nu}}{d\overset{\circ}{s}_{\nu}})}-\underline{\tilde{M}_{\nu\nu}\tilde{\vartheta}_{t}\delta(\frac{du_{\nu}}{d\overset{\circ}{s}_{t}})}]d\overset{\circ}{s}_{t}=0\,,$$

Here (see also [8])

$$Q_{\nu\nu}\approx\boldsymbol{A}^{-1}\tilde{T}_{\nu\nu}+(1+\tilde{\varepsilon})\overset{\circ}{\delta}_{t}\tilde{M}_{\nu t},\,Q_{\nu t}\approx\boldsymbol{A}^{-1}\tilde{T}_{\nu t}-(1+\tilde{\varepsilon})\overset{\circ}{\sigma}_{t}\tilde{M}_{\nu t}\,,$$

$$Q_{\nu n}\approx(1+\tilde{\varepsilon})\tilde{T}_{\nu n}+\frac{d(1+\tilde{\varepsilon})\tilde{M}_{\nu t}}{d\overset{\circ}{s}_{t}}\,, \tag{3.3}$$

$$\tilde{T}_{\nu\nu}=\tilde{T}^{\alpha\beta}\,\overset{\circ}{\nu}_{\alpha}\overset{\circ}{\nu}_{\beta},\,\tilde{T}_{\nu t}=\tilde{T}^{\alpha\beta}\,\overset{\circ}{\nu}_{\alpha}\overset{\circ}{t}_{\beta},\,\tilde{T}_{\nu n}=(\nabla_{\alpha}\tilde{M}^{\alpha\beta})\overset{\circ}{\nu}_{\beta}\,,$$

$$\tilde{M}_{\nu\nu}=\tilde{M}^{\alpha\beta}\,\overset{\circ}{\nu}_{\alpha}\overset{\circ}{\nu}_{\beta},\quad\tilde{M}_{\nu t}=\tilde{M}^{\alpha\beta}\,\overset{\circ}{\nu}_{\alpha}\overset{\circ}{t}_{\beta},\quad u_{\nu}=u_{\alpha}\overset{\circ}{\nu}^{\alpha},\quad u_{t}=u_{\alpha}\overset{\circ}{t}^{\alpha}\,,$$

$$\tilde{\varepsilon}=\varepsilon_{\alpha}^{\cdot\alpha}+\begin{vmatrix}\varepsilon_{1}^{\cdot1}&\varepsilon_{1}^{\cdot2}\\ \varepsilon_{2}^{\cdot1}&\varepsilon_{2}^{\cdot2}\end{vmatrix},\quad\tilde{\vartheta}_{1}=\vartheta_{1}+\begin{vmatrix}\vartheta_{1}&\varepsilon_{1}^{\cdot2}\\ \vartheta_{2}&\varepsilon_{2}^{\cdot2}\end{vmatrix},\quad\tilde{\vartheta}_{2}=\vartheta_{2}+\begin{vmatrix}\varepsilon_{1}^{\cdot1}&\vartheta_{1}\\ \varepsilon_{2}^{\cdot1}&\vartheta_{2}\end{vmatrix},$$

$$\varepsilon_{i}^{\cdot j}=\overset{\circ}{\nabla}_{i}u^{j}-\overset{\circ}{b}_{i}^{j}\,w,\,\vartheta_{i}=-\partial_{i}w-\overset{\circ}{b}_{i\alpha}u^{\alpha}\,. \tag{3.4}$$

Thus if *the transversal shears are absent* the boundary variational Eq. (3.1) contains variations of six independent geometrical quantities on $\partial\overset{\circ}{\Omega}$:

$$w,u_{\nu},u_{t},\frac{dw}{d\overset{\circ}{s}_{\nu}},\frac{du_{\nu}}{d\overset{\circ}{s}_{\nu}},\frac{du_{t}}{d\overset{\circ}{s}_{\nu}}\,, \tag{3.5}$$

of which only the first four ones persist in the variational equation after linearization. It is natural to suppose that in the Eq. (3.2) the influence of underlined terms is small. Nevertheless, from the point of view of the Lagrange principle the nonlinear boundary problem is generally incorrect, if one formulates the boundary conditions in terms of the quantities (3.5). The authors of nonlinear theories usually avoid this point. For example, in the monograph [2] the author obtained an expression for the work of stresses, acting on side surface, on variations of corresponding displacements (see (1.43) [2]):

$$\delta A_{1}=\oint_{\partial\overset{\circ}{\Omega}}[(T'_{\nu\nu}\mathbf{v}+T'_{\nu t}\mathbf{t}+T'_{\nu n}\mathbf{n})\cdot\delta\mathbf{r}+\underline{M_{\nu\nu}\mathbf{v}\cdot\delta\mathbf{n}}]d\overset{\circ}{s}_{t}\,. \tag{3.6}$$

Using the Eq. (3.6), the author draws only one conclusion: the set $\{T'_{\nu\nu},\ T'_{\nu t},\ T'_{\nu n},\ M_{\nu\nu}\}$ can be considered as static boundary values. But what can one say about geometrical boundary values? In the linear theory $\mathbf{v}\cdot\delta\mathbf{n}\approx\delta\vartheta_{\nu}$, so geometric boundary quantities are the first four general displacements of (3.5). In nonlinear theory the matter looks different. The integral, connected with underlined term in the Eq. (3.6), can be transformed as follows:

$$\oint_{\partial\mathring{\Omega}} M_{\nu\nu}\mathbf{v}\cdot\delta\mathbf{n}d\mathring{s}_t = -\oint_{\partial\mathring{\Omega}} \lambda_\nu^{-1} M_{\nu\nu}\mathbf{n}\cdot\delta(\frac{d\mathbf{r}}{d\mathring{s}_\nu})d\mathring{s}_t\,.$$

The variations of the last two quantities in the set (3.5) appear after expanding the last integral.

It should be noted that main boundary conditions corresponding to real strengthening of the shell edges satisfy the variational Eq. (3.2), taking into account the underlined terms. These boundary conditions (the transversal shears are absent) are conditions of clamped edges ($u_\nu = u_t = w = 0$, $\vartheta_\nu = 0$), hinged movable tangentially edges ($Q_{\nu\nu} = Q_{\nu t} = 0$, $\tilde{M}_{\nu\nu} = 0$, $w = 0$), hinged immovable tangentially edges ($u_\nu = u_t = 0$, $\tilde{M}_{\nu\nu} = 0$, $w = 0$). However in the case of smooth coupling of shells, besides the geometric conditions of the linear theory

$$u_\nu^I = u_\nu^{II}, \quad u_t^I = u_{tt}^{II}, \quad w^I = w^{II}, \quad \vartheta^I = \vartheta^{II},$$

one must check also the equalities

$$(\frac{du_\nu}{d\mathring{s}_\nu})^I = (\frac{du_\nu}{d\mathring{s}_\nu})^{II}, \quad (\frac{du_t}{d\mathring{s}_\nu})^I = (\frac{du_t}{d\mathring{s}_\nu})^{II},$$

or evaluate the errors arising from their violation.

References

1. Ainola LYa (1965) Izv AN Est SSR, Ser Fiz-Mat Tekh Nauk, 14, 3: 337-344
2. Chernykh KF (1986) Nonlinear theory of elasticity in mechanical engineering (in Russia). Mashinostroyeniye, Leningrad, p 336
3. Chernykh KF (1980) Izv AN USSR, Prikl Mat Mekh 2: 148-159
4. Galimov KZ (1976) Izv AN USSR, Prikl Mat Mekh 4: 156-166
5. Kabrits SA, Chernykh KF (1996) Izv AN USSR, Prikl Mat Mekh, 1: 124-136
6. Michailovskii EI (2001) Nonlinear problems of mechanics and physics of deformable solid bodies. St-Petersburg State University, St-Petersburg, pp 42-56
7. Michailovskii EI (1995) Izv AN USSR, Prikl Mat Mekh, 2: 109-119
8. Novozhilov VV, Chernykh KF, Mikhailovskii EI (1991) The linear theory of thin shells (in Russia). Politechnika, Leningrad, p 656
9. Pietraszkiewicz W (1989) Advances in Mechanics. 12, 1: 52-130
10. Truesdell C (1965) In: Encyclopedia of physics. Springer, Berlin New York, V: III/3
11. Zubov LM (1982) Methods of nonlinear elasticity in shell theory (in Russian). Rostov Univ Press, Rostov, p 144

Shell and Solid Type Finite Element Analysis of Creep in Thin-Walled Structures

K. Naumenko and H. Altenbach

Martin-Luther-University Halle-Wittenberg, Department of Engineering Sciences, Halle, Germany

1 Introduction

In many applications thin-walled structures (beams, plates, pipes, pipe bends, etc.) operate at elevated temperatures. Under such conditions the behavior of metals and alloys is primarily determined by irreversible time-dependent creep and material deterioration processes. In order to estimate the long-term behavior it is important to understand the mechanisms of the time-dependent stress redistribution and damage growth, particularly in the zones of nozzles, pipe connections and welds. The widely used approach in modelling the creep-damage behavior is the continuum damage mechanics, which proposes constitutive equations for the creep strain rate tensor, evolution equations for damage variables, and states nonlinear initial-boundary value problems in order to perform a structural analysis, e.g. [7]. With the progress in the material science and continuum mechanics, many new constitutive models have been developed which include physically motivated state variables and are able to consider different effects in material behavior such as the damage induced anisotropy, stress state dependence of damage evolution, etc. These models can be incorporated into a commercial finite-element code in order to analyze time-dependent deformations and stresses in a thin-walled structure under a specific mechanical loading. As usual, an important step is to select a structural mechanics model or to specify the type of finite elements. One way is a "three-dimensional approach" which is based on balance equations of three-dimensional continuum. This approach seems more preferable for creep analysis, since the existing constitutive models of creep are developed with respect to the Cauchy stress and strain (rate) tensors and the proposed measures of damage (scalars or tensors of different rank) are defined in the three-dimensional space. Another way is the use of the classical structural mechanics of beams, plates and shells and the balance equations formulated in terms of force and moment tensors. This approach often find application because of simplicity of model creation, smaller effort in solving non-linear initial-boundary value problems of creep, and easily interpretable results.

The aim of this presentation is to recall the governing equations of creep mechanics, to compare the long-term predictions based on the three-dimensional approach and a two-dimensional plate model, and to discuss the possibilities and limitations of each approach in connection with creep-damage analysis. In the first part we summarize the basic features of creep behavior in metals and alloys, introduce the widely used Kachanov-Rabotnov-Leckie-Hayhurst material model [9] and demonstrate the description possibilities for stress states which are typical for thin-walled structures. In the second part, we discuss the quasi-static initial-boundary value problems of creep. Finally, we use the ANSYS finite element code in order to simulate the time-dependent behavior of a transversely loaded plate under creep conditions. Based on the numerical example for a transversely loaded square plate with two clamped parallel edges and two simply supported parallel edges we make conclusions regarding the realization of clamped edge boundary condition as well as the applicability of the shell and the solid type finite elements to the creep analysis of engineering structures.

2 Material behavior and structural mechanics models

Creep behavior of polycrystalline metals and alloys is a phenomenon accompanied by different microstructural changes. It is known from material science that for moderate stresses (below the yield limit) and elevated temperatures above $0.4T_m$ with T_m as the melting point, the steady creep process is controlled by the climb plus glide dislocation mechanism [6]. The dependence of the creep strain rate on the applied stress can be described by the power law function. For multi-axial stress states the deviatoric part of the stress tensor and the von Mises equivalent stress are used to describe the deformation process. In addition to irreversible strains, material deterioration processes occur and lead to accelerated creep and final fracture. For polycrystalline materials, tertiary creep is accompanied by nucleation and growth of micro-voids on grain boundaries. The initially existing micro-voids have a negligible influence on the strain rate. As their number and size increase with time, they weaken the material providing the decrease in the load-bearing cross section. The nucleation kinetics can be related to the local grain boundary deformation as well as to the stress state characterized by the combination of the maximum tensile stress and the von Mises equivalent stress. The coalescence of cavities leads to the propagation of oriented micro-cracks and to final fracture. The cavities and micro-cracks nucleate on grain boundaries having different orientations. The significant influence of the damage induced anisotropy can be observed on the last stage before the creep rupture [5].

The commonly used phenomenological model characterizes the secondary creep rate by the power law stress function and includes the effect of tertiary creep by means of the single scalar valued damage parameter [9]

$$\begin{aligned} \dot{\boldsymbol{\varepsilon}}^{cr} &= \frac{3}{2}a\left(\frac{\sigma_{vM}}{1-\omega}\right)^n \frac{\mathbf{s}}{\sigma_{vM}}, \quad \dot{\omega} = b\frac{[\alpha\sigma_T + \beta\sigma_m + (1-\alpha-\beta)\sigma_{vM}]^k}{(1-\omega)^l}, \\ \sigma_{vM} &= \sqrt{\frac{3}{2}\mathbf{s}\cdot\cdot\mathbf{s}}, \quad \sigma_m = \frac{1}{3}\mathrm{tr}\,\boldsymbol{\sigma}, \quad \mathbf{s} = \boldsymbol{\sigma} - \sigma_m\mathbf{E}, \quad \sigma_T = \frac{1}{2}(\sigma_I + |\sigma_I|) \end{aligned} \qquad (1)$$

In this notation $\dot{\boldsymbol{\varepsilon}}^{cr}$ is the creep strain rate tensor, $\boldsymbol{\sigma}$ is the stress tensor, σ_I is the first principal stress, $\mathbf{E}$ is the second rank unit tensor and ω is the damage parameter. The

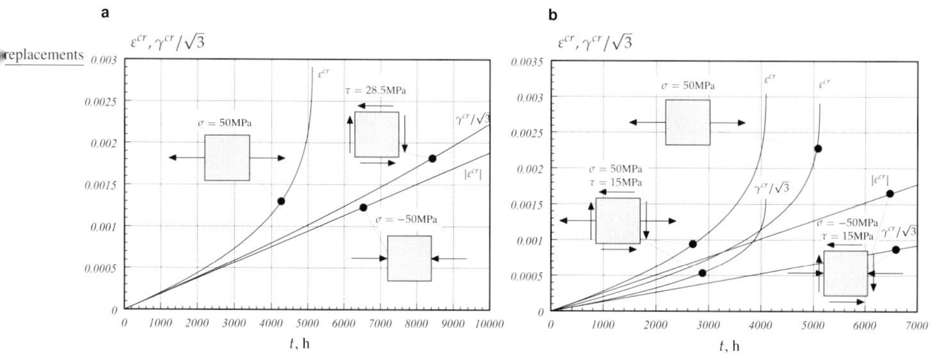

Fig. 1. Creep responses for various stress states computed using Eqs. (1): **a** responses by tension, torsion and compression; **b** responses by combined tension (compression) and torsion

weighting factor α characterizes the influence of the principal damage mechanisms (σ_T-controlled, σ_m-controlled or σ_{vM}-controlled). a, b, n, k and l are material constants, which are determined from creep tests at a constant temperature. The model (1) ignores the effects of primary creep.

As an example of application of Eqs. (1) let us introduce the 316 stainless steel at 650°C with the material constants presented in [10]: $a = 2.13 \cdot 10^{-13}$ MPa^{-n}/h, $b = 9 \cdot 10^{-10}$ MPa^{-k}/h, $n = 3.5$, $k = l = 2.8$, $\alpha = 1, \beta = 0$. The elastic constants are $E = 1.44 \cdot 10^5$ MPa as Young's modulus and $\nu = 0.314$ as Poisson's ratio. Let us illustrate the creep responses according to the constitutive model (1) and consider the following stress tensor

$$\boldsymbol{\sigma} = \sigma \mathbf{m} \otimes \mathbf{m} + \tau(\mathbf{n} \otimes \mathbf{m} + \mathbf{m} \otimes \mathbf{n}), \quad \mathbf{n} \cdot \mathbf{n} = \mathbf{m} \cdot \mathbf{m} = 1, \mathbf{m} \cdot \mathbf{n} = 0 \qquad (2)$$

Figure 1a) shows creep strain versus time curves: $\varepsilon^{cr}(t) = \mathbf{m} \cdot \boldsymbol{\varepsilon}^{cr}(t) \cdot \mathbf{m}$ under uniaxial tension $\sigma > 0, \tau = 0$; $\gamma^{cr}(t) = 2\mathbf{n} \cdot \boldsymbol{\varepsilon}^{cr}(t) \cdot \mathbf{m}$ under uniform shear $\sigma = 0$, $\tau = \sigma/\sqrt{3}$, and $|\varepsilon^{cr}(t)| = |\mathbf{m} \cdot \boldsymbol{\varepsilon}^{cr}(t) \cdot \mathbf{m}|$ under uniaxial compression $\sigma < 0, \tau = 0$. The considered stress states and the stress values provide the same secondary creep with the same minimum creep rates, Fig. 1a). The tertiary creep responses are quite different and depend significantly on the kind of the applied stress state. Such a dependence has been observed in creep tests on tubular specimen [8]. Figure 1b) shows creep curves for the combined action of the normal positive (negative) stress and the shear stress. One can observe that even a small superposed shear stress leads to a considerable shear strain response. The shear strain curve depends not only on the value of the applied shear stress, but more significantly on the value of the normal stress. The change of sign of σ leads to the considerable change of the shear strain response. The discussed effects can be explained by the dependence of damage mechanisms on the kind of the stress state.

The stress states with combined action of normal tensile (compressive) stress and small shear stress are typical for transversely loaded beams, plates and shells. For example, the stress state (2) is realized in a beam assuming that $\mathbf{m}$ is the unit vector directed along the beam axis and $\mathbf{n}$ is the unit normal to the beam axis. The creep response of transversely loaded beams is discussed in [3, 12]. It was shown that the through-the-thickness distribution of the transverse shear stresses differs from the classical parabolic one. Nevertheless, by formulation of suitable constitutive equations for the shear force, the Timoshenko type beam theory can be applied to the creep-damage analysis. In [2] a finite element analysis of a thin-walled pipe bend loaded by internal pressure is performed using the shell and the solid type finite elements. It was shown that in the steady state creep range the shell and the solid elements provide similar predictions. The results for the life-time estimation agree well only in the case of the von Mises equivalent stress σ_{vM} controlled tertiary creep. For the case of the σ_T or σ_m controlled damage evolution the shell and the solid models lead to significantly different results.

3 Equations of quasi-static creep

Let us recall the governing equations of the creep mechanics in the classical sense [11, 13]. For the sake of brevity we consider the geometrically-linear theory, assume isothermal conditions and neglect thermal strains. First, let us consider a three-dimensional solid occupying a volume V with the surface A. The solid is fixed on a part A_u of A and loaded by a constant traction vector $\mathbf{p}$ on the remaining part A_p. Let $\mathbf{r}(x^i) = \mathbf{e}_i x^i$, $i = 1, 2, 3$ and $\mathbf{R}(x^i, t)$ be the position vectors characterizing the points of the solid in the reference and actual configurations, respectively. The local form of governing equations can be summarized as follows

- Equilibrium conditions

$$\begin{aligned} &\nabla\sigma_m + \nabla\cdot\mathbf{s} + \rho\mathbf{f} = \mathbf{0}, \quad \mathbf{s}_\times = \mathbf{0}, \\ &\nabla(\ldots) = \mathbf{e}^i \otimes \frac{\partial(\ldots)}{\partial x^i}, \quad (\mathbf{a}\otimes\mathbf{b})_\times \equiv \mathbf{a}\times\mathbf{b} \end{aligned} \tag{3}$$

- Kinematical equations

$$\begin{aligned} &\boldsymbol{\varepsilon} = \frac{1}{2}\left(\nabla\mathbf{U} + (\nabla\mathbf{U})^T\right), \quad \Rightarrow \quad \nabla\times(\nabla\times\boldsymbol{\varepsilon})^T = \mathbf{0}, \\ &\boldsymbol{\varepsilon} = \frac{1}{3}\varepsilon_V\mathbf{E} + \boldsymbol{\epsilon}, \quad \varepsilon_V = \nabla\cdot\mathbf{U} \end{aligned} \tag{4}$$

- Constitutive assumptions

$$\sigma_m = \frac{2\mu}{3}\frac{1+\nu}{1-2\nu}\varepsilon_V, \quad \mathbf{s} = 2\mu(\boldsymbol{\epsilon} - \boldsymbol{\varepsilon}^{cr}), \quad \mu = \frac{E}{2(1+\nu)} \tag{5}$$

As usual we have to prescribe the boundary conditions for the displacement vector $\mathbf{U} = \mathbf{U}_*$ for $x^i \in A_u$ and/or the stress tensor $\boldsymbol{\nu}\cdot\boldsymbol{\sigma} = \mathbf{p}$ for $x^i \in A_p$ with $\boldsymbol{\nu}$ as the outer unit normal. With the constitutive and evolution equations (1) and the initial condition $\boldsymbol{\varepsilon}^{cr}|_{t=0} = 0$, the initial-boundary value problem can be solved numerically by applying a time-step algorithm and a variational method (e.g. the finite element method).

Various approaches to derive a shell theory have been developed within the assumption of elastic or viscoelastic material behavior. As far as we know, a "closed form" shell theory in the case of creep does not exist at present. The principal problem lies in establishing the constitutive equations of creep with respect to the shell type strain measures i.e. the membrane strains, changes of curvature and transverse shear strains. Although, a general structure of such equations can be found based on the direct approach, e.g. [4], the open question is the introduction of appropriate damage measures as well as the identification of damage mechanisms under the shell type stress states, i.e. under pure bending, membrane and transverse shear forces or their interactions. In this paper we follow the standard approach which can be summarized as follows:

1. Assume that Eqs. (1), (3) – (5) are applicable to the analysis of creep of thin-walled solid without any modification.
2. Formulate a variational equation of statics (e.g. the principle of virtual displacements) with the known the tensor $\boldsymbol{\varepsilon}^{cr}$ for a fixed time (time step).
3. Specify cross-section approximations for the functions to be varied (for example, the displacement vector $\mathbf{U}$).
4. Formulate and solve a two-dimensional version of Eqs. (3) – (5).
5. Recover the three-dimensional stress field $\boldsymbol{\sigma}$ from the two-dimensional model.
6. Insert $\boldsymbol{\sigma}$ into Eqs. (1) in order to calculate the time increments of $\boldsymbol{\varepsilon}^{cr}$ and ω.
7. Update the tensor $\boldsymbol{\varepsilon}^{cr}$ for the next time step and repeat the cycle from step 1.

Depending on the type of the applied variational equation (e.g. displacement type or mixed type) and the type of incorporated cross-section assumptions, different two-dimensional versions of Eqs. (3) – (5) with a different order of complexity can be obtained (i.e. models with forces and moments or models with higher order stress resultants). Note that the steps 2 and 3 can be performed numerically applying e.g. the Galerkin method to (3) – (5). Various types of finite elements which were developed for the inelastic analysis of shells are reviewed in [15]. Let us note that if studying the creep behavior coupled with damage, the type of assumed cross-section approximations may have a significant influence on the result. For example, if we use a mixed type variational equation and approximate both the displacements and stresses, a parabolic through-the-thickness approximation for the transverse shear stress or a linear approximation for the in-plane stresses is in general not suitable for the creep damage estimations [3].

4 Numerical analysis of a plate based on solid and shell elements

In order to compare the shell and the three-dimensional models let us perform a finite element analysis of a plate. As an example we selected a square plate with $l_x = l_y =$

PSfrag replacements

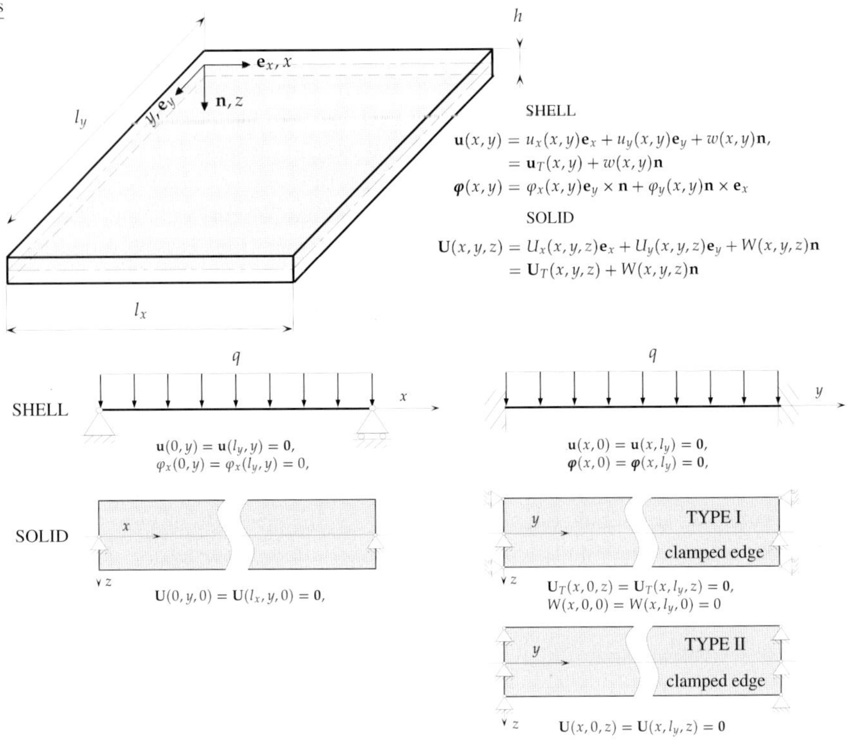

Fig. 2. Rectangular plate: geometry, loading and kinematical boundary conditions

1000 mm, $h = 100$ mm, loaded by a pressure $q = 2$ N/mm^2 uniformly distributed on the top surface, Fig. 2. The edges $x = 0$ and $x = l_x$ are simply supported and the edges $y = 0$ and $y = l_y$ are clamped. Note that applying a displacement based finite element method we can prescribe only the kinematical boundary conditions. According to the first order shear deformation type plate model we can specify the vectors of midplane displacements $\mathbf{u}(x,y) = \mathbf{u}_T(x,y) + \mathbf{n}w(x,y)$ and cross-section rotations $\boldsymbol{\varphi}(x,y)$ on the lines $x = const$ or $y = const$, Fig. 2. Applying such a model and assuming infinitesimal cross-section rotations the displacement vector $\mathbf{U}(x,y,z)$ is usually assumed to be

$$\mathbf{U}(x,y,z) \approx \mathbf{u}(x,y) + z\boldsymbol{\varphi}(x,y) \times \mathbf{n}, \quad \boldsymbol{\varphi} \cdot \mathbf{n} = 0$$

In the case of the three-dimensional model the displacement vector $\mathbf{U}$ can be prescribed on the planes x_c, y, z or x, y_c, z of the plate edges $x = x_c$ or $y = y_c$. Figure 2 illustrates the kinematical boundary conditions used for the shell and the solid models. Let us note that different boundary conditions which correspond to the clamped edge can be specified if we apply the three-dimensional model. Here we discuss two types of the clamped edge conditions. For the first type (TYPE I), see Fig. 2, we assume the vector of in-plane displacements $\mathbf{U}_T$ to be zero. The deflection W is zero

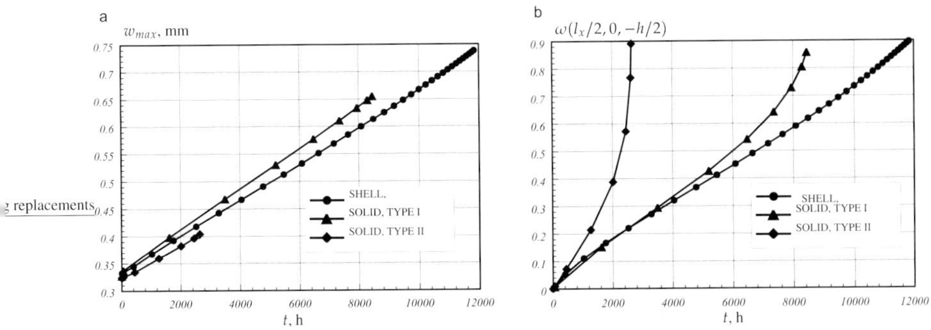

Fig. 3. Time variations: **a** maximum deflection; **b** damage parameter

only in the points of the plate mid-surface. In the second type (TYPE II) the whole displacement vector **U** is assumed to be zero in all points which belong to the plate edges. The TYPE II boundary condition is the simplest possibility with respect to the effort in the model creation and the preprocessing since all nodal displacements can be simultaneously set to zero on the whole surfaces of the edges $x = const$ and $y = const$.

The analysis has been performed using the ANSYS finite element code after incorporating the material model (1) with the help of the user defined creep material subroutine. In [1] we discussed various examples for beams and plates in bending, which verify the modified subroutine. The finite elements available in the ANSYS code for plasticity and creep analysis were used as follows: the 20 nodes solid element SOLID 95 and the 4 nodes shell element SHELL 43 [14]. 30×15 elements were used for a half of the plate in the case of the shell model and $30 \times 15 \times 3$ elements in the case of the solid model. The meshes have been justified based on the elasticity solutions and the steady state creep solutions neglecting damage. With these meshes the reference stress distributions as well as distributions of the von Mises stresses in the steady creep state were approximately the same for both the solid and the shell elements and did not change by further re-meshing. For details of time integration and equilibrium iteration methods used in ANSYS for creep calculations we refer to [14] and [16]. The time step based calculations were performed up to $\omega = \omega_* = 0.9$, where ω_* is the selected critical value of the damage parameter.

Figure 3 illustrates the results of the computations, where the maximum deflection and the maximum value of the damage parameter are plotted as functions of time. From Fig. 3a we observe that the starting values of maximum deflection as well as the starting rates of the deflection growth due to creep are approximately the same for the shell and the two solid models. Consequently the type of the elements (shell or solid) and the type of the applied boundary conditions in the case of the solid elements has a small influence on the description of the steady-state creep process. However, the three used models lead to quite different life time predictions. The difference can be clearly seen in Fig. 3b. The shell model overestimates the time to failure, while the result based on the solid model depends significantly on the type

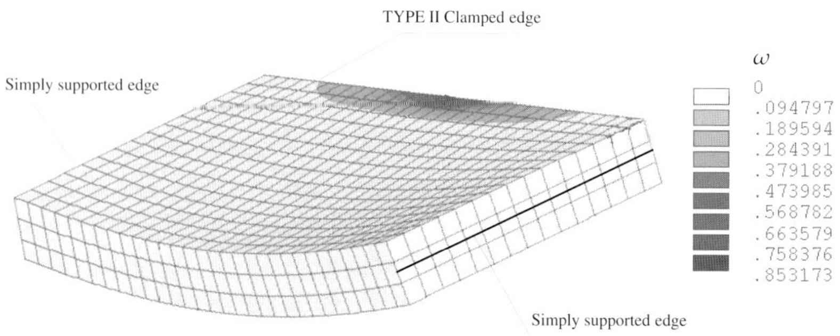

Fig. 4. Deformed shape of a half of the plate and distribution of the damage parameter in the zone of a clamped edge (SOLID elements, TYPE II boundary conditions, last time step)

of the clamped edge boundary conditions. In the case of the TYPE II clamped edge the damage parameter vs. time curve is too abrupt and the predicted time to failure is four times shorter compared to those based on the shell model. All considered models predict the zone of maximum damage to be in the midpoint of the clamped edge on the plate top surface, as shown in Fig. 4.

The creep response of a structure is connected with the time-dependent stress redistribution, that means the initial elastic stress response changes with time during the creep process. If the applied load and the boundary conditions are assumed to be constant and the effect of tertiary creep is ignored, than an asymptotic stress state exists, which is known as the state of stationary creep [11, 13]. If tertiary creep is considered, then stresses change with time up to the critical damage state. It is clear that the damage growth and the tertiary creep behavior of the considered plate is controlled by the local stress state in the vicinity of the clamped edges. Figure 5 illustrates the stress states in the midpoint of the clamped edge with the coordinates are $x = l_x/2, y = 0$. Four components of the stress tensor (two remaining components are zero due to symmetry conditions) are plotted as functions of the normalized thickness coordinate. The starting elastic distributions (solid lines) as well as creep solutions at the last time step (dotted lines) are presented. The maximum starting stresses obtained by use of three considered models are the normal in-plane stresses σ_{yy} and σ_{xx} (the stresses which results in the maximum bending and twisting moments in the clamped edge), Fig. 5. These in-plane stresses remain dominant during the whole creep process for the used shell and solid elements. Therefore, all the applied models predict the damage evolution in the zone of the clamped edge on the plate top side. However, the influence of the "second order" stresses (stresses which are usually neglected in the plate theories) is different and depends on the type of the boundary conditions. For the TYPE I clamped edge the effect of the transverse normal stress σ_{zz} decreases with time and has negligible influence on the stress state. In contrast, for the TYPE II clamped edge the initial transverse normal stress remains approximately constant, while σ_{yy} relaxes with time as consequence of creep. The

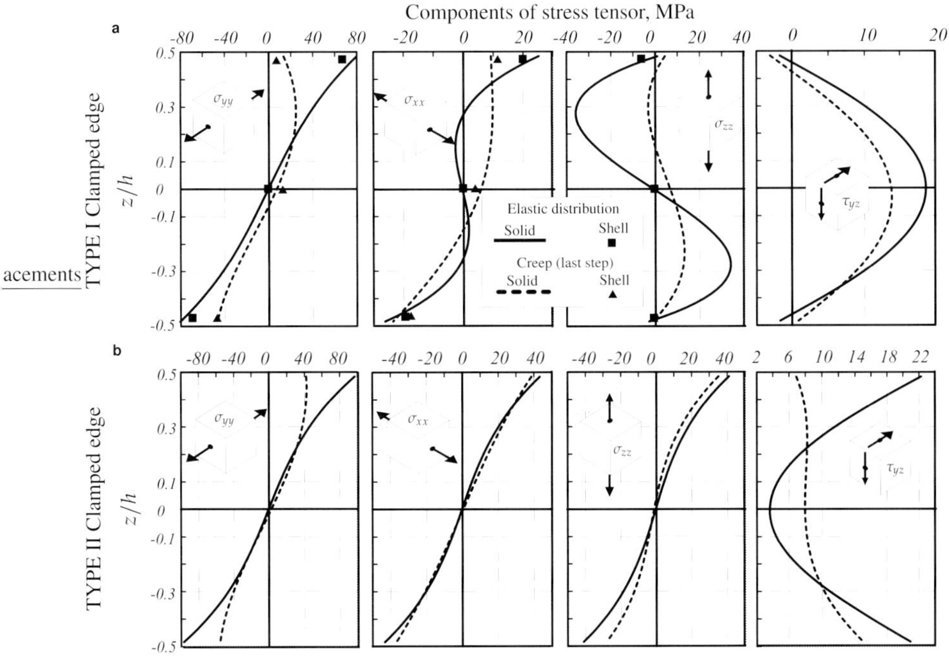

Fig. 5. Local stress state in a midpoint of the clamped edge vs. thickness coordinate: **a** TYPE I clamped edge; **b** TYPE II clamped edge

transverse normal stress becomes comparable with the bending stress and cannot be considered as the second order effect anymore.

In order to explain the difference in life-time predictions let us let us estimate the stress state for the TYPE II clamped edge $y = y_c$. In this case we have to set $\mathbf{U}(x, y_c, z) = \mathbf{0}$ on the plane x, y_c, z, Fig. 2. For $0 < x < l_x$ and $-h/2 < z < h/2$ we can write

$$\begin{aligned} \frac{\partial \mathbf{U}}{\partial x} = \frac{\partial \mathbf{U}}{\partial z} = \mathbf{0} \quad &\Rightarrow \quad \nabla \mathbf{U}(x, y_c, z) = \mathbf{e}_y \otimes \frac{\partial \mathbf{U}}{\partial y}, \\ \mathrm{tr}\boldsymbol{\varepsilon}(x, y_c, z) = \nabla \cdot \mathbf{U} = \frac{\partial U_y}{\partial y} & \end{aligned} \tag{6}$$

In addition we can set $\mathbf{e}_x \cdot \mathbf{U}(l_x/2, y, z) = 0$ due to the symmetry condition. The starting elastic stress state at $t = 0$ can be obtained from the constitutive equations (5) by setting $\boldsymbol{\varepsilon}^{cr} = \mathbf{0}$

$$
\begin{aligned}
\sigma_m|_{t=0} &= \frac{1}{3}\frac{1+\nu}{1-2\nu}\sigma_0, \quad \sigma_0 = 2\mu\frac{\partial U_y}{\partial y}|_{t=0}, \quad \tau_0 = \mu\frac{\partial W}{\partial y}|_{t=0} \\
\mathbf{s}|_{t=0} &= \frac{1}{3}\sigma_0\left[2\mathbf{e}_y\otimes\mathbf{e}_y - (\mathbf{E}-\mathbf{e}_y\otimes\mathbf{e}_y)\right] + \tau_0(\mathbf{e}_y\otimes\mathbf{n}+\mathbf{n}\otimes\mathbf{e}_y) \qquad (7)\\
\boldsymbol{\sigma}|_{t=0} &= \frac{1-\nu}{1-2\nu}\sigma_0\left[\mathbf{e}_y\otimes\mathbf{e}_y + \frac{\nu}{1-\nu}(\mathbf{E}-\mathbf{e}_y\otimes\mathbf{e}_y)\right] + \tau_0(\mathbf{e}_y\otimes\mathbf{n}+\mathbf{n}\otimes\mathbf{e}_y),
\end{aligned}
$$

From the last equation in (7) we see that $\sigma_{zz} = \sigma_{xx} = \sigma_{yy}\nu/(1-\nu)$. This well known result the linear isotropic elasticity agrees with the obtained finite element solution for $\nu = 0.314$, Fig. 5b (solid lines). Note that within the first order shear deformation type plate theory, the transverse normal stress has the order of the applied pressure q.

Let us estimate the stress redistribution in a clamped edge as a consequence of creep. For this purpose we neglect the damage evolution by setting $\omega = 0$ in (1). Because the boundary conditions and the applied pressure are independent of time, we can estimate the type of the stress state under stationary state creep by setting $\dot{\boldsymbol{\varepsilon}} \approx \dot{\boldsymbol{\varepsilon}}^{cr}$, $\dot{\varepsilon}_V \approx 0$ or

$$
\frac{1}{2}\left(\mathbf{e}_y\otimes\frac{\partial\dot{\mathbf{U}}}{\partial y} + \frac{\partial\dot{\mathbf{U}}}{\partial y}\otimes\mathbf{e}_y\right) \approx \dot{\boldsymbol{\varepsilon}}^{cr} = \frac{3}{2}a\sigma_{vM}^{n-1}\mathbf{s}, \quad \boldsymbol{\nabla}\cdot\dot{\mathbf{U}} \approx 0 \qquad (8)
$$

Consequently

$$
\frac{1}{2}\frac{\partial\dot{W}}{\partial y}(\mathbf{e}_y\otimes\mathbf{n}+\mathbf{n}\otimes\mathbf{e}_y) \approx \frac{3}{2}a\sigma_{vM}^{n-1}\mathbf{s} \qquad (9)
$$

From Eq. (9) we observe that the stress deviator in the steady-state creep has the form $\mathbf{s} \approx \tau(\mathbf{e}_y\otimes\mathbf{n}+\mathbf{n}\otimes\mathbf{e}_y)$ and is determined by the transverse shear stress. The mean stress σ_m cannot be determined from the constitutive equation, it must be found from equilibrium condition (3). The stress state in the zone of a clamped edge $(l_x/2, y, z)$ is then of the type $\boldsymbol{\sigma} \approx \sigma_m\mathbf{E} + \tau(\mathbf{e}_y\otimes\mathbf{n}+\mathbf{n}\otimes\mathbf{e}_y)$. We observe that $\sigma_{zz} \approx \sigma_{yy} \approx \sigma_{xx} \approx \sigma_m$ after the transient stress redistribution. This estimation agrees again with the obtained finite element solution Fig. 5b (dotted lines). The transverse normal stress is approximately equal to the in-plane stresses and cannot be neglected.

Let us compare the finite element results for the mean stress and the von Mises equivalent stress. Figure 6a shows the corresponding time variations in the element A of the solid model for the TYPE I and TYPE II boundary conditions. We observe that the TYPE II boundary condition leads to a lower starting value of the von Mises stress and a higher starting value of the mean stress when compared with those for the TYPE I boundary condition. In addition, for the TYPE II clamped edge we observe that the mean stress rapidly decreases within the short transition time and after that remains constant while the von Mises stress relaxes during the whole creep process. With the relaxation of σ_{vM} the stress state tends to $\boldsymbol{\sigma} = \sigma_m\mathbf{E}$. The relatively high constant value of σ_m is the reason for the obtained increase of damage and much shorter time to fracture in the case of the TYPE II clamped edge (see Fig. 3b). Note that the above effect of the mean stress has a local character and is observed only in the neighborhood of the edge. As Fig. 6b shows the value of the transverse normal stress decreases rapidly with increased distance from the boundary.

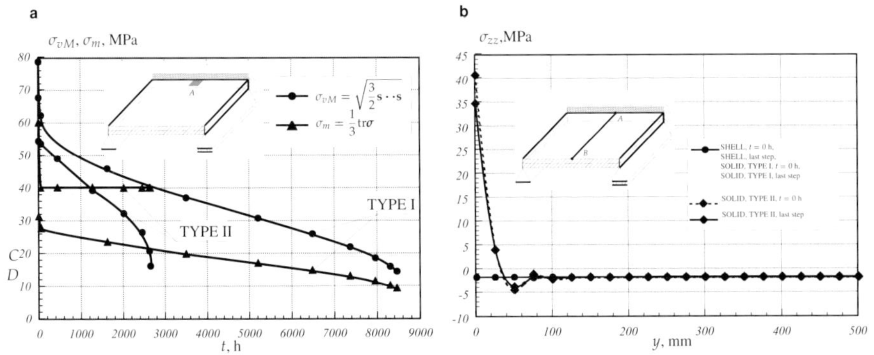

Fig. 6. Transversely loaded plate: **a** von Mises equivalent stress and hydrostatic stress vs. time, element A of the clamped edge; **b** transverse normal stress σ_{zz} in elements along the line AB;

5 Conclusions

We discussed the modelling possibilities of creep-damage behavior in thin-walled structural elements. The selected constitutive model of creep is based on the assumption that the secondary creep strain rate is determined by the deviatoric part of the stress tensor and the von Mises equivalent stress, while the increase of creep rate in the tertiary range is due to isotropic damage evolution which is controlled by the mean stress, the first principal stress and the von Mises equivalent stress. The use of this model in connection with long-term predictions of thin-walled structural elements has motivated a numerical comparative study two approaches: the three-dimensional approach and the approach based on the first order shear deformation type plate theory. The finite element results as well as some simplified estimates have shown that the approaches based on standard solid and shell finite elements provide quite different predictions. The model based on the shell elements overestimates the fracture time. The reason for the obtained differences is the local stress response in the zone of the clamped edge. In the case of linear isotropic elasticity the transverse normal and shear stresses in the zone of the clamped edge can be assumed to be the second order quantities in comparison to the dominant in-plane stresses. In the case of steady state creep, the transverse normal and shear stresses are comparable with the in-plane stresses due to the stress redistribution. If studying the creep behavior coupled with damage, the influence of these factors cannot be ignored.

If a shell or a plate theory is considered to be an approximate version of the three-dimensional equations (3)-(5) then we can conclude that "more accurate" cross-section approximations for the transverse normal and shear stresses have to be used in the case of creep. In this sense it is more reliable to solve the three-dimensional equations (3)-(5) which are "free" from ad hoc assumptions for the displacements and stresses. However, in this case the question about the relevance of "three-dimensional" constitutive assumptions (1) and (5) for the analysis of thin-walled structural elements has to be discussed.

References

1. Altenbach H, Kolarow G, Morachkovsky O, Naumenko K (2000) On the accuracy of creep-damage predictions in thinwalled structures using the finite element method. Comp Mech 25: 87–98
2. Altenbach H, Kushnevsky V, Naumenko K (2001) On the use of solid and shell type finite elements in creep–damage predictions of thinwalled structures. Arch Appl Mech 71: 164–181
3. Altenbach H, Naumenko K (2002) Shear correction factors in creep-damage analysis of beams, plates and shells. JSME International Journal, Series A 45: 77–83
4. Altenbach H, Zhilin PA (1988) Osnovnye uravneniya neklassicheskoi teorii uprugikh obolochek (Governing equations of the non-classical theory of elastic shells). Uspekhi Mekhaniki 11: 107–148
5. El-Magd E, Betten J, Palmen P (1996) Auswirkung der Schädigungsanisotropie auf die Lebensdauer von Stählen bei Zeitstandbeanspruchung. Mat-wiss u Werkstofftechn 27: 239 – 245
6. Frost HJ, Ashby MF (1982) Deformation-Mechanism Maps. Oxford et al.: Pergamon
7. Hayhurst DR (2001) Computational continuum damaged mechanics: its use in the prediction of creep in structures - past, present and future. In: Murakami S, Ohno N (Eds.) IUTAM Symposium on Creep in Structures. Dordrecht: Kluwer, pp 175–188
8. Kowalewski ZL (2001) Assessment of the multiaxial creep data based on the isochronous creep surface concept. In: Murakami S, Ohno N (Eds.) IUTAM Symposium on Creep in Structures. Dordrecht: Kluwer, pp 401–418
9. Leckie FA, Hayhurst DR (1977) Constitutive Equations for Creep Rupture. Acta Metall 25: 1059 – 1070
10. Liu Y, Murakami S, Kanagawa Y (1994) Mesh–dependence and stress singularity in finite element analysis of creep crack growth by continuum damage mechanics approach. Eur J Mech, A Solids 13: 395–417
11. Malinin NN (1981) Raschet na polzuchest' konstrukcionnykh elementov (Creep calculations of structural elements, in Russ.). Moskva: Mashinostroenie
12. Naumenko K (2000) On the use of the first order shear deformation models of beams, plates and shells in creep lifetime estimations. Technische Mechanik 20: 215–226
13. Odqvist FKG (1974) Mathematical Theory of Creep and Creep Rupture. Oxford et al.: Oxford University Press
14. ANSYS User's Manual Volume I – IV (1994) Swanson Analysis Systems, Inc.
15. Wriggers P (2001) Nichtlineare Finite-Element-Methoden. Berlin u.a.: Springer
16. Zienkiewicz OC, Taylor RL (1991) The Finite Element Method. London et al.: McGraw–Hill

An Asymptotic Model for Long-Wave Symmetric Motion of a Pre-Stressed Incompressible Plate With Fixed Faces

A. V. Pichugin[1] and G. A. Rogerson[2]

[1]Department of Mathematics, University of Manchester, UK
[2]Mathematics, School of Sciences, University of Salford UK

1 Introduction

The limitations of classical plate theories are well known and may be classified to fall into two categories. Firstly, all of these theories are only suited to describe relatively low-frequency motions, and secondly, classical theories assume that deformations of the plate's faces are not fixed. However, many modern applications have to deal with plates experiencing high-frequency loads and/or fixed on the faces, which may be illustrated by an example from geophysics of a coal layer contained within a much stiffer rock. Dynamic motion of such structures is complicated and classical *ad hoc* hypotheses are clearly not valid in this context, therefore an asymptotic analysis of three-dimensional equations of elasticity appears to be a natural approach. A total asymptotic analysis of this kind was performed for thin isotropic structures in [1]. However, in the previously mentioned example from geophysics, coal layers also usually suffer from large initial stresses, which generally requires use of a non-linear theory of elasticity.

Incremental deformation is a powerful concept for modelling of pre-stressed media. In applications it is often the case that small motions are superimposed upon large static deformation, which may be used to construct a linear theory of motions around the state of a generally non-linear deformation. We use this approach to model homogeneously pre-stressed incompressible media that is known to be a good approximation to the behaviour of rubbers and rubber-like materials, see [3]. A thin layer composed of such medium and subjected to fixed-faces (zero incremental displacement) boundary conditions is then studied. A thin rubber sheet pressed between two rigid (e.g. steel) plates may serve as a good experimental set-up to try to reproduce our results.

The contents of this paper may be outlined as follows. First the governing equations for the problem are introduced and briefly discussed. An associated symmetric dispersion relation is then analysed both numerically and asymptotically. Two important conclusions are drawn from the analysis: (a) symmetric motion in a pre-stressed layer with fixed faces does not feature any long-wave low-frequency motions (fundamental modes), but only long-wave high-frequency motions (harmonics), (b) long-wave cut-off frequencies are shown to be solutions of a transcendental equation and depend on the angle of propagation. The information obtained during this investigation is then used to obtain simplified lower-dimensional theory, describing the motion of pre-stressed incompressible plate in a vicinity of the cut-off frequencies.

2 Governing equations

We consider a finitely and homogeneously deformed layer of thickness $2h$, composed of an incompressible elastic material. The origin of a Cartesian coordinate system $Ox_1x_2x_3$ is placed in the mid-plane of the layer with Ox_2 orthogonal to the layer. Small amplitude motions of such layer may be described by the following equations of motion

$$B_{milk}u_{k,lm} - p_{t,i} = \rho \ddot{u}_i \quad , \tag{1}$$

where $\mathbf{B}$ is a fourth order elasticity tensor, $\mathbf{u}$ a particle displacement, p_t is a dynamic component of *pressure* $p = p_0 + p_t$, a reaction stress introduced to accommodate the incompressibility constraint $\operatorname{div}\mathbf{u} = 0$, ρ is a material density, commas and over-dots denote differentiation with respect to indicated space and time variables and summation is assumed over repeated suffices, see [3]. If principal axes of the initial deformation coincide with the chosen coordinate axes, non-zero components of $\mathbf{B}$ have the form B_{iiii}, B_{iijj}, B_{ijij} or B_{ijji}, $i, j = 1..3$.

Solutions of (1) are sought in form of the travelling harmonic waves

$$(\mathbf{u}, p_t) = (\mathbf{U}, kP_t)e^{kqx_2}e^{ik(x_1c_\theta + x_3s_\theta - vt)} \quad , \tag{2}$$

in which k is the wave number, v the phase speed and the vector $(c_\theta, 0, s_\theta) \equiv (\cos\theta, 0, \sin\theta)$ is the projection of wave normal onto the coordinate plane Ox_1x_3. It was shown in [5] that non-trivial solutions of the form (2) exist provided the parameter q is determined from the secular equation

$$\gamma_{21}\gamma_{23}q^6 + (c_0v_s^2 - c_1)q^4 + (v_s^4 - c_2v_s^2 + c_3)q^2 - (v_s^2 - c_4)(v_s^2 - c_5) = 0 \tag{3}$$

with $v_s^2 = \rho v^2$ and

$$c_0 = \gamma_{21} + \gamma_{23} \quad c_1 = (2\beta_{23}\gamma_{21} + \gamma_{23}\gamma_{31})s_\theta^2 + (2\beta_{12}\gamma_{23} + \gamma_{21}\gamma_{13})c_\theta^2$$
$$c_2 = (2\beta_{23} + \gamma_{21} + \gamma_{31})s_\theta^2 + (2\beta_{12} + \gamma_{23} + \gamma_{13})c_\theta^2$$
$$c_3 = (4\beta_{12}\beta_{23} + \gamma_{12}\gamma_{21} + \gamma_{23}\gamma_{32} + \gamma_{13}\gamma_{31} - \mu_{13}^2)s_\theta^2c_\theta^2 +$$
$$+(2\beta_{23}\gamma_{31} + \gamma_{21}\gamma_{32})s_\theta^4 + (2\beta_{12}\gamma_{13} + \gamma_{23}\gamma_{12})c_\theta^4$$

$$c_4 = \gamma_{32} s_\theta^2 + \gamma_{12} c_\theta^2$$
$$c_5 = \gamma_{31} s_\theta^4 + 2\beta_{13} s_\theta^2 c_\theta^2 + \gamma_{13} c_\theta^4$$

where $\mu_{13} = \beta_{13} - \beta_{12} - \beta_{23}$ and

$$\gamma_{ij} = B_{ijij}, \quad i \neq j, \quad 2\beta_{ij} = B_{iiii} + B_{jjjj} - 2B_{iijj} - 2B_{ijji}, \quad i < j \ .$$

Eq. (3) possesses six solutions $\pm q_i$, $i = 1..3$, for every angle of propagation θ. General solutions of (1) may then be formulated as superpositions of six waves of the form (2). We use these solutions to specify fixed faces (zero displacement) boundary conditions, which will comprise a system of six linear homogeneous equations. This system may further be decomposed into two homogeneous systems, associated with symmetric and anti-symmetric motions. The condition for the existence of non-trivial solutions for each of these systems is the vanishing of associated determinants, which gives so-called dispersion relations. In this paper we only consider a dispersion relation for symmetric motions, which is given by

$$\begin{gathered} q_2 q_3 (q_2^2 - q_3^2) V(q_1, v_s) \tanh(kq_1 h) - q_1 q_3 (q_1^2 - q_3^2) V(q_2, v_s) \tanh(kq_2 h) + \\ + q_1 q_2 (q_1^2 - q_2^2) V(q_3, v_s) \tanh(kq_3 h) = 0 \\ V(q_m, v_s) = (\gamma_{21} s_\theta^2 + \gamma_{23} c_\theta^2) q_m^2 + v_s^2 - c_5 \ . \end{gathered} \tag{4}$$

A more detailed derivation of the dispersion relation (4) may be found in [4].

3 Analysis of the dispersion relation

Fig. 1 depicts typical solutions of the symmetric dispersion relation (4). A particularly striking feature of both plots is the lack of any fundamental mode, which indicates the impossibility of surface wave propagation in a layer with fixed faces and the lack of classical extension-type deformations. On the other hand, the long wave high frequency behaviour of the symmetric dispersion relation is generally similar to the classical case of layer with free faces. Scaled wave speed of the branches tends to infinity as the scaled wave number kh decreases, whereas scaled frequencies tend to finite limits, so-called cut-off frequencies.

Let us investigate the long-wave high-frequency behaviour of the dispersion relation (4) in a greater detail. Our numerical solutions indicate that that the scaled wave speed $v_s \to \infty$, as $kh \to 0$, whereas the scaled frequency $\omega_s = O(1)$. It can be shown, see [5], that in this case

$$\begin{aligned} q_1^2 &= -\frac{v_s^2}{\gamma_{21}} + O(1) \\ q_2^2 &= 1 + O(v_s^{-2}) \end{aligned} \tag{5}$$

$$q_3^2 = -\frac{v_s^2}{\gamma_{23}} + O(1) \quad ,$$

which may be used to derive the leading order approximation of (4) in the form

$$\frac{ic_\theta^2}{\sqrt{\gamma_{23}}}\tanh(kq_1h) + \frac{v_s}{\sqrt{\gamma_{21}\gamma_{23}}}\tanh(kq_2h) + \frac{is_\theta^2}{\sqrt{\gamma_{21}}}\tanh(kq_3h) \approx 0 \quad . \tag{6}$$

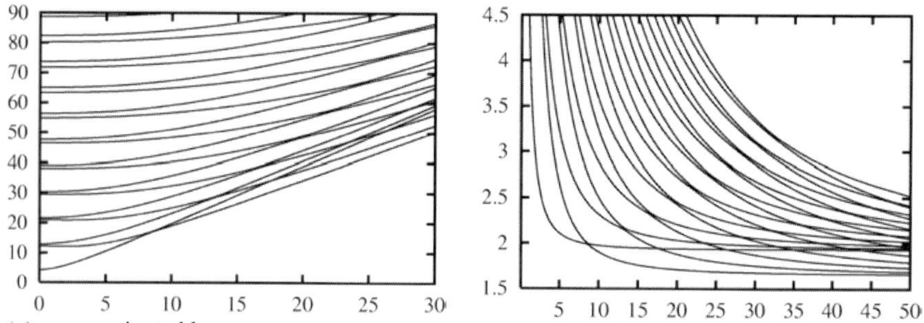

(a) ω_s against kh (b) v_s against kh

Fig. 1. (a) scaled frequency $\omega_s = v_s kh$ and (b) scaled wave speed v_s against scaled wave number kh for symmetric motion in a Mooney-Rivlin material with the angle of propagation $\theta = 45^\circ$. Principal stretches of the initial deformation are $\lambda_1 = 0.9$ and $\lambda_2 = 1.4$.

Because of $(5)_2$, it may be concluded that $\tanh(kq_2h) = O(v_s^{-1})$ and all three terms on the left-hand side of (6) appear to have the same asymptotic order. In order to determine the asymptotic structure of the solutions of (6) we introduce small parameter $\eta \equiv kh = h/l$, where l is a typical wavelength, and assume that $\omega_s = \Lambda + O(\eta^2)$. Together with (2) this yields

$$\begin{aligned} \tanh \eta q_1 &= iT_1 + O(\eta^2) \\ \tanh \eta q_2 &= \eta + O(\eta^3) \\ \tanh \eta q_3 &= iT_3 + O(\eta^2) \quad , \end{aligned} \tag{7}$$

where $T_m = \tan \Lambda_m$, $\Lambda_m = \Lambda / \sqrt{\gamma_{2m}}$. Using (6) and (7) we may obtain the leading order long-wave high-frequency approximation of (4) as

$$\sqrt{\gamma_{21}}\left(T_1 - \Lambda_1\right)c_\theta^2 + \sqrt{\gamma_{23}}\left(T_3 - \Lambda_3\right)s_\theta^2 = 0 \quad . \tag{8}$$

This form of representing the leading order dispersion relation is chosen to reflect a connection with the analogous plane strain problem analysed in [2]. When the cut-off frequency Λ is large, a solution of (8) will tend to a singularity of T_1 or T_3, which may be used to derive asymptotics of (8) for high frequencies, see [4]. We note that the leading order dispersion relation in the plane strain case is given by $\tan\Lambda = \Lambda$. It is worth remarking that the occurrence of a transcendental equation to define the cut-off frequency is unusual but not new, see e.g. [2]. However, the dependence of cut-off frequency on the angle of wave propagation, apparent from (8), is novel. Expansions (5) and (7) may also be used to estimate the relative asymptotic orders of the displacements. Due to space constraints we do not provide an explicit procedure, outlined in [5], which in our case gives

$$\begin{aligned} u_1 &= \eta^2 O(p_t) \\ u_2 &= \eta^3 O(p_t) \\ u_3 &= \eta^2 O(p_t) \quad . \end{aligned} \tag{9}$$

Long-wave motion in an incompressible elastic layer with fixed faces is very limited, which is indicated by (9). The displacements are negligible in comparison with incremental pressure. However, to leading order, the wave normal is parallel to a vector of the form $(c_\theta, O(\eta^{-1}), s_\theta)$, see (5), and the polarisation direction lies in the plane generated by vectors of the form $(O(1), O(\eta), O(1))$. We therefore observe a rapid motion of shear waves along the normal to the plane of a layer. Only physical peculiarities of motion in a layer with fixed faces, namely the lack of extension-like motions and the dependence of cut-off frequencies on the direction of wave propagation, preclude us from associating this type of motion with thickness shear resonance.

4 Derivation of an asymptotic model

We begin the derivation of an asymptotic model by choosing scales of variables, which balance x_1 and x_3 with a typical wavelength l, x_2 with a layer halfthickness h and t with a time that takes a typical wave to travel the wavelength l. The scales are therefore chosen as

$$\begin{aligned} x_1 &= l\xi_1 \\ x_2 &= h\zeta = l\eta\zeta \\ x_3 &= l\xi_3 \\ t &= l\sqrt{\rho}\eta\tau \quad . \end{aligned} \tag{10}$$

The displacements should also be re-scaled according to (9). Using star superscripts to denote non-dimensional quantities of the same order it is assumed that

$$\left(u_1, u_2, u_3, p_t\right) = \sum_{n=0,2,4\ldots} \eta^n \left(l\eta^2 u_1^{*(n)}, l\eta^3 u_2^{*(n)}, l\eta^2 u_3^{*(n)}, p_t^{*(n)}\right) \quad . \tag{11}$$

Finally, we remark that in order to produce response which is compatible with the properties of long-wave motion in a layer with fixed faces already observed in previous sections, we have to require

$$\gamma_{21} u_{m,\tau\tau}^{*(n)} + \Lambda^2 u_m^{*(n)} = O(\eta^2) \qquad m = 1..3, \quad n = 0, 2, 4..., \tag{12}$$

which follows from (2) and the form of expansion sought for $\omega_s = \Lambda + O(\eta^2)$. By introducing non-dimensional quantities (10), (11) into the equations of motion (1), incompressibility constraint $\operatorname{div} \mathbf{u} = 0$ and fixed faces boundary conditions we obtain a hierarchical system of boundary value problems at various orders. Asymptotic equalities (12) must be taken into consideration when separating boundary value problems at various orders.

At the leading order we consider the appropriate equations of motion

$$\gamma_{21} u_{1,\zeta\zeta}^{*(0)} + \Lambda^2 u_1^{*(0)} - p_{t,\xi_1}^{*(0)} = 0 \tag{13}$$

$$p_{t,\zeta}^{*(0)} = 0 \tag{14}$$

$$\gamma_{23} u_{3,\zeta\zeta}^{*(0)} + \Lambda^2 u_3^{*(0)} - p_{t,\xi_3}^{*(0)} = 0 \ , \tag{15}$$

subject to the incompressibility condition $u_{1,\xi_1}^{*(0)} + u_{2,\zeta}^{*(0)} + u_{3,\xi_3}^{*(0)} = 0$ and the leading order fixed faces boundary conditions

$$u_1^{*(0)} = u_2^{*(0)} = u_3^{*(0)} = 0 \quad \text{at} \quad \zeta = \pm 1 \ . \tag{16}$$

The general solution of (14) has the form $p_t^{*(0)} = P_t^{*(0,0)}$ within which we agree to use double superscripts for functions independent of ζ, with the first superscript referring to the order of approximation and the second denoting the degree of any possible ζ^m factor. In view of that, solutions of (13) and (15), subject to (16), may be given by

$$u_1^{*(0)} = \frac{P_{t,\xi_1}^{*(0,0)}}{\Lambda^2} \left(1 - \frac{\cos(\Lambda_1 \zeta)}{\cos \Lambda_1} \right)$$
$$u_3^{*(0)} = \frac{P_{t,\xi_3}^{*(0,0)}}{\Lambda^2} \left(1 - \frac{\cos(\Lambda_3 \zeta)}{\cos \Lambda_3} \right) \ . \tag{17}$$

It must be kept in mind that for such symmetric motion u_1^*, u_3^* and p_t^* vary as even functions and u_2^* varies as an odd function of ζ. It can be demonstrated that this property of response is ensured by the relations (8), (10) and (12), however for the sake of brevity we will simply omit terms that do not satisfy the symmetry requirements, as we did in (17). Solutions (17) may now be inserted into incompressibility condition to establish that

$$\Lambda^2 u_2^{*(0)} = \frac{P_{t,\xi_1\xi_1}^{*(0,0)} \sin(\Lambda_1 \zeta)}{\Lambda_1 \cos \Lambda_1} + \frac{P_{t,\xi_3\xi_3}^{*(0,0)} \sin(\Lambda_3 \zeta)}{\Lambda_3 \cos \Lambda_3} - \zeta \left(P_{t,\xi_1\xi_1}^{*(0,0)} + P_{t,\xi_3\xi_3}^{*(0,0)} \right) , \tag{18}$$

thus we now only need to satisfy the boundary condition $(16)_2$. Using (17) and (18) it is possible to represent that boundary condition in the form

$$\left(\sqrt{\gamma_{21}}\tan\Lambda_1-\Lambda\right)P^{*(0,0)}_{t,\xi_1\xi_1}+\left(\sqrt{\gamma_{23}}\tan\Lambda_3-\Lambda\right)P^{*(0,0)}_{t,\xi_3\xi_3}=0\ , \tag{19}$$

which together with (8) leads to an identity, automatically satisfied for any plane wave propagating along $(c_\theta,0,s_\theta)$ or any other direction whose projection on the plane of a layer is parallel to $(c_\theta,0,s_\theta)$. This is in particular true for the type of harmonic waves we consider, see (2). All of the leading order displacements are now obtained in terms of the single function $P_t^{*(0,0)}$, which is the leading order pressure. We term this function as the *long-wave pressure* and remark that an appropriate equation for it can only be determined at the next order.

The second order problem is given by the appropriate equations of motion

$$\begin{aligned}\gamma_{21}u^{*(2)}_{1,\zeta\zeta}+\Lambda^2u_1^{*(2)}&=p^{*(2)}_{t,\xi_1}-B_{1111}u^{*(0)}_{1,\xi_1\xi_1}-\gamma_{31}u^{*(0)}_{1,\xi_3\xi_3}-\\&-(B_{1122}+B_{1221})u^{*(0)}_{2,\xi_1\zeta}-(B_{1133}+B_{1331})u^{*(0)}_{3,\xi_1\xi_3}+\eta^{-2}\left(u^{*(0)}_{1,\tau\tau}+\Lambda^2u_1^{*(0)}\right)\end{aligned} \tag{20}$$

$$p^{*(2)}_{t,\zeta}=(B_{1122}+B_{1221})u^{*(0)}_{1,\xi_1\zeta}+B_{2222}u^{*(0)}_{2,\zeta\zeta}+\Lambda^2u_2^{*(0)}+(B_{2233}+B_{2332})u^{*(0)}_{3,\xi_3\zeta} \tag{21}$$

$$\begin{aligned}\gamma_{23}u^{*(2)}_{3,\zeta\zeta}+\Lambda^2u_3^{*(2)}&=p^{*(2)}_{t,\xi_3}-B_{3333}u^{*(0)}_{3,\xi_3\xi_3}-\gamma_{13}u^{*(0)}_{3,\xi_1\xi_1}-\\&-(B_{1133}+B_{1331})u^{*(0)}_{1,\xi_1\xi_3}-(B_{2233}+B_{2332})u^{*(0)}_{2,\xi_3\zeta}+\eta^{-2}\left(u^{*(0)}_{3,\tau\tau}+\Lambda^2u_3^{*(0)}\right)\ ,\end{aligned} \tag{22}$$

in which we remark that quantities in brackets are of the same asymptotic order as other terms, see (12). Eqs. (20)–(22) must be solved in conjunction with the second order incompressibility condition $u^{*(2)}_{1,\xi_1}+u^{*(2)}_{2,\zeta}+u^{*(2)}_{3,\xi_3}=0$ and second order boundary conditions

$$u_1^{*(2)}=u_2^{*(2)}=u_3^{*(2)}=0\quad\text{at}\quad\zeta=\pm1 \tag{23}$$

It is worth noting that Eqs. (20)–(22) are written in such a form that terms on the right hand side are all known at the time they are to be used. The solution procedure is quite similar at this order and for the sake of brevity we will not present intermediate steps. It is only important for our derivation that the boundary value problem (20)–(23) can only be satisfied provided a certain relation for $p\equiv P_t^{*(0,0)}$ is satisfied. In terms of original dimensional variables this relation, termed leading order governing equation, is given by

$$T_1^2\left[\rho h^2\frac{\partial^4 p}{\partial x_1^2\partial t^2}+\Lambda^2\frac{\partial^2 p}{\partial x_1^2}\right]+T_3^2\left[\rho h^2\frac{\partial^4 p}{\partial x_3^2\partial t^2}+\Lambda^2\frac{\partial^2 p}{\partial x_3^2}\right]$$
$$-h^2\left\{\left(Q_c T_1^2-\frac{2}{3}\Lambda^2\right)\frac{\partial^4 p}{\partial x_1^4}-\left(\frac{4}{3}\Lambda^2-\gamma_{31}T_1^2+\frac{E_1}{\Lambda_1}T_1\right.\right. \tag{24}$$
$$\left.\left.-\gamma_{13}T_3^2+\frac{E_3}{\Lambda_3}T_3+E_{13}\right)\frac{\partial^4 p}{\partial x_1^2\partial x_3^2}+\left(Q_s T_3^2-\frac{2}{3}\Lambda^2\right)\frac{\partial^4 p}{\partial x_3^4}\right\}=0\ ,$$

in which $Q_c=2\beta_{12}-\gamma_{12}-\gamma_{21}$, $Q_s=2\beta_{23}-\gamma_{23}-\gamma_{32}$ and

$$\frac{E_1}{4}=\frac{3}{4}(\gamma_{31}-Q_c)+\frac{\gamma_{23}}{2}-\frac{\gamma_{21}(\mu_{13}+\gamma_{21})}{\gamma_{21}-\gamma_{23}}$$
$$\frac{E_3}{4}=\frac{3}{4}(\gamma_{13}-Q_c)+\frac{\gamma_{21}}{2}+\frac{\gamma_{23}(\mu_{13}+\gamma_{23})}{\gamma_{21}-\gamma_{23}}$$
$$E_{13}=3(Q_c+Q_s-\gamma_{13}-\gamma_{31})+2(2\mu_{13}+\gamma_{21}+\gamma_{23})\ .$$

The governing Eq. (24) appears to be very complex when compared to many classical plate theories. In practical applications an additional asymptotic analysis of (24) may be performed, using the same ideas as high-frequency approximations of (8) mentioned previously. Indeed, as the harmonic number increases, terms multiplied onto one of the tangents will dominate, thus indicating the evolution of another asymptotic process. It is also worth noting that the plane strain governing equation derived in [2] may be recovered from (24) by using (8) and assuming no dependence on x_1 or x_3.

Acknowledgement

This work was performed while A.V. Pichugin was working at the University of Salford, supported by an ORS Award and a university bursary.

References

1. Kaplunov JD, Kossovich LY, Nolde EV (1998) Dynamics of thin walled elastic bodies. Academic Press, New York
2. Nolde EV, Rogerson GA (2001) Long wave asymptotic integration of the governing equations for a pre-stressed incompressible elastic layer with fixed faces. Submitted for publication
3. Ogden RW (1984) Non-linear elastic deformations. New York: Ellis Horwood
4. Pichugin AV (2002) Ph. D. Thesis. University of Salford
5. Pichugin AV, Rogerson GA (2001) A two-dimensional model for extensional motion of a pre-stressed incompressible elastic layer near cut-off frequencies. IMA J Appl Math, 66: 357-385

Construction of the Theory of Piezoelectric Shells and Plates

N. N. Rogacheva

Institute Economy Communal and Engineering, Moscow, Russia

1 Introduction

The simplest and the most used way for reducing three-dimensional equations to two-dimensional ones is the hypotheses method. The known Kirchhoff–Love theory of elastic shells and plates was obtained in this way. To derive the theory of piezoelectric shells and plates the well-known Kirchhoff's hypotheses are adopted for mechanical quantities of electroelastic state while different authors use different hypotheses for electrical quantities.

The hypotheses accepted by different author are discussed. The necessity of considering of hypotheses stems from the fact that there is no agreement among specialists regarding the validity of the hypotheses used for construction of the theory of piezoelectric shells and plates.

2 Boundary conditions in the theory of electroelasticity

In electroelasticity theory, mechanical boundary conditions are formulated just in the classical elasticity theory: the stresses, displacements, or mixed boundary conditions are specified at the boundary of the body. The type of electrical boundary conditions depends on the presence and location of electrodes and the way whereby the electrostatic states are excited in the body. We assume for simplicity that the thickness of electrodes covering a part of body surface are infinitely small compared to the smallest size of body.

A thin-walled electroelastic body of $2h$ in thickness is to be related to the three-orthogonal coordinate system $\alpha_1, \alpha_2, \lambda$, where curvilinear coordinates α_1 and α_2 coincide with the curvatures of the middle surface of the shell and a linear coordinate λ is directed along the normal to the middle surface.

We write some types of electrical boundary conditions on the shell faces. We can define electrical potential on the electrode-covered faces

$$\psi\big|_{\gamma=\pm h} = \pm V \qquad (2.1)$$

If the electrodes are short-circuited, the electrical potential on the faces is zero

$$\psi\big|_{\gamma=\pm h} = 0 \tag{2.2}$$

If the shell faces are not electrode-covered and are in contact with medium with low permittivity – say, vacuum or air – then the component of the electric induction vector, which is normal to the surface, is zero

$$D_3\big|_{\gamma=\pm h} = 0 \tag{2.3}$$

If the electrodes are shorted by a contour with known complex conductivity $Y = Y_0 + iY_1$, the electrical condition becomes

$$\int_{\Omega} \dot{D}_3 d\Omega = 2VY \tag{2.4}$$

where the integral is evaluated over the surface Ω of one of the electrodes and point denotes the time derivative, i is imaginary unit. If the electrodes are disconnected, V is calculated from the condition

$$\int_{\Omega} \dot{D}_3 d\Omega = 0 \tag{2.5}$$

3 The theory of piezoelectric shells and plates as asymptotic approach of 3D-theory

Since 2D-equilibrium equations and strain-displacement formulas in the theory of piezoelectric shells are the same as in the linear theory of elastic shells, we construct 2D-piezoelectric relations and electrostatic equations only.

The relations between the elastic field tensors (strain e_{kl} and stress σ_{kl}) and electric field vectors (induction $\boldsymbol{D}$ with components D_1, D_2, and D_3 and strength $\boldsymbol{E}$ with components E_1, E_2, and E_3) can be presented by system of piezoelectric equations

$$e_{ij} = s^E_{ijkl}\sigma_{kl} + d_{ijkl}E_k \qquad D_i = d_{ikl}\sigma_{kl} + \varepsilon^T_{ik}E_k \qquad (i, j, k, l = 1, 2, 3) \tag{3.1}$$

where s^E_{ijkl} are elastic compliances at constant electric field, ε^T_{ij} are dielectric permittivities for constant elastic stresses, d_{ikl} are piezoelectric constants that are also called piezoelectric moduli.

For this kind of materials the behavior of the electric field is described by electrostatic equations

$$div\boldsymbol{D} = 0 \qquad \boldsymbol{E} = -grad\psi \tag{3.2}$$

At first, we perform a scale extension for the electroelasticity equations with respect to the coordinates in a usual way for the asymptotic analysis:

$$\alpha_i = \eta^s L\xi_i \qquad \gamma = \eta^1 L\varsigma \qquad \eta = h/L \tag{3.3}$$

where L is the characteristic dimension of the shell and s is the index of variability of the electroelastic state. Here and below indices i and j take the values 1, 2. Every equation with subscripts i and j contains two equations: one for $i=1$ and $j=2$ and the other for $i=2$ and $j=1$. The coordinates ξ_i and ζ are chosen so that differentiation with respect to them does not change result in any substantial increase or decrease in sought-for quantities.

In (Rogacheva 1994) asymptotical analysis of piezoelectric thin-walled structures with different direction of polarization and different electrical boundary conditions is carried out in detail. Let us briefly run through construction of the theory of piezoelectric shells with thickness polarization. We define the following asymptotics for the quantities

$$\begin{aligned}
&\frac{u_i}{L} = \eta^s u_{i*}, && \frac{u_3}{L} = \eta^c u_{3*}, && \\
&s_{11}^E \sigma_{ij} = \sigma_{ij*}, && s_{11}^E \sigma_{33} = \eta^{1-c} \sigma_{33*}, && s_{11}^E \sigma_{i3} = \eta^{1-s} \sigma_{i3*}, \\
&d_{31} E_3 = E_{3*}, && d_{31} E_1 = \eta^{1-s} E_{1*}, && d_{31} \frac{\psi}{L} = \eta^1 \psi_*, \\
&\frac{d_{31}}{\varepsilon_{11}^T} D_3 = D_{3*}, && \frac{d_{31}}{\varepsilon_{11}^T} D_i = \eta^{1-s} D_{i*}, && \\
&s_{11}^E q_3^{\pm} = \eta^{1-c} q_{3*}^{\pm}, && s_{11}^E q_i^{\pm} = \eta^{1-s} q_{i*}^{\pm}, &&
\end{aligned} \tag{3.4}$$

where u_i, u_3 are the displacement vector components, $q_i^{\pm}, q_3^{\pm}$ are the surface load vector components. The introduced asymptotics are valid for shell with the faces completely covered by electrodes with given electrical potential on them (2.1). Here, quantity c is

$$c = \begin{cases} 0 \ \text{for} \ 0 \le s < 1/2 \\ -1+2s \ \text{for} \ 1/2 \le s < 1 \end{cases} .$$

For the shell without electrodes on the faces, we will take another asymptotics in the form

$$\frac{d_{31}}{\varepsilon_{11}^T} D_3 = \eta^{2-2s} D_{3*} . \tag{3.5}$$

All quantities with asterisks in Eqs. (3.4), (3.5) are of the same order and are dimensionless.

Since we deal with linear theory, there is the only asymptotic representation for internal state that leads to mathematically justified 2D-shell theory. Substituting the asymptotics (3.4), (3.5) and (3.3) in equations of electroelasticity we obtain the system of equations according to the desired quantities. Using asymptotic technique we integrate 3D-equations with respect to dimensionless thickness coordinate and present all the needed quantities (displacements, stress, components of the electric induction vector, and electrical potential) as polynominal in ζ. Each term in polynomials has a factor η in a positive power, what defines the asymptotic order of the term. We neglect the terms of order ε and less than ε in resulting asymptotic expansions

$$\varepsilon = O(\eta^1 + \eta^{2-2s}) \tag{3.6}$$

and write formulas

$$\begin{aligned}
\sigma_{ij*} &= \sigma_{ij,0} + \eta^{1-2s+c} \varsigma \sigma_{ij,1} && (\sigma_{ij*}, u_{i*}, e_{i*}, m_{i*}, E_{3*}) \\
u_{3*} &= u_{3,0} + \eta^{1-c} \varsigma u_{3,1} && (g_{i*}) \\
\sigma_{i3*} &= \sigma_{i3,0} + \varsigma \sigma_{i3,1} + \eta^{1-2s+c} \varsigma^2 \sigma_{i3,2} && (E_{i*}, D_{i*}, \psi_*) \\
\sigma_{33*} &= \sigma_{33,0} + \varsigma \sigma_{33,1} + \eta^{1-2s+c} \varsigma^2 \sigma_{33,2} + \eta^{2-4s+2c} \varsigma^3 \sigma_{33,3}
\end{aligned} \tag{3.7}$$

In the parentheses on the right of the formulas, we enumerate the quantities whose asymptotic expansions have the same form. Formula for D_{3*} for a shell with electrodes on the faces looks like

$$D_{3*} = D_{3,0} \tag{3.8}$$

D_{3*} for a shell without electrodes on the faces is a small quantity. Formula for it is given in (Rogacheva 1994).
We substitute Eqs. (3.7), (3.8) into equations of the theory of electroelasticity. Equating the coefficients at the same power of ζ we get a system 2D theory, where the number of equations is equal to the number of unknowns. The system can be integrated without resorting to the traditional notations of shell theory, i.e., forces, moments, displacements, and strains of the middle surface. Nevertheless, the transition is necessary if we make a comparison with other theories.

We apply the notations of shell theory to our equations. They are related to three-dimensional quantities by following formulas:

$$\begin{aligned}
w &= -\eta^c L u_{3,0} && v_i = \eta^s L u_{i,0} \\
T_i &= \int_{-h}^{+h} \sigma_{ii} d\gamma = \frac{2h}{s_{11}^E} \sigma_{ii,0} && S_{ij} = \int_{-h}^{+h} \sigma_{ij} d\gamma = \frac{2h}{s_{11}^E} \sigma_{ij,0}
\end{aligned} \tag{3.9}$$

$$\begin{aligned}
G_i &= -\int_{-h}^{+h} \gamma \sigma_{ii} d\gamma = -\eta^{1-2s+c} \frac{2h^2}{3 s_{11}^E} \sigma_{ii,1} \\
H_{ij} &= \int_{-h}^{+h} \gamma \sigma_{i2} d\gamma = \eta^{1-2s+c} \frac{2h^2}{3 s_{11}^E} \sigma_{ij,1},
\end{aligned} \tag{3.10}$$

where w, v_i are the middle surface displacement, T_i, S_{ij} are the forces, and G_i, H_{ij} are the moments of the shell theory. We transform Eqs. (3.10), using the foregoing equations to obtain the electroelasticity relations for shells with thickness polarization

$$T_i = \frac{2h}{s_{11}^E (1-\nu^2)} (\varepsilon_i + \nu \varepsilon_j) - \frac{2h d_{31}}{s_{11}^E (1-\nu)} E_3^{(0)} \tag{3.11}$$

$$G_i = -\frac{2h^3}{3s_{11}^E(1-\nu^2)}(\kappa_i + \nu\kappa_j) + \frac{2h^3 d_{31}}{3s_{11}^E(1-\nu)}E_3^{(1)} \tag{3.12}$$

$$S_{ij} = \frac{2h}{s_{66}^E}\omega \qquad H = \frac{4h^3}{s_{66}^E}(\tau - \frac{\omega}{2R_j}) \tag{3.13}$$

Below is given only the formulas, which are used for analysis of shell theory hypotheses. The electrical quantities vary with respect to γ according to the law

$$E_3 = E_3^{(0)} + \gamma E_3^{(1)} \qquad \psi = \psi^{(0)} + \gamma\psi^{(1)} + \gamma^2\psi^{(2)} \tag{3.14}$$

The superscripts in parentheses are coefficients at the power γ in polynomials in γ. The formulas for D_3, E_3, ψ can have different forms depending on the electrical conditions on the shell's faces. For example, if the faces are covered with electrodes and their electrical potential is given, we find the electrical quantities as

$$\psi^{(1)} = -E_3^{(0)} = \frac{V}{h}, \qquad \psi^{(0)} = -h^2\psi^{(2)}$$
$$D_3^{(0)} = -\varepsilon_{33}^T\frac{V}{h} + \frac{3d_{31}}{2h}(T_1 + T_2) \tag{3.15}$$
$$\psi^{(2)} = -\frac{1}{2}E_3^{(1)} = -\frac{3d_{31}}{4h^3\varepsilon_{33}^T}(G_1 + G_2)$$

$$G_i = -\frac{2h^3 B}{3}(\kappa_i + \sigma\kappa_j)$$
$$B = \frac{2-(1-\nu)k_p^2}{2s_{11}^E(1-\nu^2)(1-k_p^2)}, \qquad \sigma = \frac{2\nu + (1-\nu)k_p^2}{2-(1-\nu)k_p^2} \tag{3.16}$$
$$k_{31}^2 = \frac{d_{31}^2}{s_{11}^E\varepsilon_{33}^T}, \qquad k_p^2 = \frac{2k_{31}^2}{1-\nu}.$$

If electrodes are disconnected, Eqs. (3.15) still stand, but V is unknown quantity, calculated from the condition (2.5). If the faces have no electrodes we should replace Eqs. (3.15) by

$$\psi^{(1)} = -E_3^{(0)} = -\frac{3d_{31}}{2h\varepsilon_{33}^T}(T_1 + T_2). \tag{3.17}$$

We eliminate $E_3^{(0)}$ from Eq. (3.11) and get for the normal forces of the shell without electrodes on its faces the following relation:

$$T_i = 2hB(\varepsilon_i + \sigma\varepsilon_j) \tag{3.18}$$

4 Discussion of the hypotheses applied to piezoelectric shell theory

4.1 Kirchhoff's hypotheses

It is illustrated above Kirchhoff's hypotheses are true for mechanical quantities of the piezoelectric shells with thickness polarization. There are some exceptions to this rule (Rogacheva 1994), which are not treated here.

According to the first hypothesis we neglect stresses σ_{33} as compared to σ_{11} and σ_{22} in Eqs. (3.1). In different papers the use of the first hypothesis should present no problems. Things get worse with the second hypothesis. Usually it is formulated as follows: the rectilinear element that was normal to the middle surface before deformation remains straight and perpendicular to the strained surface and does not elongate. Discussions on the elastic shell theory foundation into 20th century clarify among other things (Gol'denveizer 1976), that the requirement of constant length for normal element is unnecessary. It was shown that the elongation along all axes are the same order. Really Hooke's law can be written as

$$e_3 = \frac{1}{E}\sigma_{33} - \frac{\nu}{E}(\sigma_{11} + \sigma_{22}), \qquad e_i = \frac{1}{E}\sigma_{ii} - \frac{\nu}{E}(\sigma_{jj} + \sigma_{33}) \tag{4.1}$$

Rearrangement of Eqs. (4.1) in view of the first hypothesis gives

$$\frac{e_3}{e_1 + e_2} = -\frac{\nu}{1-\nu}. \tag{4.2}$$

Thus all three strains are the same order. The first equation (4.1) is never used in construction of elastic shell theory. We can use Eq. (4.1) after solving shell problem if we need to define e_3. Because of this, inaccurate formulation of the second Kirchhoff's hypothesis does not introduce an error in the theory of elastic shells.

The situation is different if we deal with piezoelectric shells. We demonstrate it by an example. In much favour book (Tzou 1993) the assumption of constant length of normal element is used to simplify the following equation:

$$D_3 = e_{31}(e_1 + e_2) + e_{33}e_3 + \varepsilon_{33}^s E_3, \tag{4.3}$$

where ε_{kl}^s are dielectric permittivities for constant elastic stresses, e_{kl} are piezoelectric constants. The author of (Tzou 1993) neglected the strain e_3 in Eq. (4.3) and arrived at untrue formula

$$D_3 = e_{31}(e_1 + e_2) + \varepsilon_{33}^s E_3 \tag{4.4}$$

In piezoelectricity Eq. (4.3) is used to calculate important characteristic - antiresonance frequencies. The simplified Eq. (4.4) can change not only antiresonance frequencies, but also the main coefficients of 2D-electroelasticity relations (Rogacheva 1994) and electromechanical coupling coefficient. To demonstrate difference between Eq. (4.3) and Eq. (4.4) we calculate the first antiresonance frequency using equation (2.5) and electromechanical coupling coefficient k from the formula (Berlincourt et al. 1964)

$$k^2 = \frac{\omega_a^2 - \omega_r^2}{\omega_a^2},$$

where ω_r and ω_a are the resonance and antiresonance frequencies respectively. For the plate from PZT-4 of radius 1 cm with thickness polarization with electrodes on the faces and with given electrical potential we find by our theory

$$\omega_r = 114.7\text{kHz} \qquad \omega_a = 132.3\text{kHz} \qquad k^2 = 0.248$$

and by the theory (Tzou 1993)

$$\omega_a = 123.7\text{kHz} \qquad k^2 = 0.140$$

Note, that this difference does not decrease when the thickness tends to zero.

4.2 Hypotheses for electrical quantities

Asymptotical analysis proves that the electrical potential varies according to the square law at the thickness coordinate (3.14) for the considered case. But in many papers the linear law (Lee 1990; Tzou 1993) is used, that leads to the following incorrect relation for bending moment instead relation (3.16)

$$G_i = -\frac{2h^3}{3s_{11}^E(1-\nu^2)}(\kappa_i + \nu\kappa_j) \qquad (4.5)$$

We calculate the ratio of the coefficient of κ_i in the relation (3.16) to the coefficient of κ_i in the relation (4.5) for PZT-5

$$Bs_{11}^E(1-\nu^2) = 1.39$$

The ratio is much more unity, which is to say that the theory with the linear law for electrical potential can cause to a large error in case of materials with strong piezoeffect.

5 Influence of electrical conditions and directions of polarization on shell's resonance frequencies

To study the influence we calculate the following first resonance frequency for circular plate in radius 1 cm made of PZT-5 with thickness polarization and with different electrical conditions on the faces: 1) the electrical potential is given on the faces' electrodes, 2) the faces' electrodes are disconnected, 3) the plate does not have electrodes

$$1)\ \omega_r = 99.1\text{kHz}$$

$$2)\ \omega_r = 117\text{kHz}$$

$$3)\ \omega_r = 122.5\text{kHz}$$

Another example is concerned shells with different directions of polarization. We calculate the following first resonance frequency for cylindrical shell in radius 1 cm of length 3 cm, made of PZT-5 with thickness polarization and with different directions of polarization with given electrical potential on the electrodes: 1) the shell with thickness polarization with electrodes on the faces, 2) the shell polarized in lengthwise direction with electrodes on the edges, 3) the shell polarized in circular direction with electrodes on the edges

$$1)\ \omega_r = 56.7\text{kHz}$$

$$2)\ \omega_r = 39.9\text{kHz}$$

$$3)\ \omega_r = 28.7\text{kHz}$$

Our examples verify important role of electrical conditions and direction of polarization.

Acknowledgement

The work was partly supported by Russian Foundation of Basic Research (Grant No. 02-01-00843).

References

1. Berlincourt DA, Curran DR, Jaffe H (1964) Piezoelectric and piezomagnetic materials and their function as transducers. In: Mason WP (ed) Physical Acoustics vol 1A. Academic Press, New York, pp 204-326
2. Gol'denveizer AL (1976) The theory of thin elastic shells (in Russian), 2nd edn. Nauka, Moscow
3. Lee CK (1990) Theory of laminated piezoelectric plates for design of distributed sensors/actuators. Part 1. J Acoust Soc Am 87 3: 1144-1158
4. Rogacheva NN (1994) The theory of piezoelectric shells and plates. CRC Press, Boca Raton
5. Tzou HS (1993) Piezoelectric shells (Distributed Sensing and Control of Continua). Kluver Academic Publishers, Dordrecht

Boundary Layers Near the Reflected and Transmitted Dilatational Wave Fronts in a Composite Cylindrical Shell

G. A. Rogerson[1], I. V. Kirillova[2] and Ya. A. Parfenova[2]

[1]Department of Computer and Mathematical Sciences, University of Salford, UK
[2]Department of Mathematical Theory of Elasticity and Biomechanics, Saratov State University, Saratov, Russia

1 Introduction

Transient waves in composite cylindrical shells initiated by bending longitudinal edge shock loading (LM loading [1]) are considered. A version of the method of matched asymptotic expansions is used. The choice of approximations is connected with different values of variability and dynamicity indices [2,3]. Our analysis will include long-wave low-frequency approximations (tangential and transverse), the Saint-Venant quasi-static boundary layer, the quasi-plane problem of elasticity (quasi-symmetric and quasi-antisymmetric), as well as boundary layers within vicinities of the wave fronts and quasi-front. The accuracy of this separation of non-stationary stress-strain-state (SSS) has previously been established by demonstrating the existence of the overlap regions between the various approximations. In particular, it has been shown that overlap regions exist between the long-wave tangential approximation and the boundary layer in the vicinity of the quasi-front; the long-wave transverse approximation and the solution of the quasi-plane quasi-anti-symmetric problem; and solutions of the quasi-plane problems and the boundary layers in the vicinities of the dilatation wave front [2,3,4]. This paper is devoted to construction of solution for the antisymmetric plane boundary layer in the vicinity of dilatation wave front [3,5].

2 Formulation of the problem

2.1 The scheme of the shell

Consider an isotropic linear elastic composite cylindrical shell with discontinuity in thickness. The shell consists of two thin walled cylinders with common mid-surface. Elements of the shell have different thickness $h^{(i)}$ and are made of distinct materials. We use the index $^{(i)}$ $(i = 1, 2)$ to denote of mechanical and geometric parameters of corresponding cylinders.

We refer the midsurface to the coordinate system (α, θ), where α is the length of the meridian and θ is the angular coordinate. The orthogonal coordinate system is defined by the vector equality

$$\mathbf{P}(\alpha,\theta,z)=\mathbf{M}(\alpha,\theta)+z\,\mathbf{n} \tag{2.1}$$

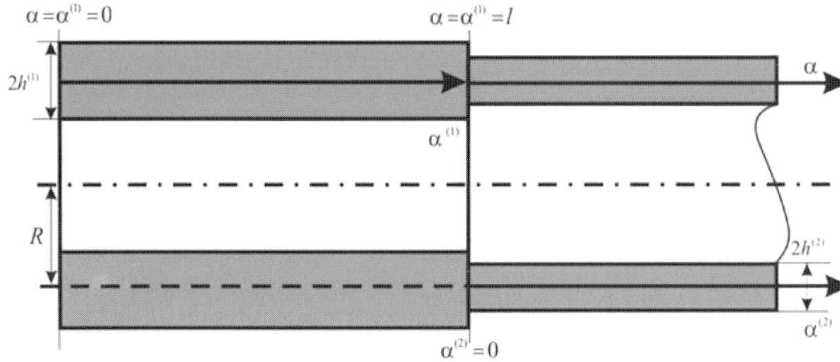

Fig. 1. Shell geometry

where $\mathbf{n}$ is the unit midsurface normal and z is the distance from midsurface along the normal. Let $\alpha = 0$ correspond to the left edge of the shell, with $\alpha = l$ corresponding to the interface. The scheme of the shell is shown in Fig.1.We will specifically consider the case in which the second section of the shell (on the right hand side of Fig. 1) is semi-infinite and thinner than the first section, i.e. $h^{(1)} > h^{(2)}$.

2.2 Boundary conditions

Let a slowly varying shock load, along the common parallel axis of both cylinders, simulated by the Heaviside unit function $H(t)$, be applied to the edge $\alpha = 0$. In the case considered, namely that of LM loading, we have a non-zero bending stress couple on the edge.

An example of the corresponding boundary conditions is

$$\sigma_{11}^{(1)} = I\,z\,H(t)f(\theta), \quad v_2^{(1)} = v_3^{(1)} = 0 \;\text{ at }\; \alpha = 0 \tag{2.2}$$

where σ_{ij} $(i, j = 1, 2, 3)$ are the stresses, v_k $(k = 1, 2, 3)$ are the displacements, t is time and I is the amplitude of the load. For simplicity throughout this paper we consider the case in which the variability of the function $f(\theta)$ is equal to zero, i.e. this function and its derivative are of the same order.

The boundary conditions on the interface have the form:

$$\sigma_{1j}^{(1)} = \begin{cases} 0, & |z| > h^{(2)}, \\ \sigma_{1j}^{(2)}, & |z| \le h^{(2)}, \end{cases} \tag{2.3}$$

$$v_j^{(1)} = v_j^{(2)}$$

$$|z| \le h^{(2)} \quad (j = 1,2,3)$$

We consider only the homogeneous initial conditions

$$v_j = \frac{\partial v_j}{\partial t} = 0 \quad (j = 1,2,3) \text{ at } t = 0 \tag{2.4}$$

Let us construct solutions for both the incident wave propagating in the first section, initiated by the edge loading, as well as for the transmitted and reflected waves initiated by the incident wave on the interface. Note that the solution for the incident wave is the same as the corresponding solution for the semi-infinite cylindrical shell.

3 Boundary layer in the vicinity of the dilatation wave front

The governing equations appropriate for the boundary layer in the vicinity of the dilatation wave front have the following form for the asymptotically leading terms $v_1, \sigma_{11}, \sigma_{33}$ [3,4,5]

$$\frac{\partial^2 v_1}{\partial \xi_h^2} + \frac{\partial^2 v_1}{\partial \zeta^2} - \kappa^2 \frac{\partial^2 v_1}{\partial \tau_h^2} = 0$$

$$\sigma_{11} = \frac{E}{2(1+\nu)\kappa^2 h} \frac{\partial v_1}{\partial \xi_h} \tag{3.1}$$

$$\sigma_{33} = \frac{E\nu}{(1-2\nu)(1+\nu)h} \frac{\partial v_1}{\partial \xi_h}$$

where $\kappa^2 = (1-2\nu)/2(1-\nu)$, $\xi_h = \alpha / h$, $\tau_h = c_2 t / h$, $\zeta = z / h$ and c_2 is the shear wave speed $c_2 = \sqrt{E/2(1+\nu)\rho}$.

The corresponding asymptotically leading boundary conditions are given by

$$\frac{E}{2(1+\nu)\kappa^2 h^2} \frac{\partial v_1}{\partial \xi_h} = I\zeta\, H(\tau_h) \text{ at } \xi_h = 0 \tag{3.2}$$

$$\frac{\partial v_1}{\partial \xi_h} = 0 \text{ at } \zeta = \pm 1$$

The analysis of the boundary layer will be in respect of bending stress couple. To begin we apply the Laplace-Fourier transform. The transformed bending stress couple is given by [4]

$$G_1^{Ls} = -\sqrt{\frac{2}{\pi}} h^3 I \frac{2\chi}{s\alpha^2}\left[\frac{\cosh\alpha}{\alpha\sinh\alpha} - \frac{1}{\alpha^2} - \frac{1}{3}\right], \tag{3.3}$$

with

$$G_1^{Ls} = -\sqrt{\frac{2}{\pi}} \int G_1^L(\xi_h, s)\sin(\chi\xi_h)d\xi_h, \tag{3.4}$$

where χ is the Fourier transform parameter, s is the Laplace transform parameter and $\alpha^2 = \chi^2 + \kappa^2 s^2$. Expanding the function G_1^{Ls} in series about its poles, we have

$$G_1^L = \frac{4h^3 I}{\pi^2}\sum_{m=1}^{\infty}\frac{1}{m^2 s}\exp\left[-\kappa s\left(1+\frac{m^2\pi^2}{\kappa^2 s^2}\right)^{1/2}\xi_h\right]. \tag{3.5}$$

Expanding the functions G_1^L as a series in negative power of the parameters s and retaining only zero-order Bessel functions we obtain [4]

$$G_1 = \frac{4h^3 I}{\pi^2}\sum_{m=1}^{\infty}\frac{1}{m^2}J_0\left(\sqrt{2}m\pi\sqrt{\frac{\xi_h}{\kappa}(t_1-\kappa\xi_h)}\right)H(t_1-\xi_h). \tag{3.6}$$

4 Solution for Laplace transforms

Let us introduce nondimensionless coordinates in the form

$$\begin{aligned} \xi_{h_1} &= \frac{\alpha_1}{h_1} \\ \xi_{h_2} &= \frac{\alpha_1 - l}{h_2} \quad (\alpha_1 \ge l) \\ \zeta_{h_1} &= \frac{\alpha_3}{h_i} \\ \tau_{h_i} &= \frac{c_2^{(i)}}{h_i} \quad (i=1,2) \end{aligned} \tag{4.1}$$

We represent the reflected wave in the first section in the form of a superposition of two waves. The first wave corresponds to the edge $\xi_{h_1} = l_{h_1}$ ($l_{h_1} = l/h_1$), which is considered as free. The solution for the second wave, together with solutions for the propagating, transmitted and first type of reflected wave, allows us to satisfy all continuity conditions on the interface. This superposition is suitable both for the formulation of the contact problem and for separating in the reflected wave the disturbance initiated by the second section with the small thickness.

The solution for the Laplace transform in the case of the propagating wave is the same as in the semi-infinite shell and is defined by the following formula

$$v_1^{(1,1)L} = \frac{4(1+\nu_1)\kappa_1^{\ 2}I}{E_1}\frac{h_1^{\ 2}}{s_1}\sum_{m=1}^{\infty}\frac{(-1)^m}{\pi m}\frac{\sin[m\pi\zeta]}{\lambda_m^{\ (1)}}\exp[-\lambda_m^{\ (1)}\xi_{h_1}] \tag{4.2}$$

$$\sigma_{11}^{(1,1)L} = -2I\frac{h_1}{s_1}\sum_{m=1}^{\infty}\frac{(-1)^m}{\pi m}\sin[m\pi\zeta]\exp[-\lambda_m^{\ (1)}\xi_{h_1}],$$

where $\lambda_m^{(i)} = \kappa_i s_i\sqrt{1+\frac{m^2\pi^2}{\kappa_i^2 s_i^{\ 2}}}$, s_i is the Laplace parameter with respect to τ_{h_i}.

The Laplace transform of the solution for the first component of the reflected wave is obtained with the help of exchanging of the sign "+" before the sum by the sign "-", $\lambda_m^{(1)}$ by (- $\lambda_m^{(1)}$) (except $\lambda_m^{(1)}$ in the argument of exponent) and $\exp[-\lambda_m^{\ (1)}\xi_{h_1}]$ by $\exp[-\lambda_m^{\ (1)}(2l_{h_1}-\xi_{h_1})]$, thus

$$v_1^{(1,2)L} = \frac{4(1+\nu_1)\kappa_1^{\ 2}I}{E_1}\frac{h_1^{\ 2}}{s_1}\cdot\sum_{m=1}^{\infty}\frac{(-1)^m}{\pi m}\frac{\sin[m\pi\zeta_{h_1}]}{\lambda_m^{\ (1)}}\exp[-\lambda_m^{\ (1)}(2l_{h_1}-\xi_{h_1})] \tag{4.3}$$

$$\sigma_{11}^{(1,2)L} = 2I\frac{h_1}{s_1}\sum_{m=1}^{\infty}\frac{(-1)^m}{\pi m}\sin[m\pi\zeta_{h_1}]\exp[-\lambda_m^{\ (1)}(2l_{h_1}-\xi_{h_1})]$$

The continuity conditions take the form

$$\sigma_{11}^{(1,3)} = \begin{cases} 0, & |\alpha_3| > h_2, \\ \sigma_{11}^{(1,2)}, & |\alpha_3| \le h_2, \end{cases} \tag{4.4}$$

$$v_1^{(1,1)} + v_1^{(1,2)} + v_1^{(1,3)} = v_1^{(2)},$$

where indices (1,3) and (2) denote the second component of the reflected wave and the transmitted wave, respectively .

Let us denote the unknown value of the normal stress in the contact region by $Z_c(\zeta_{h_1},\tau_{h_1})$. Then the Laplace transforms of the solutions for the second component of the reflected wave and the transmitted wave can be written as follows:

$$v_1^{(1,3)L} = -\frac{4(1+\nu_1)\kappa_1^{\ 2}}{E_1}\frac{h_1^{\ 2}}{s_1}\sum_{m=1}^{\infty}\frac{(-1)^m}{\pi m}\frac{D_m^{(3)L}\sin[m\pi\zeta_{h_1}]}{\lambda_m^{\ (1)}}\exp[-\lambda_m^{\ (1)}(l_{h_1}-\xi_{h_1})]$$

$$\sigma_{11}^{(1,3)L} = -\frac{2}{s_1}\sum_{m=1}^{\infty}(-1)^m D_m^{(3)L}\cos[m\pi\zeta_{h_1}]\exp[-\lambda_m^{\ (1)}(2l_{h_1}-\xi_{h_1})]$$

$$D_m^{(3)L} = \int_0^{\mu} Z_c^{L(\tau_{h_1})}(h_1\zeta_{h_1},s_1)\sin[m\pi(1-\zeta_{h_1})]d\zeta_{h_1} \tag{4.5}$$

$$v_1^{(2)L} = \frac{4(1+\nu_2)\kappa_2^{\ 2}}{E_2}\frac{h_2^{\ 2}}{s_2}\cdot\sum_{m=1}^{\infty}\frac{(-1)^m}{\pi m}\frac{D_m^{(4)L}\sin[m\pi\zeta_{h_2}]}{\lambda_m^{\ (2)}}\exp[-\lambda_m^{\ (2)}\xi_{h_2}]$$

$$\sigma_{11}^{(2)L} = -\frac{2}{s_2}\sum_{m=1}^{\infty}(-1)^m D_m^{(4)L}\sin[m\pi\zeta_{h_2}]\exp[-\lambda_m^{(2)}\xi_{h_2}]$$

$$D_m^{(4)L} = \int_0^1 Z_c^{L(\tau_{h_2})}(h_2\zeta_{h_2}, s_2)\sin[m\pi(1-\zeta_{h_2})]d\zeta_{h_2}.$$

Here $Z_c^{L(\tau_{h_1})}$ are the Laplace transforms of Z_c, $\mu = h_2 / h_1$, $Z_c^{L(\tau_{h_1})}(h_1\zeta_{h_1}, s_1) = \mu\kappa_T Z_c^{L(\tau_2)}$, $\kappa_T = c_2^{(1)} / c_2^{(2)}$.

Taking into account the continuity condition for the displacement (4.4), we obtain an equation for unknown value of normal stress Z_c, namely

$$\sum_{m=1}^{\infty}(-1)^m \frac{\sin[m\pi\zeta_{h_1}]}{\lambda_m^{(1)}}\cdot\int_0^{\mu} Z_c^{L(\tau_{h_1})}(h_1\zeta_{h_1}, s_1)\sin[m\pi(1-\zeta_{h_1})]d\zeta_{h_1}$$

$$+\kappa_T^2\mu\frac{\rho_1\kappa_2^2}{\rho_2\kappa_1^2}\sum_{m=1}^{\infty}(-1)^m\frac{\sin[m\pi\zeta_{h_2}]}{\lambda_m^{(2)}}\int_0^1 Z_c^{L(\tau_{h_2})}(h_2\zeta_{h_2}, s_2)\cdot\sin[m\pi(1-\zeta_{h_2})]d\zeta_{h_2}$$

$$= 2\mathrm{I}h_1\sum_{m=1}^{\infty}\frac{(-1)^m}{\pi m}\frac{\sin[m\pi\zeta_{h_1}]}{\lambda_m^{(1)}}\exp[-\lambda_m^{(1)} l_{h_1}]. \qquad (4.6)$$

4.1 Approximate solutions

Let us simplify Eq. (4.6) by two steps. For the first step we utilise the method of expanding the transform in a series in negative powers of transform parameter. Equation considered takes the form

$$\sum_{m=1}^{\infty}(-1)^m\sin[m\pi\zeta_{h_1}]\int_0^{\mu} Z_c^{L(\tau_{h_1})}(h_1\zeta_{h_1}, s_1)\sin[m\pi(1-\zeta_{h_1})]d\zeta_{h_1} +$$

$$+\kappa_T\frac{\rho_1}{\rho_2}\frac{\kappa_2}{\kappa_1}\sum(-1)^m\sin[m\pi\zeta_{h_2}]\int_0^1 Z_c^{L(\tau_{h_2})}(h_2\zeta_{h_2}, s_2)\sin[m\pi(1-\zeta_{h_2})]d\zeta_{h_2} \qquad (4.7)$$

$$= 2Ih_1\sum_{m=1}^{\infty}\frac{(-1)^m}{\pi m}\sin[m\pi\zeta_{h_1}]\exp[-\lambda_m^{(1)} l_{h_1}],$$

where $\lambda_m^{(1)} = \kappa_1 s_1 + \dfrac{m^2\pi^2}{2\kappa_1 s_1}$.

For the second step, we use an assumption about the smallness of the thickness h_2 with respect to thickness h_1 and expand the contact stress Z_c into a series in powers of normal coordinate ζ_{h_1}. We now consider the leading part of this expansion:

$$Z_c(\zeta_{h_1}, \tau_{h_1}) = \zeta_{h_1} Z_0(\tau_{h_1}) \qquad (4.8)$$

where Z_0 is unknown function. Substituting (4.8) into (4.7) we obtain the following expression for the transform of the function considered

$$Z_0^{L(\tau_{h_1})} = 4I\frac{\rho_2\kappa_1}{\rho_2\kappa_1+\rho_1\kappa_2}\sum_{m=0}^{\infty}(-1)^{m-1}\cos[m\pi\zeta_1]\exp[-\lambda_m^{(1)}l_{h_1}] \tag{4.9}$$

Taking into account (4.9) we can now write approximate expressions for solutions for all waves in the forms

$$G_1^{(1,1)} = \frac{4Ih_1^3}{\pi^2}\sum_{m=1}^{\infty}\frac{1}{m^2}J_0\left[\sqrt{2}m\pi\sqrt{\frac{\xi_{h_1}}{\kappa_1}(\tau_{h_1}-\kappa_1\xi_{h_1})}\right]H(\tau_{h_1}-\kappa_1\xi_{h_1})$$

$$G_1^{(1,2)} = \frac{4Ih_1^3}{\pi^2}\sum_{m=1}^{\infty}\frac{1}{m^2}J_0\left[\sqrt{2}m\pi\sqrt{\frac{2l_{h_1}-\xi_{h_1}}{\kappa_1}(\tau_{h_1}-\kappa_1(2l_{h_1}-\xi_{h_1}))}\right]\cdot$$
$$\cdot H(\tau_{h_1}-\kappa_1(2l_{h_1}-\xi_{h_1}))$$

$$\begin{aligned}G_1^{(1,3)} &= \frac{16Ih_1^3}{\pi^2}\frac{\rho_2\kappa_1}{\rho_2\kappa_1+\rho_1 k_2}\sum_{m=1,n=1}^{\infty}\frac{(-1)^m(-1)^{n-1}}{m}\left(\cos(m\pi\mu)-\frac{\sin(m\pi\mu)}{\pi}\right)\\ &\times J_0\left[\sqrt{2}\pi\sqrt{\frac{1}{\kappa_1}(m^2(l_{h_1}-\xi_{h_1})+nl_{h_1})(\tau_{h_1}-\kappa_1(2l_{h_1}-\xi_{h_1}))}\right]\cdot\\ &\cdot H(\tau_{h_1}-\kappa_1(2l_{h_1}-\xi_{h_1}))\\ G_1^{(2)} &= \frac{16Ih_2^3}{\pi^2}\frac{\rho_2\kappa_1}{\rho_2\kappa_1+\rho_1 k_2}\sum_{m=1,n=1}^{\infty}\frac{(-1)^m(-1)^{n-1}}{m}\\ &\times J_0\left[\sqrt{2}\pi\sqrt{\left(\frac{m^2}{\kappa_2}\xi_{h_2}+\frac{n^2}{\kappa_1}\mu^2\kappa_T l_{h_2}\right)\left(\tau_{h_2}-\left(\kappa_2\xi_{h_2}+\kappa_1\frac{l_{h_1}}{\kappa_T}\right)\right)}\right]\cdot\\ &\cdot H(\tau_{h_2}-(\kappa_1 l_{h_2}+\kappa_2\xi_{h_2}))\end{aligned} \tag{4.10}$$

Acknowledgement

The work of the author (YP) is supported by a grant N YSF 01/1-19 from the INTAS. This award is very gratefully acknowledged.

References

1. Kaplunov JD, Kossovich LYu, Nolde EV (1998) Dynamics of thin walled elastic bodies. Academic Press, San Diego
2. Kirillova IV Ranges of applicability of boundary layers in vicinities of wave fronts in shells of revolution of zero Gaussian curvature. – Izv. RAN Ser. Mekhanika Tverdogo Tela (accepted for publication)
3. Kirillova IV, Kossovich LYu (2001) Boundary layers in shells of revolution with a discontinuity in the thickness. – Proceedings of the international conference ICOVP Moscow, Russia, pp 263-266
4. Kossovich LYu (1986) Non-stationary Problem in Theory of Elastic Thin Shells (in Russian). Saratov University Press, Saratov
5. Nigul U (1969) Regions of effective application of the methods of three-dimensional and two-dimensional analysis of transient stress waves in shells and plates. Int J Solid Structures 5: 607-629

On Some Interior and Boundary Effects in Thin Plates Based on the Asymmetric Theory of Elasticity

S. H. Sargsyan

Gyumri State Pedagogical Institute, Armenia

1 Main equations and boundary conditions of three-dimensional asymmetric theory of elasticity

For plates with microstructure, the classical theory of elasticity considering only force interactions between the material points results in some well-known errors. In this case, the plate or shell theories must be based on the asymmetric theory of elasticity (ATE) [3]. Below, several mechanical phenomena in the theory of micro-polar elastic thin plates [4-10] are stated and discussed in connection with the use of the asymptotic method [1, 11]. Let us consider a plate of *2h* constant thickness, bounded with the cylindric lateral surface as a three–dimensional problem of asymptotic theory of elasticity (ATE) with the independent fields of displacements and rotations [2], which will present in x_k (k = 1,2,3) Cartesian coordinate system. The principal equations are:

$$\sigma_{ij,j} = 0, \qquad \mu_{ij,j} + e_{ijk}\sigma_{jk} = 0. \tag{1.1}$$

Physical relations in reverse form

$$\begin{cases} \gamma_{ji} = (\mu' + \alpha')\sigma_{ji} + (\mu' - \alpha')\sigma_{ij} + \lambda'\sigma_{kk}\delta_{ij}, \\ \chi_{ji} = (\gamma' + \varepsilon')\mu_{ji} + (\gamma' - \varepsilon')\mu_{ij} + \beta'\mu_{kk}\delta_{ij}. \end{cases} \tag{1.2}$$

Geometric relations

$$\gamma_{ij} = u_{j,i} - e_{kij}\omega_k, \qquad \chi_{ij} = \omega_{j,i}. \tag{1.3}$$

where σ_{ij}, μ_{ij} are components of force and moment tensors of stress, γ_{ij}, χ_{ij} are components of asymmetric deformation tensor and bending-rotation tensor, $\overline{u}$ is the vector of displacements; $\overline{\omega}$ is the vector of independent rotations of body's points; λ, μ, α, β, γ, ε (or with primes) are elastic constants of the plate material, e_{kij} are symbols of Levi-Cevita ; the comma following an index means differentiation.

On $x_3 = \pm h$ planes of plate the force and momental boundary conditions are considered as given

$$\sigma_{3i}\Big|_{x_3=\pm h} = p_i^{\pm}, \ \mu_{3i}\Big|_{x_3-\pm h} = m_i^{\pm}, \quad (i=1,2,3) \ . \tag{1.4}$$

On lateral cylindric surface of plate $(\Sigma = \Sigma_1 \cup \Sigma_2)$ the boundary conditions of mixed type are considered as given

$$\begin{cases} \sigma_{ji} n_j = P_{*i}, \ \mu_{ji} n_j = m_{*i} & \text{on } \Sigma_1 \\ \vec{u} = \overrightarrow{u_*}, \ \vec{\omega} = \overrightarrow{\omega_*} & \text{on } \Sigma_2 \end{cases} . \tag{1.5}$$

The solution of the defined by Eqs. (1.1)-(1.5) boundary value problem for plate forms from the sum of the solutions of symmetric by x_3 (plane stress state on ATE) and asymmetric (bending of plate on ATE) problems. We'll state that the asymptotic theory of bending of plates on ATE (the analog theory is constructed for plane stress state of plate on ATE as well).

It is assumed that the thickness of plate is small in comparison with its characteristic size in midplane ($2h << a$, $\delta = h/a << 1$ is a small parameter).

Taking into account the principle of asymptotic theory of integration of singular disturbance system of differentional equations, as the base of the reasoning we'll put the property of stress-deformable state (SDS) on (ATE), testing statistical influence which is expressed by the structural formulae

$$(\text{SDS})_{\text{comp}} = (\text{SDA})_{\text{int}} + (\text{SDA})_{\text{bound}} \tag{1.6}$$

In this equation by the lower indexes the complete, interior (attenuating) and boundary (SDS) of plate on ATE are marked. The latter origins near the Σ lateral surface of plate and quickly (exponentially) attenuates in the case of taking away from the lateral surface of thin three-dimensional body of plate. The conception of dismemberance of SDS of this plate on ATE into interior (ISDS) and boundary (BSDS) and the fact of radical difference among their physical properties will bring to their diverse studies. From this point of view, the main subject of research will be the problem on approximate methods of interior account of plate (the definition of ISDS) and boundary account of plate (the definition of BSDS).

2 The definition of interior stress-deformable state of plate on ATE

In Eqs. (1.1)-(1.3) of three-dimensional ATE we'll come to dimensionless coordinate system

$$\xi = \frac{x_1}{a}, \ \eta = \frac{x_2}{a}, \ \zeta = \frac{x_3}{h} \ . \tag{2.1}$$

After substitution of independent variables by Eq. (2.1), on the basis of Eqs. (1.1)-(1.3) of ATE , the singular disturbance system of differential equations with δ

small parameter is obtained. The solution of interior problem will be presented by the asymptotic expansions

$$Q = \delta^{-q} \sum_{s=0}^{S} \delta^{s} Q^{(s)} \quad , \tag{2.2}$$

where Q is any of stresses (force and moment), displacements and independent rotations; q is a natural number, which is different for different values and is defined from the condition of obtaining of noncontradictory recurrent system of equations in asymptotic approaches. After putting of Eqs. (2.1) and (2.2) in Eqs. (1.1)-(1.3) for asymmetric by ζ problem (the bending of plate on ATE), where σ_{13}, σ_{23}, σ_{31}, σ_{32}, μ_{11}, μ_{22}, μ_{33}, μ_{12}, μ_{21}, ω_1, ω_2, u_3, are even functions by ζ, and σ_{11}, σ_{22}, σ_{33}, σ_{12}, σ_{21}, μ_{31}, μ_{32}, μ_{13}, μ_{23}, u_1, u_2, ω_3 are odd functions by ζ, the noncontradictory system of equations is obtained relatively to $Q^{(s)}$ coefficients of Eq. (2.2) only for

$$\begin{aligned} &q = 1 \quad \text{for } \sigma_{13}, \sigma_{23}, \sigma_{31}, \sigma_{32}, \mu_{11}, \mu_{22}, \mu_{33}, \mu_{12}, \mu_{21}, \omega_1, \omega_2, u_3 \\ &q = 0 \quad \text{for } \sigma_{11}, \sigma_{22}, \sigma_{33}, \sigma_{12}, \sigma_{21}, \mu_{31}, \mu_{32}, \mu_{13}, \mu_{23}, \omega_3, u_1, u_2 \ . \end{aligned} \tag{2.3}$$

Specific properties of interior iterational process on ATE for thin three-dimensional plate are

- BSDS in the case of taking away from the lateral surface of plate changes by the thickness of plate (by ζ variable) by the simple law;
- In the description of BSDS the role of (ξ, η) variables, which give the position of a point on the plate's midplane, remains to elusidate.

From this point of view it is expedient to substitute force and moment stresses with statically equivalent forces, moments and hypermoments

$$\begin{cases} N_{nm} = \int_{-h}^{h} \sigma_{nm} dx_3, \quad \Lambda_{nm} = \int_{-h}^{h} \mu_{nm} x_3 dx_3 \ (nm = 13; 31; 23; 32), \\ L_{pq} = \int_{-h}^{h} \mu_{pq} dx_3, \quad \mathrm{M}_{pq} = \int_{-h}^{h} \sigma_{pq} x_3 dx_3 \ (pq = 11; 22; 33; 12; 21) . \end{cases} \tag{2.4}$$

On the base of these integrate by thickness of plate concepts, BSDS on ATE, in reference ($s = 0$) asymptotic approach, is described with two-dimensional equations by (ξ, η) variables (by (x_1, x_2)):

- Balance equations

$$\begin{cases} \dfrac{\partial N_{13}^{(0)}}{\partial x_1} + \dfrac{\partial N_{23}^{(0)}}{\partial x_2} = -(p_3^+ + \mathrm{p}_3^-) , \\ \dfrac{\partial L_{11}^{(0)}}{\partial x_1} + \dfrac{\partial L_{21}^{(0)}}{\partial x_2} + N_{23}^{(0)} - N_{32}^{(0)} = -(m_1^+ + m_1^-) , \\ \dfrac{\partial L_{12}^{(0)}}{\partial x_1} + \dfrac{\partial L_{22}^{(0)}}{\partial x_2} + N_{31}^{(0)} - N_{13}^{(0)} = -(m_2^+ + m_2^-) . \end{cases} \tag{2.5}$$

It should be taken into consideration also

$$N_{31}^{(0)} = h(p_1^+ - p_1^-),\ N_{32}^{(0)} = h(p_2^+ - p_2^-),\ L_{33}^{(0)} = h(m_3^+ - m_3^-)\ . \tag{2.6}$$

- Elastic relations

$$\begin{cases} N_{13}^{(0)} = 2h\left[(\mu+\alpha)\Gamma_{13}^{(0)} + (\mu-\alpha)\Gamma_{31}^{(0)}\right], & 1 \to 2 \\ N_{31}^{(0)} = 2h\left[(\mu+\alpha)\Gamma_{31}^{(0)} + (\mu-\alpha)\Gamma_{13}^{(0)}\right], & 1 \to 2 \\ L_{11}^{(0)} = 2h\left[(2\gamma+\beta)K_{11}^{(0)} + \beta(K_{22}^{(0)} + K_{33}^{(0)})\right], & 1 \to 2 \\ L_{33}^{(0)} = 2h\left[(2\gamma+\beta)K_{33}^{(0)} + \beta(K_{11}^{(0)} + K_{22}^{(0)})\right], & \\ L_{12}^{(0)} = 2h\left[(\gamma+\varepsilon)K_{12}^{(0)} + (\gamma-\varepsilon)K_{21}^{(0)}\right]. & 1 \leftrightarrow 2\ . \end{cases} \tag{2.7}$$

- Geometric relations

$$\begin{cases} \Gamma_{13}^{(0)} = \dfrac{\partial w^{(0)}}{\partial x_1} + \Omega_2^{(0)}, \quad \Gamma_{23}^{(0)} = \dfrac{\partial w^{(0)}}{\partial x_2} - \Omega_1^{(0)}, \quad \Gamma_{31}^{(0)} = \beta_1^{(0)} - \Omega_2^{(0)} \\ \Gamma_{32}^{(0)} = \beta_2^{(0)} + \Omega_1^{(0)}, \quad K_{11}^{(0)} = \dfrac{\partial \Omega_1^{(0)}}{\partial x_1} \quad (1 \to 2), \quad K_{12}^{(0)} = \dfrac{\partial \Omega_2^{(0)}}{\partial x_1} \\ K_{33}^{(0)} = O_3^{(0)} \end{cases} \ . \tag{2.8}$$

Here $w^{(0)}(x_1, x_2)$, $\Omega_1^{(0)}(x_1, x_2)$, $\Omega_2^{(0)}(x_1, x_2)$ are displacements and independent rotations around x_1 and x_2 axes of points of plate's midplane; $\omega_3^{(0)} = x_3 O_3^{(0)}(x_1, x_2)$ is independent rotation around x_3 axis; $\Gamma_{13}^{(0)}$, $\Gamma_{31}^{(0)}$ $(1 \to 2)$ are components of deformation, and K_{11}, K_{12}, K_{21}, K_{22}, K_{33} are components of bending-rotation in the points of plate's midplane.

Eqs. (2.5)-(2.8) are the resolving system of differential equations of interior problem in reference asymptotic approach. The system consists of 24 equations relative to 24 unknowns

$$\begin{gathered}(N_{13}^{(0)}, N_{23}^{(0)}, N_{31}^{(0)}, N_{32}^{(0)}, L_{11}^{(0)}, L_{22}^{(0)}, L_{33}^{(0)}, L_{12}^{(0)}, L_{21}^{(0)} \\ \Gamma_{13}^{(0)}, \Gamma_{23}^{(0)}, \Gamma_{31}^{(0)}, \Gamma_{32}^{(0)}, K_{11}^{(0)}, K_{22}^{(0)}, K_{33}^{(0)}, K_{12}^{(0)}, K_{21}^{(0)} \\ w^{(0)}, \Omega_1^{(0)}, \Omega_2^{(0)}, \beta_1^{(0)}, \beta_2^{(0)}, O_3^{(0)})\ .\end{gathered}$$

The rest values, characterizing the defined problem, are defined by corresponding formulas after solution of the main system of equations. For example,

$$\begin{cases} M_{11}^{(0)} = \dfrac{2Eh^3}{3(1-\upsilon^2)}(K_{11}^{(0)} + K_{22}^{(0)}) + \dfrac{\upsilon}{1-\upsilon} M_{33}^{(0)}, & 1 \leftrightarrow 2 \\ M_{12}^{(0)} = \dfrac{2h^3}{3}\left[(\mu+\alpha)K_{12}^{(0)} + (\mu-\alpha)K_{21}^{(0)}\right], & 1 \leftrightarrow 2 \\ \Lambda_{13}^{(0)} = \dfrac{2h^3}{3}\dfrac{4\gamma\varepsilon}{\gamma+\varepsilon} L_{13}^{(0)} + \dfrac{\gamma-\varepsilon}{\gamma+\varepsilon}\Lambda_{31}^{(0)}, & 1 \leftrightarrow 2 \end{cases} \tag{2.9}$$

$$M_{33}^{(0)} = \frac{h^2}{3}(P_3^+ + P_3^-), \qquad \Lambda_{31}^{(0)} = \frac{h^2}{3}(m_1^+ + m_1^-) \ , \tag{2.10}$$

$$\begin{cases} K_{11}^{(0)} = \dfrac{\partial \beta_1^{(0)}}{\partial x_1} \ (1 \leftrightarrow 2), \quad K_{12}^{(0)} = \dfrac{\partial \beta_2^{(0)}}{\partial x_1} - O_3^{(0)}, \quad K_{21}^{(0)} = \dfrac{\partial \beta_1^{(0)}}{\partial x_2} + O_3^{(0)}, \\ L_{13}^{(0)} = \dfrac{\partial O_3^{(0)}}{\partial x_1} \ (1 \leftrightarrow 2) \ . \end{cases} \tag{2.11}$$

Resolving system of Eqs. (2.5)-(2.8) simultaneously is the system of equations of applied two-dimensional theory of bending of plate on ATE. The general variational functional of applied two-dimensional theory of plate's bending on ATE is constructed from the three-dimensional general variational functional by the asymptotic method.

3 The construction of theory of boundary layer for thin plate on ATE

Now, we'll consider the study of the boundary account of plate on ATE. We consider Σ lateral surface of plate as a closed face and choose three-orthogonal system of equations of (α_1, α_2, α_3) curvilinear coordinates in such a way that it combines with $\alpha_1 = \alpha_{10}$ coordinate surface.

The asymptotic properties of $(SDS)_{\text{bound}}$ in structural Eq. (1.6) will be noticed by the following way. In three-dimensional Eqs. (1.1)-(1.3) of ATE, written in α_i independent variables in this case, we'll do the following substitutions

$$\begin{aligned} \alpha_1 - \alpha_{10} &= h\xi_1 \\ \alpha_2 = a\eta, \ \alpha_3 &= h\zeta \end{aligned} \tag{3.1}$$

The solution of transformed Eqs. (1.1)-(1.3) of ATE we'll present in this form

$$R = \sum_{s=0}^{S} \delta^{\chi_R + s} R^{(s)} \ , \tag{3.2}$$

where $R^{(S)}$ is any of force and momental stresses, linear displacements and independent rotations. Because Eq.(1.4) force and momental nonhomogeneous conditions, given on ($\zeta = \pm 1$) facial planes, were satisfied with the solution of the interior problem, then Eq. (3.2) solution must satisfy to homogeneous boundary conditions on $\zeta = \pm 1$ planes.

After substitution of Eq. (3.2) in system of Eqs. (1.1)-(1.3) of ATE transformed with the help of Eq.(3.1), we'll obtain noncontradictory system of educations relative to $R^{(s)}$, if

$$\sigma_{ij} = \delta^{\chi} \sum_{s=0}^{S} \delta^{s} \sigma_{ij}^{(s)}, \; \mu_{ij} = \delta^{\chi} \sum_{s=0}^{S} \delta^{s} \mu_{ij}^{(s)}$$
$$v_i = \delta^{\chi+1} \sum_{s=0}^{S}{}' \delta^{s} v_i^{(s)}, \; \omega_i = \delta^{\chi+1} \sum_{s=0}^{S}{}' \delta^{s} \omega_i^{(s)} \quad . \tag{3.3}$$

Here χ whole number, characterizing the intensity of boundary layer, should be defined in the case of considering the problem of interaction of boundary layer with the interior SDS of plate to satisfy Eq. (1.5) defined three-dimensional boundary conditions of ATE on Σ lateral surface of plate.

The obtained system of equations in s approach transforms in four groups of equations (they are independent when $s = 0$ and interconnected when $s > 0$). The first time were discovered that on ATE besides the force (plane and antiplane) boundary layers there exist also moment (plane and antiplane) ones.

The equations of corresponding boundary layers possess important property of the type

$$\int_{-1}^{1} p^{s}(t=0)d\zeta = \int_{-1}^{1} d\zeta \int_{0}^{\infty} T^{(s-1)} dt \quad (t = \frac{1}{H_{10}} \xi_1) \; , \tag{3.4}$$

where as $P^{(s)}(t=0)$ it should be $\sigma_{1m}^{(s)}(t=0)$ or $\mu_{1m}^{(s)}(t=0)$ $(m = 1,2,3)$ and also $v_2^{(s)}(t=0)$ or $\omega_2^{(s)}(t=0)$.

To construct the functions of boundary layer type the systems of equations of all four types of boundary layers should be considered in reference ($s = 0$) asymptotic approach. Each of such systems is homogeneous, independent as was mentioned and we can obtain the attenuating solution by method of dismembering of variables representing it in this way

$$\overset{p}{R}{}^{(0)} = \sum_{(\lambda_n)} e^{-\lambda_n t} \overset{p}{R}{}_n^{(0)}(\zeta) \quad (p \leftrightarrow a) \; , \tag{3.5}$$

where $\overset{p}{R}{}^{(0)}$ $(p \leftrightarrow a)$ is any of force and moment stresses, displacements and rotations of plane (p) and antiplane (a) force and momental boundary layers.

Let's mention very important from our point of view circumstance that exists the full analogy between force plane and momental antiplane boundary layers, as well as between antiplane force and plane momental boundary layers

$$(\overset{p}{v}_i \leftrightarrow \overset{a}{\omega}_i, \overset{p}{\sigma}_{ij} \leftrightarrow \overset{a}{\mu}_{ij}, \; \mu \leftrightarrow \varepsilon, \, \alpha \leftrightarrow \gamma, \, \lambda \leftrightarrow \beta) \; .$$

The solution (eigenfunctions) of plane boundary layer relatively to force stresses and linear displacements, and in analog way, the solution (eigenfunctions) of antiplane boundary layer relatively to momental stresses and independent rotations, in the result, it's come to the solution of linea, homogeneous, depending on the parameter (eigenvalues) of ordinary differentional equations of the fourth order with the homogeneous and also depending on λ_n boundary conditions.

$$\frac{d^4\Phi_n}{d\zeta^4}+2\lambda_n^2\frac{d^2\Phi_n}{d\zeta^2}+\lambda_n^4\Phi_n^{(0)}=0 \tag{3.6}$$

$$\overset{p}{\sigma}{}_{31}^{(0)}\Big|_{\zeta=\pm1}=\overset{p}{\sigma}{}_{33}^{(0)}\Big|_{\zeta=\pm1}=0 \quad or\ \mu_{31n}^{(0)}\Big|_{\xi=\pm1}=\overset{p}{\mu}{}_{33n}^{(0)}\Big|_{\zeta=\pm1}=0\ . \tag{3.7}$$

Here in the case of force plane boundary layer $\Phi_n = \overset{p^{(0)}}{v}_{3n}$ should be accepted and in the case of antiplane momental boundary layer $\Phi_n = \overset{a^{(0)}}{\omega}_{3n}$.

The solution of (Eqs. (3.6), (3.7)) boundary value problem (in the case of bending problem) is expressed by the following formulae

$$\Phi_n(\lambda_n,\zeta)=-(\mathrm{ctg}\lambda_n-\frac{1}{\lambda_n}\frac{b_2}{b_1})\cdot\sin\lambda_n\zeta+\zeta\cos\lambda_n\zeta\ , \tag{3.8}$$

where λ_n is the root of the equation

$$k\cdot\sin 2\lambda_n-2\lambda_n=0\ , \tag{3.9}$$

b_1, b_2, k values in the case of force plane boundary layer are expressed with λ, μ, α elastic constants, and in the case of momental antiplane boundary layer with β, γ, ε elastic constants.

There were proved that eigenfunctions of boundary value problem for force plane and momental antiplane boundary layers for different eigenvalues satisfy correspondingly to the following formulas of generalized orthogonality.

$$\begin{gathered}\int_{-1}^{1}(\overset{p^{(0)}}{\sigma}_{11m}\overset{p^{(0)}}{v}_{1n}-\overset{p^{(0)}}{\sigma}_{13n}\overset{p^{(0)}}{v}_{3m}\,d\zeta=0\\ \int_{-1}^{1}(\overset{a^{(0)}}{\mu}_{11m}\overset{a^{(0)}}{\omega}_{1n}-\overset{a^{(0)}}{\mu}_{1n3}\overset{a^{(0)}}{\omega}_{3m}d\zeta=0\ .\end{gathered} \tag{3.10}$$

The solution (eigenfunctions) of antiplane force boundary layer, and in analog way, the solution (eigenfunctions) of plane moment boundary layer, in the result, it brings to the to the solution of linear, homogeneous, depending on λ_n parameter (eigenvalues) of ordinary differentional equations of the second order with the homogeneous boundary conditions,

$$\frac{d^2F_n}{d\zeta^2}+\lambda_n^2F_n=0,\ \ \overset{a\,(0)}{\sigma}_{32n}\Big|_{\zeta=\pm1}=0,\ \text{or}\ \ \overset{p\,(0)}{\mu}_{32n}\Big|_{\zeta=\pm1}=0\,. \tag{3.11}$$

Here in the case of force plane boundary layer $F_n\equiv\overset{p^{(0)}}{v}_{2n}$ and in the case of plane moment boundary layer $F_n\equiv\overset{p^{(0)}}{\omega}_2$.

The solution of (Eq.(3.11)) boundary value problem (in the case of bending problem) can be presented in this form

$$F_n(\lambda_n,\zeta)=\sin\lambda_n\zeta\ , \tag{3.12}$$

where λ_n is the root of the following equation.

$$\cos\lambda_n=0,\ \ \lambda_n=(2n+1)\frac{\pi}{2}\quad (n=0,1,2...)\ . \tag{3.13}$$

The general solution of all four types of mentioned boundary layers (each separately) can be presented as the series with independent coefficients (arbitraries of boundary layers) by the corresponding eigenfunctions. When $s\ >=1$ the systems of equations of all four types of boundary layers (each of which will be with "p" index as well as with "a" index) become nonhomogeneous (with the known right sights). In this work the general solution of all mentioned four types boundary layers is constructed for $s>=1$ case as well.

4 The matching of asymptotic expansion of interior iterational process and boundary layers for thin plate on ATE

So, the problem of separate construction of differentional equations of interior problem and boundary layers in the theory of thin plates on ATE can be considered as solved and the general solution of defined (1.1)-(1.5) boundary value problem on ATE can be presented in this form

$$J=Q^{bound}+\delta^{\chi+s}\overset{a^{(s)}}{R}+\delta^{\mu+s}\overset{p^{(s)}}{R}\ , \tag{4.1}$$

where χ and μ whole numbers characterize the intensity of boundary layers. They must be chosen in such a way that the satisfaction of (1.5) boundary conditions on Σ lateral surface of plate will be possible.

Now let's consider the problem of dismembering of three-dimensional boundary conditions of ATE on Σ, i.e. in the given stage of asymptotic research it is necessary to find out what boundary conditions should be refer to the interior differentional equations and what boundary conditions to the boundary layer differentional equations on ATE. To do the matching of the solution of interior problem with the solutions of all four types of boundary conditions, the three-dimensional boundary conditions on ATE should be concrete. For this purpose we'll consider the case when the three-dimensional boundary conditions of the first boundary value problem on ATE (i.e. in Eq. (1.5) $\Sigma_1=\Sigma,\ \Sigma_2=0$) are given on lateral surface ($\alpha_1=\alpha_{10}$) of plate.

Let's put Eq. (4.1) in the (1.5) boundary conditions (for the mentioned case). When satisfying this boundary conditions, in order to obtain the iterational process, for intensities of boundary layers χ and μ should be taken $\chi=\mu=-1$. Using Eq. (3.4) property of boundary layers from the obtained iterational process in reference asymptotic approach (when $s = 0$) we obtain the boundary conditions of interior problem:

$$N_{13}^{(0)}\Big|_{\alpha_1=\alpha_{10}} = \int_{-h}^{h} P_3^* d\alpha_3 ,$$
$$L_{11}^{(0)}\Big|_{\alpha_1=\alpha_{10}} = \int_{-h}^{h} m_1^* d\alpha_3 , \tag{4.2}$$
$$L_{12}^{(0)}\Big|_{\alpha_1=\alpha_{10}} = \int_{-h}^{h} m_2^* d\alpha_3 .$$

From the mentioned systems of equations of iterational process of boundary conditions taking into account Eq. (4.2) we obtain the boundary conditions in reference asymptotic approach ($s = 0$) corresponding to boundary conditions of all four types of boundary layers:

- For plane boundary layer

$$\overset{p^{(0)}}{\sigma_{11}}\Big|_{\xi_1=0} = 0 \qquad \overset{p^{(0)}}{\sigma_{13}}\Big|_{\xi_1=0} = f_3(\zeta) , \qquad \text{(for force boundary layer)} \tag{4.3}$$

$$\overset{p^{(0)}}{\mu_{12}}\Big|_{\xi_1=0} = \Psi_2(\zeta) ; \qquad \text{(for moment boundary layer)} \tag{4.4}$$

- For antiplane boundary layer

$$\overset{a^{(0)}}{\sigma_{11}}\Big|_{\xi_1=0} = 0 , \qquad \text{(for force boundary layer)} \tag{4.5}$$

$$\overset{a^{(0)}}{\mu_{11}}\Big|_{\xi_1=0} = \Psi_1(\zeta) \qquad \overset{a^{(0)}}{\mu_{12}}\Big|_{\xi_1=0} = 0 \qquad \text{(for moment boundary layer)} \tag{4.6}$$

Here the following abbreviaties are used .

$$f_3(\zeta) = P_3^* - \frac{1}{2}\int_{-1}^{1} P_3^* d\zeta, \qquad \psi_i(\zeta) = m_i^* - \frac{1}{2}\int_{-1}^{1} m_3^* d\zeta, \ (i = 1,2) .$$

So, in the reference asymptotic approach ($s = 0$), the interior problem with its differentional Eqs. (2.5)-(2.8) and (4.2) boundary conditions separates completely as independent boundary value problem in the bending problem of plate on ATE. In analog way (4.3)-(4.6) boundary conditions should be refer to the corresponding differentional equations which define boundary conditions separately.

When s >= 1 the worth side of the obtained result remains. In the work in high asymptotic approaches, the specifying results are obtained both for interior boundary value problem and for boundary value problems of all four types of boundary layers. The mechanisms of influence (by the boundary conditions) of boundary layers on interior SDS of plate on ATE and the contrary case are elucidated.

It should be mentioned that the matching of asymptotic expansions of interior problem and problems of boundary layers can be realized in the variational way as well (using the corresponding variational principles of ATE). In this case, for the interior problem we'll obtain the same Eq. (4.2) boundary conditions, and for definition of arbitraries of all four types of boundary conditions we'll obtain the linear algebraic system of equations. The other variants of three-dimensional boundary conditions of ATE on Σ are considered as well.

References

1. Kaplunov I, Kossovich L, Nolde E (1998) Dynamics of thin walled elastic bodies. Academic press, San Diego, London, Boston, New York, Sydney, Tokyo, Toronto
2. Nowacki W (1975) Theory of elasticity. Mir , Moscow (in Russian)
3. Palmov V, Altenbach H (1982) Über eine Cosseratsche Theorie für elastische Platten. In: Technische Mechanik, 3: 5-9
4. Sargsyan S (1999) Asymptotic theory and variational equation of plane problem of elastic thin plate on momental theory of elasticity. Reports of NSA of Armenia, vol 99, N2, pp 138-147 (in Russian)
5. Sargsyan S (1999) Asymptotic theory and variational equation of bending problem of elastic thin plate on momental theory of elasticity. Reports of N SA of Armenia, vol 99, N3, pp 216-225 (in Russian)
6. Sargsyan S (2002) On some results of interior and boundary accounts of thin plate on asymmetric theory of elasticity. In: Problems of mechanics of thin deformable bodies. Publ. NSA of Armenia, Yerevan, pp 285-307 (in Russian)
7. Sargsyan S, Farmanyan A (2000) On boundary layer and character of its interaction with interior stress-deformable state of plate on asymmetric theory of elasticity. In: Actual problems of mechanics of shells, International Conference, Novo Znanie, Kazan, pp 362-367 (in Russian)
8. Sargsyan S, Farmanyan A (2001) Generalized orthogonality in plane problem of asymmetric theory of elasticity. Reports of NSA of Armenia, vol 101, N2, pp 128-134 (in Russian)
9. Sargsyan S, Farmanyan A (2001) Conditions of existence of attenuating solutions for plates on asymmetric theory of elasticity. Reports of NSA of Armenia, vol 101, N1, pp 55-60 (in Russian)
10. Sargsyan S, Farmanyan A (2002) Matching of asymptotic expansions of the main iterational process and boundary layer in boundary value problems of thin plates on asymmetric theory of elasticity. In: Izvestiya of Establishment of Higher Education, North Caucase region, Natural Sciences, N2, pp 36-38 (in Russian)

On the Numerical Simulation of Thin Textile Reinforced Concrete Layers for the Strengthening of Shell Structures Using a Surface-Related Shell Theory

R. Schlebusch, J. Matheas and B. W. Zastrau

Institute for Mechanics, Technische Universität Dresden, Germany

1 Introduction

This contribution deals with the development of a surface-related shell theory and its conversion into the finite element method. In particular, composites made of thin textile reinforced concrete layers are examined for the strengthening of existing shell structures. The theoretical foundation of the shell theory is a modified form of the principle of Hu-Washizu functional. This variational principle permits the adapted formulation of the enhanced assumed strain method, Simo and Rifai [12]. Special attention is paid to the interface between the existing structure and the thin strengthening layer. It is modeled as a contact problem, whereby the contact conditions are introduced by following the continuum based formulation as suggested in Laursen and Simo [8] augmented to problems involving adhesion.

The authors contribution to the development of the new composite material, textile reinforced concrete, is focused on the derivation of a suitable material description and a surface-related shell theory. A surface-related shell theory is a natural approach for several reasons, i.e. a more accurate evaluation of the interface stresses between the existing structure and the strengthening layer and a finite element approach that is numerically more efficient. The idea of surface-related shell theories was pursued by one of the authors already many years ago, Rothert and Zastrau [10], Zastrau [14] and recently in Zastrau et al. [15]. Unlike the standard shell theories, that refer to the middle surface of the shell, the shell theory introduced here with an outer shell surface as the reference surface offers among others the following advantages:

- Contact problems can be considered without the usual difficulty of mapping the middle surface data onto the outer shell surface.
- The discretization of the contact surface and therewith the complete discretization may remain unchanged, contrary to the classical approach. In an optimization of e.g. the utilization of material, only the elements of

those matrices belonging to elements coordinate to the surface have to be changed iteratively.

- Discontinuities of the stress field, caused e.g. by concentrated loads occurring on the surface, which is peculiar to contact problems, are directly determined where they occur.
- The detection of compound failure and delamination becomes easier.

2 Continuum mechanics

Starting-point for the mechanical description of the thin strengthening structure is the augmented Hu-Washizu functional. It describes the following saddle point problem:

$$\begin{aligned}\Pi_{\mathrm{HW}}(\mathbf{E},\mathbf{U},\mathbf{S},\mathbf{t}_0) = &\int_{\mathcal{B}} \rho_0\, \mathrm{f}\, \mathrm{dV} + \int_{\mathcal{B}} \mathbf{S} :(\mathbf{E}^{\mathbf{u}} - \mathbf{E})\, \mathrm{dV} - \int_{\mathcal{B}} \rho_0 \mathbf{f}\cdot \mathbf{U}\, \mathrm{dV} \\ &- \int_{\partial_\sigma \mathcal{B}} \hat{\mathbf{t}}_0 \cdot \mathbf{U}\, \mathrm{dA} - \int_{\partial_\sigma \mathcal{B}} \mathbf{t}_0 \cdot (\hat{\mathbf{U}} - \mathbf{U})\, \mathrm{dA} \quad \rightarrow \quad \text{stat}\,. \end{aligned} \tag{1}$$

Herein is: $\mathbf{E}$ – the Green-Lagrange strain tensor, $\mathbf{U}$ – the displacement field, $\mathbf{S}$ the second Piola-Kirchhoff stress tensor, $\mathbf{t}_0$ – the surface traction per area of the undeformed boundary, ρ_0 – the mass density in the reference configuration, f – the Helmholtz free energy, $\mathbf{E}^{\mathbf{u}}$ – the displacement compatible Green-Lagrange strain tensor, $\mathbf{f}$ – the body force per unit mass. Additionally $\partial_\sigma \mathcal{B}$ is the part of the surface of the undeformed body where the surface traction per area of the boundary $\hat{\mathbf{t}}_0$ is prescribed and $\partial_u \mathcal{B}$ is the part of the boundary of the undeformed body where the displacements $\hat{\mathbf{U}}$ are prescribed. The superposed hat indicates prescribed quantities. In addition $\mathbf{N}$ is the outer normal vector to the surface of the body in the reference configuration.

To simplify the four field functional Eq. (1) and to avoid an interpolation for the stress field $\mathbf{S}$ the second term on the right-hand side is forced to vanish. This procedure characterizes the above mentioned EAS method that was first presented in the context of a variational formulation of continuum problems in Simo and Rifai [12] and demands that the additional strain tensor is orthogonal to the stress tensor. From this results that the orthogonality condition:

$$\int_{\mathcal{B}} \mathbf{S} : (\mathbf{E} - \mathbf{E}^{\mathbf{u}})\, \mathrm{dV} = \int_{\mathcal{B}} \mathbf{S} : \tilde{\mathbf{E}}\, \mathrm{dV} = 0 \tag{2}$$

in the discretized form must be fulfilled. The defined strain tensor $\tilde{\mathbf{E}}$ enriches the displacement compatible strain tensor $\mathbf{E}^{\mathbf{u}}$ avoiding locking phenomena in the shell theory and in the finite element formulation, respectively. This point will be discussed later in conjunction with the formulation of the shell theory. In contrast to the hybrid stress concept the EAS concept preserves the basic features of pure displacement finite elements, because the resulting additional strain parameters need not be compatible across the element boundary. They can be eliminated on the element level by condensation. The fulfillment of the orthogonality condition reduces the augmented Hu-Washizu functional Eq. (1) to a modified three field functional being the underlying principle for the surface-related shell theory and the finite element formulation:

$$\Pi_{HW}^{mod}(\tilde{\mathbf{E}}, \mathbf{U}, \mathbf{t}_0) = \int_{\mathcal{B}} \rho_0 f \, dV - \int_{\mathcal{B}} \rho_0 \mathbf{f} \cdot \mathbf{U} \, dV - \int_{\partial_\sigma \mathcal{B}} \hat{\mathbf{t}}_0 \cdot \mathbf{U} \, dA - \int_{\partial_u \mathcal{B}} \mathbf{t}_0 \cdot (\hat{\mathbf{U}} - \mathbf{U}) \, dA \quad \rightarrow \quad \text{stat.} \tag{3}$$

2 Surface-related shell theory

Since every shell theory is an approximation of a three dimensional continuum theory, basic assumptions have to be made, Basar and Krätzig [1], Naghdi [9]. The most serious assumption is related to the kinematics. Before the chosen kinematics is discussed in detail, the differential geometry of the shell continuum has to be described. Achieving this a general curvilinear, convected coordinate system Θ^1, Θ^2 and Θ^3 is introduced in the shell continuum, whereby Θ^1 and Θ^2 parameterize one of the outer surfaces, i.e. the reference surface (see Fig. 1). With the aid of this parameterization every point in the shell continuum is uniquely identified by the position vector **X**. Due to the displacement field **U** the shell continuum is deformed from the reference configuration into the actual configuration. This general deformation is restricted by a kinematical assumption.

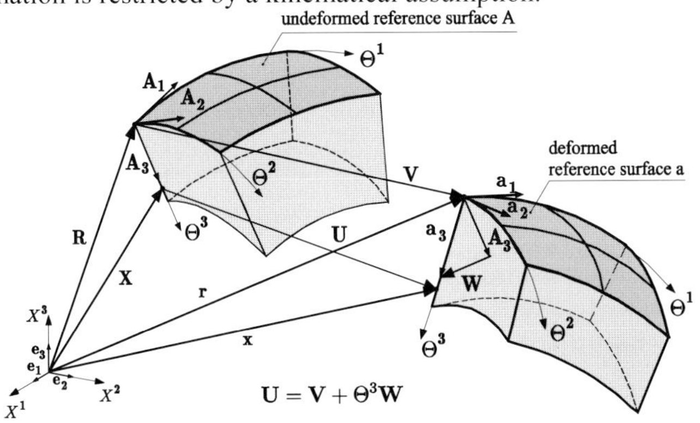

Fig. 1. Undeformed and deformed part of the shell continuum and shell kinematics

Within the scope of the presented shell formulation it is assumed that the displacement field is a sum of two parts:

$$\mathbf{U} = \mathbf{V} + \Theta^3 \, \mathbf{W} \tag{4}$$

first, the displacement field **V** of the reference surface and, second, a part in the direction of **W** being linearly dependent from the normal coordinate Θ^3. Herein **W** is a vector field describing the displacement field of points along the normal vector. The disadvantage of such a 6-parameter shell kinematics is that it suffers from so-called Poisson thickness locking. From this results a relative error of the order of ν^2 in bending dominated cases, Büchter [5]. It is obviously not adequate to approximate the deformation of a shell with unchanging thickness by kinematics leading to constant transverse normal strains. To avoid this locking effect an enhancement of the Green-Lagrange strain tensor is made using the EAS method. For this the displacement gradients are calculated from the linear approach Eq. (4) and introduced into the definition of the strain tensor:

$$\mathbf{E} = \mathbf{E}^{\mathbf{u}} + \tilde{\mathbf{E}} = \boldsymbol{\alpha} + \Theta^3\,\boldsymbol{\beta} + (\Theta^3)^2\,\boldsymbol{\gamma} \quad . \tag{5}$$

Herein $\boldsymbol{\alpha}$, $\boldsymbol{\beta}$ and $\boldsymbol{\gamma}$ are so-called sub-strain tensors. If the index notation is used, the displacement compatible strains E^u_{ij} and the components of the enhancement $\tilde{E}_{ij}$ can be identified as:

$$\begin{aligned} E^u_{ij} &= \alpha^u_{ij} + \Theta^3\,\beta^u_{ij} + (\Theta^3)^2\,\gamma^u_{ij} \\ \tilde{E}_{ij} &= \begin{cases} 0 & (i,j) \neq (3,3) \\ (\frac{H}{2}\text{-}\Theta^3)\,\tilde{\beta}_{33} & (i,j) = (3,3) \end{cases} \end{aligned} \tag{6}$$

According to this the enrichment of the strain field has only one component, namely $(\frac{H}{2}\text{-}\Theta^3)\,\tilde{\beta}_{33}\mathbf{G}^3 \otimes \mathbf{G}^3$, unequal to zero. It is important to note, that this single component is sufficient to avoid the Poisson thickness locking, because it introduces the missing, in Θ^3 linearly varying expression into the transverse normal strain. The determination of the additional strain $\tilde{\beta}_{33}$ is achieved by exploiting the orthogonality condition Eq. (2):

$$\int_B \mathbf{S} : \tilde{\mathbf{E}}\,\mathrm{dV} = \int_B S^{33}\,\tilde{E}_{33}\,\mathrm{dV} = 0 \quad \Rightarrow \quad m^{33} - n^{33}\,\frac{H}{2} = 0 \tag{7}$$

The coupling of the stress (n^{33}) and stress moment (m^{33}) resultant tensor is characteristic for surface-related shell theories. With Eq. (7) the value of the additional strain $\tilde{\beta}_{33}$ can finally be determined from:

$$\tilde{\beta}_{33} = \frac{\begin{bmatrix} D_1^{33kl} & D_2^{33kl} & D_3^{33kl} \end{bmatrix} - \frac{H}{2}\begin{bmatrix} D_0^{33kl} & D_1^{33kl} & D_2^{33kl} \end{bmatrix}}{-\frac{H^2}{4}\,D_0^{3333} + H\,D_1^{3333} - D_2^{3333}} \begin{bmatrix} \alpha^u_{kl} \\ \beta^u_{kl} \\ \gamma^u_{kl} \end{bmatrix} \tag{8}$$

The value of $\tilde{\beta}_{33}$ depends on the components C^{ijkl} of the material tensor:

$$D_K^{ijkl} = \int_0^H (\Theta^3)^K\,C^{ijkl}\,\mu\,\mathrm{d}\Theta^3 \qquad \text{with} \qquad K \in \{0,\ 1,\ 2,\ 3,\ 4\} \tag{9}$$

and the displacement compatible sub-strain measures α^u_{kl}, β^u_{kl} and γ^u_{kl}.

Interpreting the additional strain likewise as a kinematical variable of the shell theory one can speak of a 7-parameter shell theory. This is valid, because the strain parameter can be determined in any place of the reference surface. It should however be mentioned that this seventh parameter can be eliminated on the element level. This results again in a 6-parameter shell theory. Hence an expansion of the element stiffness matrix and the system stiffness matrix is prevented by the aforementioned elimination.

3 Locking phenomena of the finite element

Among others, the kinematical assumption Eq. (4) and the interpolation of the degrees of freedom are the source of several further locking phenomena, Bischoff [2], Bischoff and Ramm [3]. In particular the following phenomena are stated in this context:

- Poisson thickness locking
- Membrane locking
- Volume locking
- Curvature thickness locking
- Shear locking

Hence arrangements have to be made to reduce or avoid these locking effects, because all these artificial stiffening effects cause a significant deviation from the continuum mechanical solution. In this contribution the appropriate concepts can only be stated. As described before in connection with the formulation of the shell theory the EAS method is used to prevent Poisson thickness locking. The described specialization is in close relation to Büchter and Ramm [6], Büchter et al. [7]. Furthermore the EAS method is used to prevent or reduce membrane and volume locking. An effective concept against curvature thickness locking is the ANS method, Simo and Hughes [11], whereas the discrete shear gap (DSG) method possesses high efficiency and flexibility to avoid shear locking, Bletzinger et al. [4]. With a combination of the aforementioned concepts a very efficient finite volume shell element has been developed.

4 Contact mechanics

Investigating the behavior of textile reinforced concrete the contact mechanics described extensively in Laursen and Simo [8], Wriggers [13] is utilized to treat the compound failure. In this context it is necessary to build the tangential stiffness matrix for normal contact and tangential contact with help of contact segments, which were placed on the outer surface of the shell element. These segments can be used easily with a body situated in contact that is discretized by volume elements as well. The difficulty in middle surface related shell elements is the determination of the stiffness for tangential and normal displacement degrees of freedom determined by the segments, because they directly have effect on the stiffness affiliated to the rotational degrees of freedom of the shell element. Using shell theories with higher order kinematics, this mapping from surface data on the middle surface is connected with considerable effort. Furthermore, it is necessary to know the geometry of the finite volume element connected with the contact segment. A combination of contact segment and shell element becomes unavoidable. Using surface related shell theories the assignment of the segments can performed in the same manner like using volume elements, because only the stiffness corresponding to the three displacements of the surface has to be calculated. The coordinate systems of the contact segment as well as of the shell element are both situated in the contact area.

The principle of the virtual work can be utilized to derive the tangential stiffness matrix of the shell element because of the normal and the tangential contact conditions in the case of sticking contact partners. If the tangential contact stress exceeds a static limit that can either be dependent on the contact normal pressure p in the form $\tau_{max} = \mu\, p$ with μ as coefficient of friction or independent with $\tau_{max} = \tau_{stick}$ a shear stress:

$$\tau\left(g_N, \mathbf{g}_T\right) = -\frac{\mathbf{g}_T}{\left|\mathbf{g}_T\right|}\left(\mu\, p\left(g_N\right) + k\left(\mathbf{g}_T\right)\right) \qquad (10)$$

is used in the case of slipping. It depends on the gap function g_N and the tangential relative displacements represented by the slip vector $\mathbf{g}_T$. The first part in Eq. (10) states Coulomb's friction law. The resulting stiffness $k\left(\mathbf{g}_T\right)$ can be regarded as tangential compound stiffness, e.g. adhesion within the contact area. This offers the possibility to describe compound problems within the scope of contact mechanics.

5 Numerical example

The example shows the application of contact mechanics in the context of strengthening and retrofitting of structural members. We chose a square plate with the dimensions 7.20×7.20×0.20 m³ with build-in edges and a concentrated central loading. The plate is made of St. Venant-Kirchhoff material with a Young's modulus $E = 2.5 \cdot 10^7$ kN/m² and a Poisson's ratio $\nu = 0.2$. The plate is strengthened by a symmetrically applied textile reinforced concrete ply with the dimensions 6.30×6.30×0.02 m³ that is pressed against the plate by a contact pressure of 0.1 kN/m² to establish contact. For the maximum shear stress a friction coefficient $\mu_0 = \mu = 0.15$ is used. The Young's modulus of the strengthening layer is $E = 2.0 \cdot 10^7$ kN/m² and the Poisson's ratio is $\nu = 0.2$. Making use of the symmetry only one quarter of the plate is discretized with 4-node shell elements.

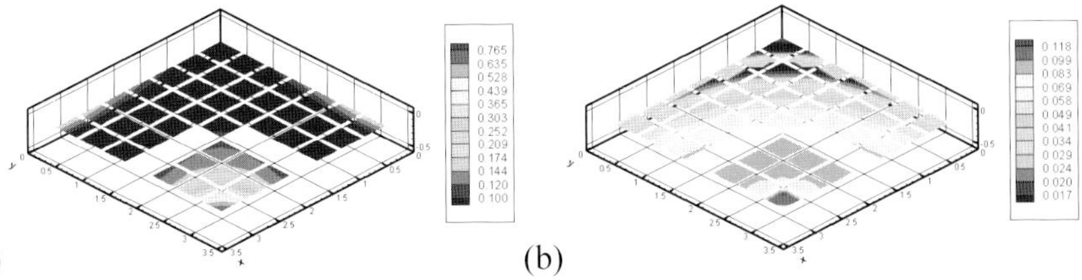

(a) (b)

Fig. 2. Contact area and contact pressure (a) and contact shear stress distribution (b) both in kN/m²; $F = 3.75$ kN

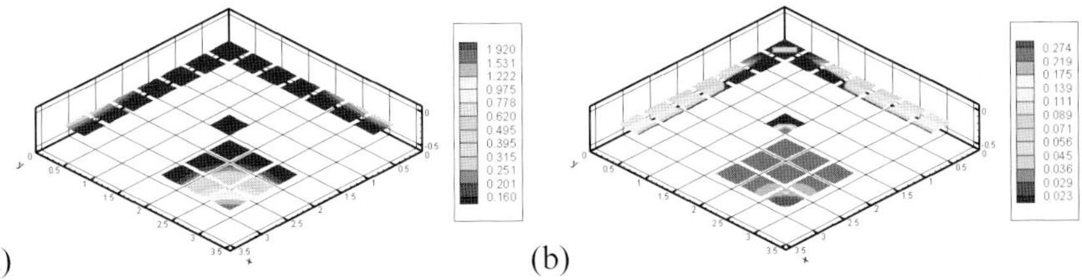

(a) (b)

Fig. 3. Contact area and contact pressure (a) and contact shear stress distribution (b) both in kN/m²; $F = 10.0$ kN

In a first step a resultant load of $F = 3.75$ kN is applied (see Fig. 2) and in a second step a resultant load of $F = 10$ kN is applied (see Fig. 3). The white areas in the contact domain represent delaminated regions of the strengthening layer. The size of the delaminated regions grows extensively with an increase of the load F.

6 Conclusion

The formulation of surface-related shell theories allows the efficient simulation of the compound behavior of thin textile reinforced concrete layers for the strengthening of structural members. Due to the particular position of the reference surface the well established concepts against locking phenomena are extended and implemented which leads to a reliable surface-related finite volume shell element. With the aid of a nonlinear example the efficiency of the developed shell and contact formulation is demonstrated.

Acknowledgement

The authors gratefully acknowledge the financial support of this research from Deutsche Forschungsgemeinschaft DFG within the Sonderforschungsbereich SFB 528 "Textile Reinforcement for Structural Strengthening and Retrofitting" at Technische Universität Dresden.

References

1. Başar Y, Krätzig WB (1985) Mechanik der Flächentragwerke. Friedrich Vieweg & Sohn Verlag mbH, Braunschweig
2. Bischoff M (1999) Theorie und Numerik einer dreidimensionalen Schalenformulierung. PhD thesis, Universität Stuttgart, Institut für Baustatik
3. Bischoff M, Ramm E (1997) Shear deformable shell elements for large strains and rotations. Int J Num Meth Engng 40: 4427-4449
4. Bletzinger KU, Bischoff U, Ramm E (1998) A unified approach for shear-locking-free triangular and rectangular shell finite elements. In: Computational Mechanics, New Trends and Applications, CD-Rom, Proceedings of the Fourth Congress on Computational Mechanics, pp 1-22
5. Büchter N (1999) Zusammenführung von Degenerationskonzept und Schalentheorie bei endlichen Rotationen. Ph.D. thesis, Universität Stuttgart, Institut für Baustatik
6. Büchter N, Ramm E (1992) 3d-extension of nonlinear shell equation based on the enhanced assumed strain concept. Computational Mechanics in Applied Mechanics, First European Conference on Numerical Methods in Engineering, pp 39-59
7. Büchter N, Ramm E, Roehl D (1994) Three-dimensional extension of nonlinear shell formulation based on the enhanced assumed strain concept. Int J Num Meth Engng 37: 2551-2568
8. Laursen TA, Simo JC (1993) A continuum-based finite element formulation for the implicit solution of multibody, large deformation frictional contact problems. Int J Num Meth Engng 36: 3451-3485
9. Naghdi PM (1963) Foundations of elastic shell theory. Progress in Solid Mechanic 4 pp 1-90
10. Rothert H, Zastrau BW (1981) Herleitung und Anwendung einer kontaktproblemorientierten Schalentheorie. Konstruktiver Ingenieurbau-Berichte (KIB) 61/39: 76-84
11. Simo JC, Hughes TJR (1986) On the variational foundations of assumed strain methods. J Appl Mech 63: 224-225
12. Simo JC, Rifai MS (1990) A class of mixed assumed strain methods and the method of incompatible modes. Int J Num Meth Engng 29: 1595-1638
13. Wriggers P (1995) Finite element algorithms for contact problems. Arch Comp Meth Engng 2(4): 1-49
14. Zastrau BW (1983) Ein Beitrag zur Modifikation der klassischen Schalentheorie für die unmittelbare Berechnung von Kontaktproblemen. ZAMM 63: 224-225
15. Zastrau BW, Schlebusch R, Matheas J (2000) Surface-related shell theories for the treatment of composites and contact problems. In: Proceedings of the Fourth Colloquium on Computation of Shells & Spatial Structures, pp 1-18

Refined Wave Theories of Plates and Shells

I. T. Selezov

Department of Wave Processes, Institute of Hydromechanics, National Academy of Sciences, Kiev, Ukraine

The paper is devoted to construction of refined hyperbolic models of plates and shells predicting the wave propagation with the finite velocities unlike the classical Kirchhoff plate theory and Kirchhoff-Love shell theory and other known modifications of parabolic type. It is shown that these refined models can be obtained as mathematical approximations of the original hyperbolic model of elastodynamics without introducing some additional corrections of physical kind unlike widely used refined phenomenological theories of Timoshenko-Mindlin plates (1951), Herrmann-Mirsky, Lin-Morgan, Naghdi-Cooper shells (1956) and others. From the point of view of the spectral theory, any theories of plates and shells are of approximate kind and, as a result, deal with a finite number of eigenvalues and wave modes instead of infinite numbers of eigenvalues corresponding to exact elastodynamic problem.

In general, the construction of degenerated models of such a kind can be developed by using various methods. The most popular of them is the phenomenological approach – the method of hypotheses and averagings over the thickness coordinate $x^s \in \left[-h^s, h^s\right]$. As far as analytical methods are concerned, the asymptotic approaches and the power series method can be noted. Asymptotic methods approximate the operator P as the expansion in small parameter ξ while the power series method represents the operator P in the vicinity $x^s = 0$ .The analytical approaches find there beginning with the works of Lagrange (1781) who derived the shallow water wave equation on the basis of asymptotic expansions. The power series method finds its beginning with the works of Cauchy (1828) and Poisson (1829) derived the equation of plate vibration. In the theory of plates and shells the power series method has been used by Krauss (1929), Kilchevsky (1939), Selezov (1960, 1962, 1994) and others [1-4].

It should be noted that above-mentioned approaches originate new simpler models which, however, can be of different type compared to the original model. In this paper the problem of degeneration of the original IBV hyperbolic problem with a spatial coordinate is considered as a tool to construct the simpler hyperbolic models of less dimensionality, that is to realise hyperbolic degeneration.

The initial boundary value problem in R^n for the system of hyperbolic equations of arbitrary order in a hyperlayer $\Omega \subset R^n$ is stated. On the basis of the power series method this model is reduced to the infinite system of equations and recurrence relationships when $\xi = h^s / l \to 0$ (h^s is the hyperlayer thickness, l is a horizontal length scale). This infinite system can be truncated by different manner allowing to obtain different approximations. In this connection three cases are possible: degeneration to models of hyperbolic, parabolic or mixed type. The question is what is correct? From the point of view of finite velocity of disturbance propagation only the case of hyperbolic to hyperbolic is correct and having the physical sence. Necessary and sufficient conditions of hyperbolic degeneration for selection of hyperbolic approximations are presented below.

Let us consider the functions $u_r(x)\left(r=\overline{1,k}\right)$ in $n-$dimensional Euklidean space R^n the point of which $(x)=x^1,...,x^n$, where $x^1=t$ is the time coordinate. Consider the region $\Omega \subset R^n$ bounded hypersurfaces $x^s = \pm h^s$, $h^s > 0$ (index s is fixed):

$$\Omega = \left\{ x \in R^n : -\infty < \left(x^1, x^2, ..., x^{s-1}, x^{s+1}, ..., x^{n-1}\right) < \infty, x^n \geq 0, -h^s \leq x^s \leq h^s \right\}.$$

Consider a finite hyperbolic system of linear partial differential equations in R^n, for which a boundary value problem or initial boundary value problem is stated with given real-valued data and real-valued solutions belonged C^k. The differential equations and boundary conditions are written as follows (Kythe,1996) [3]

$$L_i \equiv a_{ilq} \frac{\partial^q u_l}{\partial x^{1(\alpha_1)} ... \partial x^{n(\alpha_n)}} + F_i = P_i \tag{1}$$

In Ω, $t \geq 0$, $(i,l)=\overline{1,\ k}$, $q = \alpha_1 + \alpha_2 + ... + \alpha_n$, $q = \overline{1,\ p}$

$$\left\{ b_{jlq} \frac{\partial^q u_l}{\partial x^{1\alpha} ... \partial x^{n(\alpha_n)}} + f_j \right\}_{x^s = \pm h^s} = Q_j^{\pm},\ j = p \cdot k,\ q = \overline{1,(p-1)},\ l = \overline{1,k}, \tag{2}$$

where (1) is the system of k equations of the p-th order with k unknown functions to be determined as the solutions of this system satisfying the boundary conditions (2) and the initial conditions (not presented here), so that a well-posed statement is guaranteed. The first term in (1) is the principal part of the operator, the second term F_i is the remained part of the operator. In (2) the term f_j is the operator of lower order than the first term. The summation convention over repeated subscripts is implied and subscripts preceded by a comma denote partial differentiation with respect to the corresponding coordinate. It is assumed that coefficients a_{ilq} and b_{jlq} are constant but they can depend on small parameter $\varepsilon \ll 1$.

The system (1) can be reduced to the system of equations of the first order (Misokhata, 1965; Bers et al.,1964) [4]

$$\frac{\partial v}{\partial t} - A_v \frac{\partial v}{\partial x_v} - Bv = p, \quad v = 2,3,...,n, \tag{3}$$

where v is the vector having m components $v_1(x,t),...,v_m(x,t)$, $m > k, p$ is the vector of similar kind, $\boldsymbol{A}$ and $\boldsymbol{B}$ are matrices of the order m. It is assumed that the characteristic equation (without lowest terms)

$$\det\left(\lambda I - A_v \xi_v\right) = 0 \tag{4}$$

has the real roots $\lambda_i(\xi)$ and matrix $A_v \xi_v$ is reduced to diagonal kind for each $\xi \in R^{n-1}$ that is the necessary condition is satisfied for system (3) to be of hyperbolic type.

Hyperbolicity is necessary but not enough condition for propagation of disturbances with finite velocities. Principle of finite velocity can be violated by boundary conditions including boundary and initial data, and variable coefficients corresponding to inhomogeneity and /or anisotropic properties of media. For example, Hersh (1964) has shown that there are boundary conditions for which the initial problem will be incorrect posed and hence there are no exist solutions with the finite velocities. We consider the boundary conditions for which the IBV problem is correct posed.

The original model (1), (2) is reduced to degenerated model when $\varepsilon \to 0$ or $\xi = h^s / l \to 0$. Our aim is to derive hyperbolic approximations, i.e. to construct a mapping of the reference space $R^n(\varepsilon)$ or $R^n(\xi)$ into the degenerated space, $R^n(\varepsilon) \to R^{n-1}$ or $R^n(\xi) \to R^{n-1}$ satisfying the condition of limiting correctness to be of hyperbolic type, that is the condition of the finite velocity of disturbance propagation (Hersh, 1964; Selezov, 1969) [2]. At that, in degenerated space instead of the functions $u(x)$ we have new functions $\widehat{u}(x)$. In further, we shall consider the case of coordinate degeneration $\xi \to 0$.

Expanding field functions in power series by a degenerated coordinate s with respect to a middle hypersurface yields the degenerated problem to determine the series coefficients depending now only on *n-1* coordinates

$$u_i\left(t, x^2, ..., x^{n-1}, x^n\right) = \sum_{k=1}^{\infty} u_{ik}\left(t, x^2, ..., x^{s-1}, x^{s+1}, ..., x^n\right)\left(x^s\right)^k \quad . \tag{5}$$

Substituting (5) into PDE Eqs. (1) and boundary conditions, (2) yields recurrence relationships from Eqs. (1) and several systems of differential equations of infinite order from (2) in R^{n-1}. The next principal step is the reduction of these infinite systems which is possible in different ways keeping the terms according to different regulations. It is possible from recurrence relationships to express all u_{ik} in terms of minimal finite number of desired functions corresponding to a number of the systems of differential equations. Substituting these functions into reduced equations yields the resolving equations. It allows us to obtain different approximations, i.e. simplified models. Regulations of keeping the terms should be of such a kind that this system will be of hyperbolic type. The necessary condition is to obtain a closed system to determine new unknown functions. The sufficient condition of hyperbolic degeneration is announced as follows: in order to obtain the hyperbolic approximation it is enough to keep in infinite systems all space-time differential operators up to definite order.

On this basis refined hyperbolic models of plates and cylindrical shells have been obtained as mathematical approximations of corresponding problems of elastodynamics for the layer and hollow cylinder. These models include all known phenomenological models as particular cases.

Let us consider the elastodynamic problem for an infinite layer of the thickness $2\,h$ occupied the region

$$\Omega=\left\{ (x_1,\, x_2,\, x_3) \in R^3 : x_1,\ \ x_2 \in (-\infty,\, \infty),\quad x_3 \in \left[-\xi/2,\ \xi/2\right]\right\}$$

in a Cartesian coordinate system $(x_1,\, x_2,\, x_3)$. The corresponding IBV-problem for the displacement vector $\vec{u} = (u_1,\, u_2,\, u_3 = w)$ is presented as follows: to find the functions $u_k = u_k(x_1,\, x_2,\, x_3,\, t)$ as the solutions of the hyperbolic equations in $\Omega \times [0,\, T]$, $T>0$

$$\nabla^2 u_k + (1 + \lambda / G)\, \partial_k (\vec{\nabla} \cdot \vec{u}) = \partial_{tt}\, u_k, \qquad k=1,2,3 \tag{6}$$

satisfying the boundary conditions for the stress-tensor components

$$\sigma_{33}\Big|_{x_3=\pm\xi/2} = q^{\pm}(x_1,x_2,t), \quad \sigma_{3i}\Big|_{x_3=\pm\xi/2} = p_i^{\pm}(x_1,x_2,t), \quad i=1,2 \tag{7}$$

and the initial conditions:

$$u_k\Big|_{t=0} = 0, \qquad \partial_t\, u_k\Big|_{t=0} = 0, \qquad k=1,2,3. \tag{8}$$

Here dimensionless quantities are used according to the formulae (asteriscs are suppressed throughout)

$$\left(x_i^*,\, u_i^*\right)=\frac{1}{l}\left(x_i,\, u_i\right), \qquad t^* = \frac{c_s}{l}\, t, \qquad c^* = \frac{c}{c_s}$$

$$\left(\sigma_{kl}^*,\, q^*,\, p_i^*\right)=\frac{1}{G}\left(\sigma_{kl},\, q,\, p_i\right), \qquad \frac{2h}{l}=\xi\ .$$

The following expressions for stress-tensor components are used
$\sigma_{kk} = (\lambda / G)(\nabla \cdot u) + 2\,u_{k,k}$ (no summation over k),

$$\sigma_{kp} = G(u_{k,p} + u_{p,k}) \qquad (k,\ p=1,2,3;\ k \neq p)$$
$$\vec{\nabla} \cdot \vec{u} = u_{k,k} = e + u_{3,3} \qquad e = u_{1,1} + u_{2,2}\ .$$

It is assumed that $2h/l = \xi \ll 1$ (l is a horizontal length scale), λ and G are constants. The functions u_k are presented in the form of power series in x_3

$$u_k\,(x_1,x_2,x_3,t) = \sum_{\nu=0}^{\infty} u_{k\nu}\,(x_1,\, x_2,t)\, x_3^{\nu}, \qquad k=1,2,3\ . \tag{9}$$

The functions $u_{k\nu}$ are assumed to be differentiated so many times as it is required, and all the derivatives of $u_{k\nu}$ are continuous, and series (9) converge uniformly. The convergence of these series depends only on the value ξ characterising changes of the fields in the coordinates x_1, x_2, t. The smoother these changes are, the greater is l and the smaller ξ and the faster series (9) are converging. Substituting (9) into (6)-(8) leads after long transformations to an infinite system which is decomposed onto two independent subsystems for symmetric (planar) and asymmetric (bending) deformations with respect to the mid–surface. The above-mentioned approach for the construction of hyperbolic approximations allows to obtain both the known models obtained earlier on the basis of phenomenological approach involving physical assumptions and the new more exact models.

In the case of symmetric deformation with respect to the middle surface we obtain from Eqs. (6)-(8) (when q_1, $q_2 = 0$), as the second approximation – the new more exact two-mode hyperbolic equation

$$\left(-\xi a_1' \nabla^2 + \xi a_2' \frac{\partial^2}{\partial t^2} + \xi^3 b_1' \nabla^2 \nabla^2 - \xi^3 b_2' \nabla^2 \frac{\partial^2}{\partial t^2} + \xi^3 b_3' \frac{\partial^4}{\partial t^4}\right) e_0 = \\ = \left\{ d_1' + \xi^2 d_2' \nabla^2 + \xi^2 d_3' \frac{\partial^2}{\partial t^2} \right\} \left(\frac{\partial}{\partial x_2} \frac{p_1 - p_2}{2} + \frac{\partial}{\partial x_1} \frac{p_3 - p_4}{2} \right) \tag{10}$$

Coefficients in (10) are of the form

$$a_1' = \frac{4}{1+2\nu}, \qquad a_2' = 2\frac{1-\nu}{1-2\nu}, \qquad b_1' = \frac{8}{1-2\nu}\frac{1}{3!},$$
$$b_2' = \left(\frac{6}{1-2\nu} + 2\frac{2-\nu}{1-\nu} + 4\frac{1-\nu}{1-2\nu}\right)\frac{1}{3!} \qquad b_3' = \frac{5-8\nu}{1-2\nu}\frac{1}{3!}. \tag{11}$$

The Eq. (10) includes as a particular case the one-mode hyperbolic approximation (generalized stress state): two the first terms in the left hand side and the first term in the right hand side.

In the case of asymmetric deformation, keeping all the operators up to seventh order inclusively, we obtain from (6)-(8) (when $p_1, p_2, p_3, p_4 = 0$) the new three-mode hyperbolic approximation which can be reduced to the following generalized resolving hyperbolic equation

$$\left\{ \left[\left(\xi \frac{\partial^2}{\partial t^2} + \xi^2 a_1 \nabla^2 \nabla^2 \right)_K - \xi^3 a_2 \frac{\partial^2}{\partial t^2} \nabla^2 + \xi^3 a_3 \frac{\partial^4}{\partial t^4} \right]_{TM} - \right. \\ \left. - \xi^5 b_1 \nabla^2 \nabla^2 \nabla^2 + \xi^5 b_2 \frac{\partial^2}{\partial t^2} \nabla^2 \nabla^2 - \xi^5 b_3 \frac{\partial^4}{\partial t^4} \nabla^2 + \xi^5 b_4 \frac{\partial^6}{\partial t^6} \right\}_{TMS} w_0 \\ = \left\{ \left[1 - \xi^2 d_1 \nabla^2 + \xi^2 d_2 \frac{\partial^2}{\partial t^2} \right]_{TM} + \xi^4 d_3 \nabla^2 \nabla^2 - \right. \\ \left. - \xi^4 d_4 \frac{\partial^2}{\partial t^2} \nabla^2 + \xi^4 d_5 \frac{\partial^4}{\partial t^4} \right\}_{TMS} \left(q^+ - q^-\right) \tag{12}$$

Coefficients in (12) are determined by formulae

$$a_1=\frac{1}{3}\frac{\lambda+G}{\lambda+2G},\quad a_2=\frac{1}{3!}\frac{3\lambda+4G}{\lambda+2G},\quad a_3=\frac{1}{4!}\frac{3\lambda+7G}{\lambda+2G}$$

$$b_1=\frac{1}{10\cdot 3!}\frac{\lambda+G}{\lambda+2G},\qquad b_2=\frac{1}{5!}\frac{4\lambda^2+14G\lambda+11G^2}{(\lambda+2G)^2}$$

$$b_3=\frac{1}{2^2\cdot 5!}\frac{9\lambda^2+39G\lambda+3G^2}{(\lambda+2G)^2},\qquad b_4=\frac{1}{2^4\cdot 5!}\frac{5\lambda^2+30G\lambda+41G^2}{(\lambda+2G)^2}$$

$$d_1=\frac{1}{8}\frac{3\lambda+4G}{\lambda+2G},\quad d_2=\frac{1}{8!},\qquad d_3=\frac{1}{2^4\cdot 5!}\frac{5\lambda^2+16G\lambda+12G^2}{(\lambda+2G)^2}$$

$$d_4=\frac{1}{8\cdot 4!}\frac{2\lambda^2+8G\lambda+7G^2}{(\lambda+2G)^2},\qquad d_5=\frac{1}{16\cdot 4!} \tag{13}$$

The hyperbolic operator TMS of the sixth order includes the Timoshenko-Mindlin hyperbolic operator TM of the fourth order and the parabolic Kirchhoff operator ***K***.

It should be noted that coefficients (11) and (13) depend only on the Poisson number. As a result, the exact analytical formula for the shear coefficient corresponding to two-mode approximation (TM-operator) can be obtained

$$k^2=2/\left(2-\nu+\sqrt{0.5+\nu^2}\right) \tag{14}$$

where ν is the Poisson's number. In conclusion it can be noted that unlike phenomenologocal approaches, for realization of the purely analytical approach presented in this paper it is not necessary to attract some physical prerequisities to construct the hyperbolic models.

Acknowledgement

This research is supported by SFFR Project of Ukraine (Grant №01.07/00079).

References

1. Grigoluk EI, Selezov IT (1973) Nonclassical theories of vibrations of bars, plates and shells (in Russian). Acad Sci, Moscow, USSR
2. Selezov IT (1999) Some hyperbolic models for wave propagation. Hyperbolic problems: Theory, numerics, applications. In: Fey M, Jeltsch R (eds) Int Series of Numerical Mathematics. Birkhäuser Verlag, Basel, Switzerland, Vol. 130: 833-842
3. Selezov IT (1994) Hyperbolic models of wave propagation in bars, plates and shells (in Russian). Izvestia Akedemii Nauk Rossii, Mekhanika tverdogo tela 2: 64-77
4. Selezov IT (2000) Degenerated hyperbolic approximations of the wave theory of elastic plates. Operator theory. advances and application. Birkhäuser Verlag, Basel, Switzerland, 117: 339-353

On Dynamics of Thin Plates With a Periodic Structure

Cz. Woźniak and E. Wierzbicki

Institute of Mathematics and Informatics, Częstochowa University of Technology, Poland

1 Introduction

The subject of investigations is elastodynamics of periodic plates like that shown in Fig.1. So far, two general approaches to the modelling of these plates leading to the effective 2D-models (i.e. represented by equations with constant coefficients) have been proposed. The first is based on the asymptotic analysis and multi-scale expansions, cf. [4], and yields what are called homogenized plate equations which have the same form as equations of homogeneous plates, [5,9]. Hence, the main attention in this approach is concentrated on calculations of constant coefficients in a homogenized equation, which are referred to as the effective plate stiffness (cf. [5,9] and [10] for an overview of papers on this subject). The second approach applies the tolerance averaging technique as a tool of modelling, [14]. We can mention here a series of papers [1-3,7,8,11-13] related to dynamics and stability of elastic plates and beams with a periodic structure. The tolerance averaging approach, in contrast to homogenization, makes it possible to describe the local dynamic plate behaviour (e.g. to determine higher order vibration frequencies caused by the plate periodic structure) and takes into account the rotational inertia effect on the plate dynamics. However, until now the effective 2D-models for dynamics of periodic plates, based on the tolerance averaging technique, have been restricted to situations in which the period length is sufficiently large when compared to the plate thickness.

The main aim of this contribution is to propose a new effective 2D-model of periodic plates which, in contrast to the aforementioned models, at the same time describes the local dynamic plate behaviour and can be applied to plates with the mean thickness of an order of the period lengths. We shall deal with the bending of plates for which the midplane is a symmetry plane and loadings are transverse to the midplane.

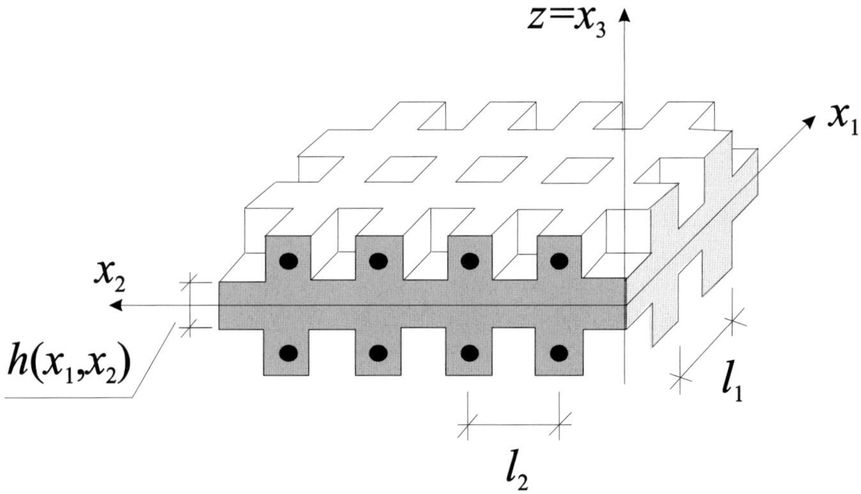

Fig. 1. Geometry of a periodic plate

As a mathematical tool of modelling we shall apply the tolerance averaging technique to the equations of the linear elastodynamics. To make this paper self-consistent in the subsequent section we shall outline the foundations of the tolerance averaging technique; for details the readers is referred to [14].

1.1 Notations

By x_1, x_2, x_3 we denote the orthogonal Carthesian coordinates in the physical space, by $\mathbf{x} = (x_1,x_2)$ a position vector of points on the Ox_1x_2-plane and by $\mathbf{n}$ the unit vector related to the Ox_3-axis. We also define $z = x_3$. The distance between points $\mathbf{x}$, $\mathbf{y}$ on the Ox_1x_2-plane will be denoted by $\|\mathbf{x}-\mathbf{y}\|$. We shall use the gradient operators $\nabla=(\partial_1,\partial_2,\partial_3)$, $\overline{\nabla}=(\partial_1,\partial_2,0)$, where ∂_k, $k = 1,2,3$, is a partial derivative with respect to x_k. We also introduce notation $\Delta=(0,l_1)\times(0,l_2)$, $\Delta(\mathbf{x})=\mathbf{x}+\Delta$ and $\Delta(\mathbf{x},z)=\Delta(\mathbf{x})+\mathbf{n}z$. By a Δ–periodic function we mean a function of $\mathbf{x} = (x_1,x_2)$ which has periods l_1, l_2 in arguments x_1, x_2, respectively, by l we define the length of the diameter of Δ. Superscripts A, B run over 1,...,N and subscripts α, β, γ, δ over 1,2; summation convention holds.

2 Preliminaries

Let the plate under consideration in its undeformed configuration occupies in the physical space a region

$$\Omega = \{(\mathbf{x}, z) : |z| \leq h(\mathbf{x}), \mathbf{x} \in \Pi\}$$

where Π is a region on the Ox_1x_2-plane (representing the plane midplane) and $h(\cdot)$ is a positive valued Δ–periodic function which may be non-continuous. The maximum plate thickness will be denoted by δ. We assume that the plate is thin, i.e., $\delta \ll L$ where L stands for the smallest characteristic length dimension of Π, and that the plate thickness δ is of an order of periods l_1, l_2. Moreover, all material plate properties are assumed to be Δ–periodic functions of $\mathbf{x} \in \Pi$ and even functions of z. Setting $\Pi_o = \{\mathbf{x} \in \Pi : \Delta(\mathbf{x}) \subset \Pi\}$ we conclude that every region

$$V(\mathbf{x}) := \{(\mathbf{y}, z) : z \in (-h(\mathbf{y}), h(\mathbf{y})), \mathbf{y} \in \Delta(\mathbf{x})\}, \quad \mathbf{x} \in \Pi_o$$

is a three dimensional periodicity cell with a center at $\mathbf{x}$.

Let $g(\cdot)$ be an arbitrary integrable function defined in Ω. In the subsequent considerations we shall apply two kinds of averaging operators, given by

$$\begin{aligned}
\langle g\rangle(\mathbf{x}; z) &= \frac{1}{|\Delta(\mathbf{x}; z) \cap \Omega|} \int_{\Delta(\mathbf{x}; z)\cap\Omega} g(\mathbf{y}, z) d\mathbf{y}, \quad \mathbf{x} \in \Pi_o, \quad |z| \leq \delta / 2, \\
\langle g\rangle(\mathbf{x}) &= \frac{1}{\delta} \int_{-\delta/2}^{\delta/2} \langle g\rangle(\mathbf{x}; z) dz, \quad \mathbf{x} \in \Pi_o
\end{aligned} \tag{1}$$

where $|\Delta(\mathbf{x}; z) \cap \Omega|$ is the area of the plane element $\Delta(\mathbf{x}; z) \cap \Omega$ (this area is independent of $\mathbf{x}$) and $d\mathbf{y}= dy_1dy_2$. Let us observe that $\langle g\rangle(\mathbf{x})$ is the value of function $g(\cdot)$ averaged over a periodicity cell $V(\mathbf{x})$. If $g(\cdot, z)$ is a Δ–periodic function then $\langle g\rangle(\mathbf{x})$ is independent of $\mathbf{x}$ and will be denoted by $\langle g\rangle$.

Let $F(\cdot)$ be an integrable function defined on Ω such that for an arbitrary integrable Δ–periodic function $g(\cdot, z)$, $|z| < \delta / 2$, the following approximation takes place

$$\langle gF\rangle(\mathbf{x}; z) \cong \langle g\rangle F(\mathbf{x}, z), \quad \mathbf{x} \in \Pi_o, \quad |z| < \delta / 2. \tag{2}$$

Here and subsequently an accuracy of approximation $\cong$ also depends on the form of function g, cf. [14]. If the above approximation holds also for all derivatives of F which occur in the problem under consideration then $F(\cdot, z)$ will be referred to as a *slowly-varying function*. It means that in the course of modelling increments $DF(\mathbf{x}, z) - DF(\mathbf{y}, z)$, where DF stands for F and all its derivatives, can be neglected provided that condition $\|\mathbf{x} - \mathbf{y}\| \leq l$ holds.

Let $f(\cdot)$ be an integrable function defined on Ω such that for every $\mathbf{x} \in \Pi_o$ there exists a Δ–periodic function $f_{\mathbf{x}}(\cdot)$ which for an aforementioned function g satisfies the approximation

$$\langle gf \rangle(\mathbf{x}; z) \cong \langle gf_{\mathbf{X}} \rangle(\mathbf{x}; z), \quad \mathbf{x} \in \Pi_o, \quad |z| < \delta / 2. \tag{3}$$

If the above condition holds for every Δ–periodic integrable function g then $f(\cdot, z)$ will be referred to as *a periodic-like function*. Moreover, a Δ–periodic function $f_{\mathbf{x}}(\cdot, z)$ will be called *a local periodic approximation of f* near $\mathbf{x}$.

Formulae (2), (3) represent what is called the tolerance approximation of averages $\langle gF \rangle(\mathbf{x}; z)$, $\langle gf \rangle(\mathbf{x}; z)$, respectively, [14]. Symbol $\cong$ stands for a certain tolerance relation, i.e., a binary relation defined on R^2 which is symmetric and reflexive but not transitive. To be more exact, if φ_1, φ_2 are two values of a certain quantity expressed in the fixed unit measure, then $\varphi_1 \cong \varphi_2$ holds if $|\varphi_1 - \varphi_2| \leq \varepsilon_\varphi$ where ε_φ stands for a tolerance parameter related to measurements and/or calculations of φ. It follows that in order to specify approximations (2) and (3) we have to assign to every quantity under consideration a certain tolerance parameter; this mapping is referred to as a tolerance system and denoted by T. Let us observe that in order to introduce concepts of slowly-varying and periodic-like function we have to specify T and a cell Δ. That is why the collections of these functions will be denoted by $SV_\Delta(T)$ and $PL_\Delta(T)$, respectively. For a detailed discussion of the aforementioned concepts the reader is referred to [14]; the notion of the tolerance parameter was introduced in [6] as an "upper bound for negligibles". Following results given in [14] we recall that if $f(\cdot, z) \in PL_\Delta(T)$ and g is a Δ–periodic integrable function then $\langle gf \rangle(\cdot, z) \in SV_\Delta(T)$. This assertion will be used in the course of the modelling procedure in the subsequent section.

3 Modelling approach

The starting point for the proposed modelling procedure are the known equations of the linear elastodynamics

$$\nabla \cdot (\mathsf{C} : \nabla \mathbf{u}) - \rho \ddot{\mathbf{u}} + \rho \mathbf{b} = \mathbf{0} \tag{4}$$

where $\mathbf{u}(\cdot, t)$ is a displacement field at time t defined on Ω and $\mathbf{b}$ is a constant body force; in the class of problems under considerations the tensor of elastic moduli $\mathsf{C}(\cdot, z)$ and the mass density $\rho(\cdot, z)$ are for every $|z| < \delta / 2$ the known Δ–periodic functions. We shall also assume that $z = const.$ are elastic symmetry planes, $\mathsf{C}(\mathbf{x}, \cdot)$, $\rho(\mathbf{x}, \cdot)$ are even functions of z and that the plate is subjected to the same distributions of loadings at boundaries $z = \pm h(\mathbf{x}), \mathbf{x} \in \Pi$; these loadings are assumed to be transversal with respect to the midplane. Eqs. (4) have to be satisfied together with the stress continuity conditions and the initial conditions as well as boundary conditions on $\partial '\Omega = \{(\mathbf{x}, z) : |z| \leq h(\mathbf{x}), \mathbf{x} \in \partial\Pi\}$.

Obviously, only a certain class of dynamic problems related to Eqs. (4) can be described in the framework of the specified below 2D-model represented by equations with constant coefficients. This class of problems will be restricted by the following heuristic assumption.

Conformability Assumption (CA). *The displacement field u=u(·,z,t), for every z and t, has to conform to the Δ-periodic plate structure, i.e. u(·,z,t) has to be a periodic-like function; this condition can be violated only near boundary* $\partial '\Omega$*.*
The tolerance averaging of equations (4) will be carried out in five steps.

1. We define for every $z \in (-\delta/2, \delta/2)$ the averaged displacement field **w**(·,z) by means of

$$\mathbf{w}(\mathbf{x}, z, t) = (\langle\rho\rangle(z))^{-1} \langle\rho\mathbf{u}\rangle(\mathbf{x}, z, t), \quad \mathbf{x} \in \Pi_o, \quad |z| \leq \delta/2.$$

Setting **v**=**u**-**w**, we conclude that

$$\langle\rho\mathbf{v}\rangle(\mathbf{x}, z, t) = \mathbf{0}. \tag{5}$$

Thus, we have introduced the decomposition of displacements

$$\mathbf{u}(\mathbf{x}, z, t) = \mathbf{w}(\mathbf{x}, z, t) + \mathbf{v}(\mathbf{x}, z, t), \quad (\mathbf{x}, z) \in \Omega \tag{6}$$

into the averaged part **w** and the fluctuating part **v**. By means of CA the fluctuating part satisfies conditions $\mathbf{v}(\bullet, z, t) \in PL_\Delta(T)$. It can be shown that the averaged part of displacement is slowly varying, $\mathbf{w}(\bullet, z, t) \in SV_\Delta(T)$ (cf. assertion at the end of Section 2).

2. Let us observe that the slowly varying averaged displacement field $\mathbf{w}(\cdot, t)$ is defined on $\Pi \times (-\delta/2, \delta/2)$ and hence, from a formal point of view, can be treated like a displacement field in a certain virtual plate with a constant thickness δ. Thus, it is reasonable to introduce a 2D-model of this plate by means of any from the known 2D-modelling hypotheses. In this contribution it will be done by introducing the Kirchhoff-type constrains

$$w_3 = w(\mathbf{x}, t), \quad w_\alpha = -z\partial_\alpha w(\mathbf{x}, t) \tag{7}$$

and by the plane stress assumption imposed on stresses $\bar{\sigma}_{\alpha\beta}$ given by

$$\bar{\sigma}_{\alpha\beta} = \bar{C}_{\alpha\beta\gamma\delta}\partial_\gamma w_\delta(\mathbf{x}, t), \quad \bar{C}_{\alpha\beta\gamma\delta} = C_{\alpha\beta\gamma\delta} - C_{\alpha\beta 33}C_{\gamma\delta 33}(C_{3333})^{-1} \tag{8}$$

In general, instead of (7), (8) we can also impose on the averaged displacement field arbitrary restrictions leading to any other 2D-plate theory.

3. Let $\mathbf{v}_\mathbf{x}(\cdot,z,t)$ be a local periodic approximation of $\mathbf{v}(\cdot,z,t)$ and let $\tilde{\mathbf{v}}=\tilde{\mathbf{v}}(\mathbf{y},z)$ be a test function which is Δ-periodic in **y**, odd in z, and satisfies condition of the form (5). It can be shown that multiplying (4) by $\tilde{\mathbf{v}}$, substituting into (4) the decomposition (6) and taking into account (7),(8), after tolerance averaging the resulting equations by using (2),(3), we shall arrive at the following variational equation

$$\langle\rho\tilde{\mathbf{v}}\cdot\ddot{\mathbf{v}}_\mathbf{x}\rangle(\mathbf{x},t)+\langle\nabla\tilde{\mathbf{v}}:\mathsf{C}:\nabla\mathbf{v}_\mathbf{x}\rangle(\mathbf{x},t)=\langle\bar{\nabla}\tilde{\mathbf{v}}:\bar{\mathsf{C}}z\rangle:\bar{\nabla}\otimes\bar{\nabla}w(\mathbf{x},t),\quad \mathbf{x}\in\Pi_o, \tag{9}$$

which has to be satisfied by every aforementioned test function $\tilde{\mathbf{v}}$. The components of the fourth-order tensor $\bar{\mathsf{C}}$ are defined by (8). Eq. (9) together with condition

$$\langle\rho\mathbf{v}_\mathbf{x}\rangle(\mathbf{x},t)=\mathbf{0},\quad \mathbf{x}\in\Pi_o,$$

which is implied by (5), represents what is called a periodic fluctuation 3D-problem. Let us observe that every $\mathbf{x}\in\Pi_o$ can be treated in (9) as a parameter.

4. We look for the approximate solution to (9) in the form

$$\mathbf{v}_\mathbf{x}(\mathbf{y},z,t)=\mathbf{g}^A(\mathbf{y},z)\upsilon^A(\mathbf{x},t),\quad (\mathbf{y},z)\in V(\mathbf{x}), \tag{10}$$

where $\mathbf{g}^A$, $A=1,\ldots,N$ are postulated *a priori* linear independent vector functions which are Δ–periodic in **y**, odd with respect to z and satisfy conditions

$$\langle\rho\mathbf{g}^A\rangle=\mathbf{0},\quad \mathbf{g}^A(\mathbf{y},z)\in O(l),\quad l\bar{\nabla}\mathbf{g}^A(\mathbf{y},z)\in O(l).$$

Scalar functions υ^A in (10) are new unknowns which will be referred to as the fluctuations variables; it can be shown that $\upsilon^A(\cdot,t)$ have to be slowly-varying functions.

In general, functions $\mathbf{g}^A$ can be treated as the local mode-shape functions, i.e. functions describing the form of free periodic vibrations of every periodic cell $V(\mathbf{x})$, $\mathbf{x}\in\Pi_o$. Thus, the above functions can be specified as eigenvectors related to the eigenvalue periodic problem for a vector funtion **g** given by the variational condition

$$\langle\bar{\nabla}\tilde{\mathbf{v}}:\mathsf{C}:\nabla\mathbf{g}\rangle-\omega^2\langle\rho\tilde{\mathbf{v}}\cdot\mathbf{g}\rangle=0 \tag{11}$$

which has to hold for every aforementioned test function $\tilde{\mathbf{v}}$, where ω^2 is an eigenvalue and $\langle\rho\mathbf{g}\rangle=\mathbf{0}$. However, in many special problems functions $\mathbf{g}^A$ can be taken as certain approximations of the expected free periodic vibrations of an arbitrary cell $V(\mathbf{x})$, $\mathbf{x}\in\Pi_o$.

5. Substituting the decomposition (6) into Eq. (4) and averaging the obtained result over an arbitrary cell $V(\mathbf{x})$, $\mathbf{x} \in \Pi_o$, after taking into account (2),(3),(7),(8) and (10), we arrive at the equation

$$\begin{aligned} &\langle \rho z^2 \rangle \bar{\nabla} \cdot \bar{\nabla} \ddot{w} - \langle \rho \rangle \ddot{w} - \bar{\nabla} \otimes \bar{\nabla} : \langle \mathsf{C} z^2 \rangle : \bar{\nabla} \otimes \bar{\nabla} w + \\ &+ \langle \nabla \mathbf{g}^A : \mathsf{C} z \rangle : \bar{\nabla} \otimes \bar{\nabla} \mathsf{u}^A + \langle p \rangle = 0 \end{aligned} \tag{12}$$

where p are the total external loadings transverse to the plate midplane. Substituting (10) into (9) and setting $\tilde{\mathbf{v}} = \mathbf{g}^A$ we obtain the system of N equations

$$\langle \rho \mathbf{g}^A \cdot \mathbf{g}^B \rangle \ddot{\mathsf{u}}^B + \langle \nabla \mathbf{g}^A : \mathsf{C} : \nabla \mathbf{g}^B \rangle \mathsf{u}^B - \langle \nabla \mathbf{g}^A : \mathsf{C} z \rangle : \bar{\nabla} \otimes \bar{\nabla} w = 0 \tag{13}$$

It can be seen that (12),(13) represent the system of N+1 equations for N+1 scalar unknowns w, υ^A defined on Π for every time t. The above equations have constant coefficients and hence represent the proposed effective 2D-model of periodic plates under consideration.

4 Conclusions

Among the new informations on the effective 2D-modelling of dynamic problems for periodic plates, provided by this contribution, the following seems to be the most important.

1. Independently of model Eqs. (12), (13) the tolerance averaging approach also yield conditions

$$w(\cdot,t) \in SV_\Delta(T) \qquad \upsilon^A(\cdot,t) \in SV_\Delta(T)$$

which can be used as certain *a posteriori* estimations of the obtained solutions, [14].

2. Coefficients $\langle \rho \mathbf{g}^A \cdot \mathbf{g}^B \rangle$ in (13) are of an order $O(l^2)$ and hence the model equations describe also the local dynamic plate behaviour (e.g. vibration frequencies depending on the periodicity cell size), in contrast to the homogenized model which, after neglecting the rotational inertia term, is independent of l.
3. Eq. (12) involve the rotational inertia $\langle \rho z^2 \rangle \bar{\nabla} \cdot \bar{\nabla} \ddot{w}$ which plays an important role in dynamics of ribbed plates.
4. For quasi-stationary problems Eq. (13) lead to the system of linear algebraic equations for υ^A. In this case unknowns υ^A can be eliminated from the model equations and we obtain a single equation for the averaged midplane deflection w which, after neglecting the rotational inertia term, is similar to the know equation representing the homogenized plate model.
5. The tolerance averaging technique makes it possible to formulate the effective 2D-model for periodic plates on different levels of accuracy; to this end we have to replace (7),(8) by more general 2D-modelling assumptions.
6. The main disadvantage in the formulation of model equations lies in a proper choice of the mode-shape functions $\mathbf{g}^A$. For more complicated structures of a

3D-periodicity cell the above functions have to be derived from (11) by using numerical methods.

Some applications of the proposed model to the analysis of selected dynamic problems are going to appear in a separate paper.

References

1. Baron E (2002) On modelling of medium thickness plates with an uniperiodic structure. J Theor Appl Mech 40: 7-22
2. Baron E, Jędrysiak J (1998) On vibrations of plates under periodically distributed inertial loadings. J Theor Appl Mech 36: 1001-1020
3. Baron E, Woźniak Cz (1995) On microdynamics of composites plates. Arch Appl Mech 66: 126-133
4. Bensoussan A, Lions JL, Papanicolaou G (1978) Asymptotic analysis for periodic structures. North Holland, Amsterdam
5. Caillerie D (1984) Thin elastic and periodic plates. Math Meth Appl Sci 6: 159-191
6. Fichera G (1992) Is the theory of heat propagation paradoxical? Rendiconti del Circ. Mat. Palermo Ser II, XLI: 5-28
7. Jędrysiak J (1999) Dynamics of thin periodic plates resting on a periodically inhomogeneous Winkler foundations. Arch Appl Mech 69: 345-356
8. Jędrysiak J (2000) On the stability of thin periodic plates. Eur J Mech A / Solids 19: 487 -502
9. Kohn RV, Vogelius M (1984) A new model for thin plates with rapidly varying thickness. Int J Solids Structures 20: 333-350
10. Lewiński T, Telega JJ (2000) Plates, laminates and shells. Asymptotic analysis and homogenization. World Sci Publ Co
11. Mazur-Śniady KV (2001) A kinematic internal variable approach to dynamics of beams with a periodic structure. J Theor Appl Mech 89
12. Michalak B (1998) Stability of elastic slightly-wrinkled plates. Acta Mech 130: 111-119
13. Michalak B (2000) Vibrations of plates with initial geometrical imperfections interacting with a periodic elastic foundation. Arch Appl Mech 70: 508-518
14. Woźniak Cz, Wierzbicki E (2000) Averaging techniques in thermomechanics of composite solids. Tolerance averaging versus homogenization. Wyd Politechniki Częstochowskiej, Częstochowa

Critical Review of the Contemporary State of the Theory and Design Methods of Ribbed Plates and Shells

V. A. Zarutsky

S. P. Timoshenko Institute of Mechanics, National Academy of Sciences Kiev, Ukraine

As a result of analysis of the three-dimensional problem of elasticity theory for cylindrical shell, reinforced by annular ribs, it is shown, that the well-behaved problem formulation on the stress-strain state of the ribbed shell is possible only in the case, when at the initial relations deduction the ground width, on which the sheathing (the shell itself) and the rib are contacting, is taken into account. It is shown as well that in the framework of the applied shell theory it is impossible to study the stress distribution near the rib.

The different variants of the ribbed shell (plate) theory, taking into account the discrete ribs placement, listed in the scientific literature, are compared. At the equilibrium (motion) equations derivation the attention is accented on the difficulties, concerned with the equilibrium (motion) equations formulation taking into the account the ribs torsion at the shells deformation. The possibility of the complete determination of the stress-strain state on their base (in particular, in view of the theory appropriateness for the description of the beam-type state and the shell displacement as a rigid body) is analyzed.

General solution of the equilibrium (motion) equations is obtained for the cylindrical shell, reinforced by the regular system of longitudinal and annular ribs. The analysis of the named solution allowed to propose the number of the approximate approaches to the solution of problem on the boundary load action on the mentioned shells, as well as to formulate the conditions of applicability of the constructively orthotropic shell theory for their calculation.

The methods of the determination of stress-strain state of shells, reinforced by the cross system of ribs, are proposed and substantiated.

The comparison of the results of calculations, realized on the base of the numerical and analytical methods, showed that in the first case the results receipt, close to the accurate ones, is possible to obtain only at the dense mesh (finite elements, finite differences), and in the second case the sufficiently trustworthy results can be obtained only for simple problems, in particular, for the cylindrical and the sloping shells with the rectangular plane and the rectangular shells, reinforced by the cross framework of ribs, as well as the shells of revolution, reinforced by the annular ribs only.

At solving the problems on the ribbed shells dynamics it is shown that, as a rule, at the real relations of shell and ribs rigidities for the determination of the minimal vibration eigenfrequencies and close to them ones the approximate approaches and, in particular, the theory of the constructively orthotropic shells can be used. The determination of more high eigenfrequencies demands the utilization of relations of the theory, which takes into account the discrete ribs placement. The forced vibrations cannot be studied using the reductive methods. The same conclusion follows also from the realized investigation of the harmonic waves propagation along the longitudinally reinforced cylindrical shell. The natural and forced vibrations of the ribbed conical and spherical shells, as well as rectangular plates, are also considered.

The problems of the equilibrium of the ribbed plates and shells at the static loading can be solved only by using the theory, which takes into account the discrete ribs placement. The ribbed cylindrical shell stability at the axial compression, joint action of the axial compression and the internal pressure, torsion, as well as other loads action, are considered in details. The stability of the ribbed conical shells at the external pressure and the ribbed rectangular plates at the longitudinal compression is studied also. It is shown that at the dynamical loading of the ribbed shells the influence of the discrete ribs placement grows.

The listed above questions are considered in detail in the resumptive papers of author [2,3,4].

The accurate general solution of the uniform equations system of the circular closed cylindrical shell reinforced by the regular system of the longitudinal ribs and some most substantial results of its analysis are given below. The mentioned solution has the form [1]:

$$
\begin{aligned}
u = {} & A_1\xi + A_0 + \left[C_1^{'} + 2C_2^{'}\xi + 6(2+\nu)C_3^{'} + 3C_3^{'}\xi^2 \right]\cos\theta + \\
& + \frac{12C_3^{'}}{1-\nu}(\gamma_c - \delta_c)\Phi_1^1\left(b_{n_1}\frac{\cos n_1\theta}{n_1^2} \right) + \\
& + \left[C_1^{''} + 2C_2^{''}\xi + 6(2+\nu)C_3^{''} + 3C_3^{''}\xi^2 \right]\sin\theta + \\
& + \frac{12C_3^{'}}{1-\nu}(\gamma_c - \delta_c)\Phi_2^1\left(b_{n_1}\frac{\sin n_1\theta}{n_1^2} \right) + \\
& + \sum_{s=1}^{\infty} e^{\chi_{0s}^1\xi} C_0^s \Phi_1^0\left(A_{1n_1}^s \cos n_1\theta \right) + \sum_{s=1}^{\infty} e^{\chi_{0s}^2\xi} D_0^s \Phi_1^0\left(B_{1n_1}^s \sin n_1\theta \right) + \\
& + \sum_{n=1}^{n_2}\sum_{s=1}^{\infty} e^{\chi_{ns}\xi}\Big\{ C_n^s \Big[\Phi_1^n\left(E_{n_1}^s \cos n_1\theta \right) \\
& + \Phi_2^n\left(B_{n_1}^s \cos n_1\theta \right) \Big] D_n^s + \Big[\Phi_2^n\left(E_{n_1}^s \sin n_1\theta \right) - \Phi_1^n\left(B_{n_1}^s \sin n_1\theta \right) \Big] \Big\},
\end{aligned}
\qquad (1a)
$$

$$v = B_1\xi + B_0 + \left[C_0^{'} + 2\nu C_2^{'} + \left(C_1^{'} - 6\nu C_3^{'} \right)\xi + C_2^{'}\xi^2 + C_3^{'}\xi^3 \right]\sin\theta -$$
$$-\left[C_0^{''} - 2\nu C_2^{''} + \left(C_1^{''} - 6\nu C_3^{''} \right)\xi + C_2^{''}\xi^2 + C_3^{''}\xi^3 \right]\cos\theta +$$
$$+\sum_{s=1}^{\infty} e^{\chi_{0s}^{1}\xi} C_0^s \Phi_1^0 \left(A_{2n_1}^s \cos n_1\theta \right) + \sum_{s=1}^{\infty} e^{\chi_{0s}^{2}\xi} D_0^s \Phi_1^0 \left(B_{2n_1}^s \sin n_1\theta \right) + \tag{1b}$$
$$+\sum_{n=1}^{n_2}\sum_{s=1}^{\infty} e^{\chi_{ns}\xi} \left\{ C_n^s \left[\Phi_1^n \left(M_{n_1}^s \sin n_1\theta \right) + \Phi_2^n \left(N_{n_1}^s \sin n_1\theta \right) \right] + \right.$$
$$\left. + D_n^s \left[\Phi_2^n \left(M_{n_1}^s \cos n_1\theta \right) - \Phi_1^n \left(N_{n_1}^s \cos n_1\theta \right) \right] \right\},$$

$$w = \nu A_1 + \left(C_0^{'} + C_1^{'}\xi + C_2^{'}\xi^2 + C_3^{'}\xi^3 \right)\cos\theta +$$
$$+\left(C_0^{''} + C_1^{''}\xi + C_2^{''}\xi^2 + C_3^{''}\xi^3 \right)\sin\theta +$$
$$\sum_{s=1}^{\infty} e^{\chi_{0s}^{1}\xi} C_0^s \Phi_1^0 \left(A_{3n_1}^s \cos n_1\theta \right) +$$
$$+\sum_{s=1}^{\infty} e^{\chi_{0s}^{1}\xi} D_0^s \Phi_1^0 \left(B_{3n_1}^s \sin n_1\theta \right) + \left(B_{3n_1}^s \sin n_1\theta \right) + \tag{1c}$$
$$+\sum_{n=1}^{n_2}\sum_{s=1}^{\infty} e^{\chi_{ns}\xi} \left\{ C_n^s \left[\Phi_1^n \left(O_{n_1}^s \cos n_1\theta \right) - \Phi_2^n \left(Q_{n_1}^s \cos n_1\theta \right) \right] + \right.$$
$$\left. + D_n^s \left[\Phi_2^n \left(O_{n_1 s} \sin n_1\theta \right) - \Phi_1^n \left(Q_{n_1}^s \sin n_1\theta \right) \right] \right\}$$

Here: A_0, A_1, B_0, B_1, $C_2^{'}$, $C_3^{'}$, $C_0^{'}$, $C_1^{''}$, $C_2^{''}$, $C_3^{''}$, $C_0^{''}$, C_n^s, D_n^s are an arbitrary constants; $n_2 = \dfrac{k}{2}$ when k is even, or $\dfrac{k-1}{2}$ when k is odd, k is the number of ribs; χ_{ns} is the s-th nonzero root of the equation

$$\left| \delta_{i_1,i_2} + L_{i_1,i_2}^n \right| = 0 \qquad (i_1, i_2 = 1, 2, 3, 4) \tag{2}$$

χ_{0s}^1 is the s-th nonzero root of the equation

$$1 + A_{11}^0 + A_{22}^0 + A_{11}^0 A_{22}^0 - A_{12}^0 A_{21}^0 = 0 \tag{3}$$

χ_{0s}^2 is the s-th nonzero root of the equation

$$1 + B_{11}^0 + B_{22}^0 + B_{11}^0 B_{22}^0 - B_{12}^0 B_{21}^0 = 0 \tag{4}$$

$A_{s_1 n_1}^s$, $B_{s_1 n_1}^s$ (s_1 = 1, 2, 3), $E_{n_1}^s$, $B_{n_1}^s$, $F_{n_1}^s$, $M_{n_1}^s$, $N_{n_1}^s$, $O_{n_1}^s$, $Q_{n_1}^s$, L_{i_1,i_2}^n, A_{i_3,i_4}^0, B_{i_3,i_4}^0 (i_3, i_4 = 1, 2) are the known functions of the parameters of the sheathing and ribs, characteristics of the numbers χ_{0s}^1, χ_{0s}^2, or χ_{ns}, as well as n_1, ν is a Poisson's ratio, γ_c, δ_c are the dimensionless rigidity on the strain and the dimensionless eccentricity of ribs (it is assumed, that the ribs are reinforced inside the shell), b_{n_1} is the parameter, that takes into account the area width, along which the sheathing and ribs are contacting;

$$\Phi_1^n\left(X_{n_1}\right)=X_n+\sum_{l=1}^{\infty}\left(X_{lk+n}+X_{lk-n}\right)$$

$$\Phi_2^n\left(X_{n_1}\right)=\sum_{l=1}^{\infty}\left(X_{lk+n}-X_{lk-n}\right)$$

Substituting the items, that are in (1) out of the sum, to the formulas for the calculation the forces and moments, we come to the conclusion, that in the longitudinal sections of sheathing they become zero. Determining after that the forces and the moments in the cross-sections of the sheathing and ribs by the mentioned formulas, as well as the transverse shearing forces and torque strengths, we make certain of the mentioned forces and moments do not depend on the arbitrary constants $A_0', B_0', C_0', C_0'', C_1', C_1''$ (it could be shown, that their choice determines the shell placement in the space), hence, the considered stress state of the shell is described by the arbitrary constants $A_1, B_1, C_2', C_3', C_2'', C_3''$. The latter are chosen in such a way, that the static boundary conditions take place. It is assumed, that on the butt of sheathing the axial forces (T_l), shear forces (S_l), bending moments (M_l) and transverse shear forces (Q_l), and on the ribs butts the mentioned forces and moments (T_j, S_j, M_j, Q_j), and the torsion moments $M_{\kappa pj}$ can be set.

Further we determine the relations of the boundary forces, at which the stress-strain states, determined by the arbitrary $A_1, B_1, C_2', C_2'', C_3', C_3''$, are realized.

Stress-strain. Let's substitute the solution, determined by the constant A_1 ($u=A_1\xi;\ v=0;\ w=\nu A_1$), to the formulas, that describe static boundary conditions. They are true if:

$$M_l=Q_l=S_l=M_j=Q_j=S_j=M_{\kappa pj}=0;\quad T_j=T_l\frac{E_cF_c}{Eh} \tag{5}$$

In this case the shell is in the one-axial stress-strain state (bending moment in the cross-sections of the sheathing and ribs, transverse shearing force and torsion moment in the sheathing are equal to zero).

Pure shear. If to substitute the solution, determined by the constant B_1, to the boundary conditions, the latter is satisfied if:

$$T_l=Q_l=M_l=T_j=M_j=S_j=Q_j=0,\quad M_{\kappa pj}=-S_1\frac{2G_cJ_{\kappa p.c}}{rh\left(1+\nu\right)E\pi} \tag{6}$$

In this state the normal forces, bending moments are lacking in the cross-sections of the sheathing and ribs, as well as a torsion moment in the sheathing.

Pure bending is described by the arbitrary constants C_2' and C_2''. Assuming, that it is caused by the boundary forces $\boldsymbol{T_1}=\begin{cases} T_1^* \cos\theta \\ T_1^{**} \sin\theta \end{cases}$, we obtain such conditions of its realization:

$$S_1 = Q_1 = S_j = Q_j = M_{\kappa pj} = 0,\ T_j = \frac{E_c F_c}{Eh} T_1 \Big|_{\theta=\theta_j} \tag{7}$$

At this in the cross-sections of the shell the bending moments, determined by the formulas $M_x = -ra^2 T_1$ (in sheathing) and $M_j^* = \frac{E_c J_{yc}}{Ehr} T_1 \Big|_{\theta=\theta_j}$ (in ribs), appear.

Maximum stresses, caused by bending moments in the sheathing and ribs, are small comparatively with the stresses caused by forces, that's why the stress state of the ribbed shell at the pure bending should be assumed to be momentless. Transverse shear forces and torsion moments in the body are absent.

Bending by butt transverse forces. Shell condition in this case is determined by the arbitrary constants C_3' and C_3''; this is possible such a relation between Q_1 and Q_j:

$$Q_j = \frac{E_c\left(J_{yc} - h_c r F_c\right)}{Ehr^2\left(1-\nu\right)} Q_1 \Big|_{\theta=\theta_j} \quad \left(Q_1 = \begin{cases} Q_1^* \cos\theta \\ Q_1^{**} \sin\theta \end{cases} \right)$$

Under the action of such load in the shell cross-sections the following forces and moments appear

$$N_x = -\frac{1+\nu}{1-\nu}\frac{\xi Q_1}{a^2}, \qquad M_x = \frac{1+\nu}{1-\nu}\xi r Q_1$$

$$N_{xy} = \frac{1+\nu}{1-\nu}\frac{1}{a^2}Q_1 + \frac{E_c F_c\left(1+\nu\right)}{2\pi Ehra^2\left(1-\nu\right)} \begin{cases} Q_1^* \Phi_1^1 \left(b_{n_1} \dfrac{\sin n_1\theta}{n_1} \right) \\ Q_1^{**} \Phi_2^1 \left(b_{n_1} \dfrac{\cos n_1\theta}{n_1} \right) \end{cases}$$

$$N_{xj} = -\frac{E_c F_c}{Eh}\frac{1+\nu}{1-\nu}\frac{\xi}{a^2} Q_1 \Big|_{\theta=\theta_j} \qquad M_{xj} = \frac{E_c J_{yc}}{Eh^2}\xi Q_1 \Big|_{\theta=\theta_j} \tag{8}$$

It is not difficult to verify that this state as well as previous one is momentless.

In the mentioned above formulas, which describe the momentless state of the longitudinally reinforced shell, such additional denotes are assumed: $a^2 = \frac{h^2}{12r^2}$, h, r are the width and radius of sheathing, E is the elasticity material modulus, of which the plate is constructed. F_c, I_{yc}, I_{kpc} are the ribs cross-section area and its inertia moments at the bending in the radial plane and at the torsion, h_c

is the eccentricity of rib, $h_c > 0$, if the rib is reinforced on the inside sheathing surface, E_c, G_c are the elasticity modulus and shear modulus of ribs.

We notice, that the considered above momentless shell states correspond to its deformation at the boundary loads action as a hollow beam.

Returning now to the states of the longitudinally reinforced shell, described by the items, lied in (1) under the sum signs, it is necessary to notice, that the realized investigations [1] showed, that the transcendental Eqs. (2)-(4) do not have the purely imaginary roots. From this it follows, that at the action on the shell of the boundary loads, which are different of ones, that cause the considered above momentless conditions, the stress-strain state appears, that changes at the moving away of the loaded edge by the exponential function. Extension of "boundary effects", caused by the boundary loads, depends on both character of the load changing in the annular direction and on the geometrical and mechanical characteristics of the sheathing and ribs, as well as on the number of ribs. Having consider the half-infinite shell, loaded by the self-balanced boundary loading, we obtain, that there are three characteristic zones, on which it can be divided: in the zone most distant of the edge the so-called "half-momentless state" of shell is realized, which is well described by the theory of constructively orthotropic shells; in the middle zone the state, analogue to the "simple boundary effect" in the constructively orthotropic shell, is realized, however its extension depends substantially on the ribs number and their discrete placement (the approximate formulas for the determination of this state characteristics are listed in [1]); in the shell zone, immediately adjacent to its edge, the building of the approximate calculation formulas is not possible and the calculation taking into account the discrete placement of ribs must be fulfilled in the complete volume. This fact shows, that at the numerical methods application for the determination of stress-strain state characteristics for the ribbed shells, it is necessary to use the dense mesh.

Reference

1. Amiro IYa, Zarutsky VA (1980) Theory of ribbed shells. In: Methods of Shell Design Vol.2 [in Russian], Naukova Dumka, Kiev
2. Zarutskii VA (2000) Features of the buckling of ribbed shells. Int Appl Mech 36, 5: 559-585
3. Zarutskii VA (2000) The theory and methods of the stress-strain analysis of ribbed shells. Int Appl Mech 36, 10: 1259-1289
4. Zarutskii VA (2001) Integrated stability-and-vibration study of structurally inhomogeneous shells. Int Appl Mech 37, 8: 998-1022

Printing: Saladruck, Berlin
Binding: Stein+Lehmann, Berlin